PARTICLE SYSTEMS IN NANO- AND BIOTECHNOLOGIES

PARTICULATE
SYSTEMS
IN
NANO- AND
BIOTECHNOLOGIES

Edited by
Wolfgang Sigmund • Hassan El-Shall
Dinesh O. Shah • Brij M. Moudgil

CRC Press
Taylor & Francis Group
Boca Raton London New York

CRC Press is an imprint of the
Taylor & Francis Group, an **informa** business

CRC Press
Taylor & Francis Group
6000 Broken Sound Parkway NW, Suite 300
Boca Raton, FL 33487-2742

First issued in paperback 2019

© 2009 by Taylor & Francis Group, LLC
CRC Press is an imprint of Taylor & Francis Group, an Informa business

No claim to original U.S. Government works

ISBN-13: 978-0-8493-7436-4 (hbk)
ISBN-13: 978-0-3670-40346-1 (pbk)

Library of Congress Cataloging-in-Publication Data

Particulate systems in nano- and biotechnologies / editors, Wolfgang Sigmund ... [et al.].
    p. cm.
Includes bibliographical references and index.
ISBN 978-0-8493-7436-4 (alk. paper)
    1. Nanoparticles. 2. Biotechnology. 3. Nanotechnology. I. Sigmund, Wolfgang. II. Title.

TP248.25.N35P54 2008
660.6--dc22
                                    2008043505

Visit the Taylor & Francis Web site at
http://www.taylorandfrancis.com

and the CRC Press Web site at
http://www.crcpress.com

Contents

v

Preface

With the explosion of opportunities in nano-biotechnology, the scope of particle technology has acquired a new focus on soft particulate (bacteria, viruses, cells, droplets) systems and adsorbed films. Considering that these (soft) particles are highly sensitive to temperature and shear forces, new protocols for their synthesis, characterization, processing and handling must be developed to enable timely applications. Complexity and lack of understanding of multiphase systems have led to inefficient utilization of particulate materials, especially for nano-bio applications. Nano-bio advances are further complicated by the lack of adequate education of engineers and scientists in particle science and technology, especially at the nano-bio interface including the field of adsorbed films.

This volume embodies in part the advances presented at the International Symposium on the Role of Adsorbed Films and Particulate Systems in Nano and Biotechnologies held in Gainesville, Florida, August 24–26, 2005. This symposium was organized to celebrate the successes of the Particle Engineering Research Center (PERC) during its 11 year funding by the National Science Foundation (NSF) under grant numbers EEC-94-02989 and BES/9980795. This book summarizes the most exciting advances in adsorbed films and particulate systems over the last decade. Select researchers including Plenary and Invited symposium speakers were asked to write review papers on the advances that were achieved worldwide in this exciting field. These experts represent the European, Austral-Asian as well as the Americas leading Particle and Nano-Bio research centers.

Among the topics covered in this volume are: overview of particle technology now and then, production of nanoparticles by top-down approach, self-assembled surfactants for nonmaterial synthesis, engineering of polymeric latex particles for hemodialysis, product engineering of nanoscaled materials, surface engineering of quantum dots, fundamental forces in powder flow, pharmaceutical aerosols, inhalers for pulmonary drug delivery, imaging of particle size and concentration in heterogeneous scattering media, surfactants/hybrid polymers and their nanoparticles for personal care applications, FloDots for bioimaging and bioanalysis, particles for biocidal applications, iron nanoparticles for abatement of environmental pollutants, functionalized magnetite nanoparticles, multimodal nanoparticles as contrast agents in bioimaging.

We trust that both the established as well as those embarking on their maiden research voyage in this important field will find this treatise valuable.

We are grateful to the authors for their interest, enthusiasm, and contributions, without which this book would not have been possible. We are thankful to faculty members, postdoctoral associates, graduate students and administrative staff of both the Particle Engineering Research Center (PERC) and the Center for Surface

Science and Engineering, University of Florida for their valuable assistance during the symposium and in finalizing the drafts. We acknowledge the generous support of the following organizations: the National Science Foundation, the University of Florida, and the Industrial Partners of the Particle Engineering Research Center. Finally, we thank the editing and production staff at CRC/Taylor and Francis for their valued assistance.

Wolfgang Sigmund
Hassan El-Shall
Dinesh O. Shah
Brij M. Moudgil

Editors

Dr. Sigmund is Professor of Materials Science and Engineering at the University of Florida. He has been at UF since 1999 coming from the Max-Planck Institute of Metals Research and University of Stuttgart. He is editor-in-chief of the journal *Critical Reviews of Solid State and Materials Sciences* and an elected member of the European Academy of Sciences.

Wolfgang Sigmund
University of Florida
Department of Materials Science & Engineering
wsigm@mse.ufl.edu

Dr. El-Shall is Associate Professor of Materials Science and Engineering and Associate Director for Research at The Particle Engineering Research Center of University of Florida. He has been at UF since 1994. He also served as Associate

Director for Beneficiation Research at the Florida Institute of Phosphate Research from 1986-1992. He co-edited six books related to phosphate processing. He serves as an editor and reviewer for several archival Journals.

Hassan El-Shall
University of Florida
Department of Materials Science & Engineering
helsh@mse.ufl.edu

Dinesh O. Shah is Professor Emeritus and Founding Director (1984–2008) of the Center for Surface Science and Engineering at University of Florida. He was the First Charles Stokes Professor in Chemical Engineering Department. He served as a joint faculty member in the Department of Chemical Engineering and Department of Anesthesiology, working on interfacial phenomena in engineering and biomedical systems. He has been a recipient of several awards for excellence in teaching, research and scholarship over the past four decades. He is among the top one percent of most frequently cited scientists in the world. He has published widely in the areas of monolayers, foams, macro-and microemulsions, and surface phenomena in contact lenses, membranes, lungs and anesthesia. He was a thrust leader for nano-bio systems in the particle engineering research center. He has edited 11 books and monographs and has published about 250 papers.

Dinesh O. Shah
University of Florida
Department of Chemical Engineering and
Department of Anesthesiology
shah@che.ufl.edu

Brij M. Moudgil is a Distinguished Professor and Alumni Professor of Materials
Science and Engineering at the University of Florida. He is also serving as the
Director of the Particle Engineering Research Center. His research interests are
in polymer and surfactant adsorption, dispersion and aggregation of fine particles,
nanotoxicity, multifunctional nanoparticles for bioimaging, diagnosis, and therapy,
nanoparticulate processing and separation technology for enhanced performance
in mineral, chemical, microelectronics, pharmaceutics, advanced materials, and
resource recovery & waste disposal applications. He received his B.E degree in
Metallurgical Engineering from the Indian Institute of Science, Bangalore, India
and his M.S and Eng.Sc.D degrees from the Columbia University, New York, NY.
He has published more than 300 technical papers and has been awarded 14 patents.
He has been recognized by his peers with several professional awards. In 2002 he
was elected as a member of the U.S National Academy of Engineering. He can be
reached at bmoudgil@perc.ufl.edu.

Brij M. Moudgil
University of Florida
Department of Materials Science & Engineering
bmoudgil@perc.ufl.edu

Contributors

R. Davies
Particle Engineering Research Center
University of Florida
Gainesville, Florida

J.C.M. Marijnissen
Delft University of Technology
Department of Chemical Technology
Particle Technology Group
Delft, The Netherlands

J. van Erven
Delft University of Technology
Department of Chemical Technology
Particle Technology Group
Delft, The Netherlands

K.-J. Jeon
Department of Environmental
  Engineering Sciences
University of Florida
Gainesville, Florida

M. Andersson
Department of Applied Surface Chemistry
Chalmers University of Technology
Göteborg, Sweden

A.E.C. Palmqvist
Department of Applied Surface Chemistry
Chalmers University of Technology
Göteborg, Sweden

K. Holmberg
Department of Applied Surface Chemistry
Chalmers University of Technology
Göteborg, Sweden

S. Kim
Materials Science and Engineering
University of Florida
Gainesville, Florida

H. El-Shall
Particle Engineering Research Center
Materials Science and Engineering
University of Florida
Gainesville, Florida

R. Partch
Department of Chemistry
Clarkson University
Potsdam, New York

B. Koopman
Environmental Engineering Sciences
University of Florida
Gainesville, Florida

T. Morey
Anesthesiology, College of Medicine
University of Florida
Gainesville, Florida

W. Peukert
Institute of Particle Technology
University of Erlangen
Germany

A. Voronov
Coatings and Polymeric Materials
North Dakota State University
Fargo, North Dakota

J. Orbulescu
Department of Chemistry
University of Miami
Coral Gables, Florida

R.M. Leblanc
Department of Chemistry
University of Miami
Coral Gables, Florida

N. Stevens
Particle Engineering Research Center
Materials Science and Engineering
University of Florida
Gainesville, Florida

S. Tedeschi
Particle Engineering Research Center
Materials Science and Engineering
University of Florida
Gainesville, Florida

M. Djomlija
Department of Chemical Engineering
University of Florida
Gainesville, Florida

B. Moudgil
Particle Engineering Research Center
Materials Science and Engineering
University of Florida
Gainesville, Florida

M.S. Coates
Faculty of Pharmacy
School of Chemical and Biomolecular
  Engineering
University of Sydney
Australia

P. Tang
Faculty of Pharmacy
University of Sydney
Australia

H.-K. Chan
Faculty of Pharmacy
University of Sydney
Australia

D.F. Fletcher
School of Chemical and Biomolecular
  Engineering
University of Sydney
Australia

J.A. Raper
Department of Chemical & Biological
  Engineering
University of Missouri-Rolla
Rolla, Missouri

C. Li
Department of Biomedical Engineering
University of Florida
Gainesville, Florida

H. Jiang
Department of Biomedical Engineering
University of Florida
Gainesville, Florida

P. Somasundaran
Center for Advanced Studies in Novel
  Surfactants
Langmuir Center for Colloid
  and Interfaces
Columbia University
New York, NewYork

P. Deo
Center for Advanced Studies in Novel
  Surfactants
Langmuir Center for Colloid
  and Interfaces
Columbia University
New York, New York

G. Yao
Center for Research at the Bio/nano
  Interface
Department of Chemistry and Shands
  Cancer Center
University of Florida
Gainesville, Florida
Life Science Inc.
St. Petersburg, Florida

Y. Wu
Center for Research at the Bio/nano
 Interface
Department of Chemistry and Shands
 Cancer Center
University of Florida
Gainesville, Florida

D.L. Schiavone
Center for Research at the Bio/nano
 Interface
Department of Chemistry and Shands
 Cancer Center
University of Florida
Gainesville, Florida

L. Wang
Center for Research at the Bio/nano
 Interface
Department of Chemistry and Shands
 Cancer Center
University of Florida
Gainesville, Florida

W. Tan
Center for Research at the Bio/nano
 Interface
Department of Chemistry and Shands
 Cancer Center
University of Florida
Gainesville, Florida

G. Pyrgiotakis
Particle Engineering Research Center
Materials Science and Engineering
University of Florida
Gainesville, Florida

W. Sigmund
Materials Science and Engineering
University of Florida
Gainesville, Florida

X.-Q. Li
Center for Advanced Materials and
 Nanotechnology
Department of Civil and
 Environmental Engineering
Lehigh University
Bethlehem, Pennsylvania

D.W. Elliott
Center for Advanced Materials
 and Nanotechnology
Department of Civil and
 Environmental Engineering
Lehigh University
Bethlehem, Pennsylvania

W.-X. Zhang
Fritz Engineering Laboratory
Lehigh University
Bethlehem, Pennsylvania

P. Majewski
Ian Wark Research Institute
University of South Australia
Adelaide, SA, Australia

B. Thierry
Ian Wark Research Institute
University of South Australia
Adelaide, SA, Australia

P. Sharma
Particle Engineering Research Center
Materials Science and Engineering
University of Florida
Gainesville, Florida

A. Singh
Particle Engineering Research Center
Materials Science and Engineering
University of Florida
Gainesville, Florida

S.C. Brown
Particle Engineering Research Center
Materials Science and Engineering
University of Florida
Gainesville, Florida

G.A. Walter
Physiology and Functional Genomics
University of Florida
Gainesville, Florida

S. Santra
Nanoscience Technology Center
Department of Chemistry and
  Biomolecular Science Center
University of Central Florida
Orlando, Florida

S.R. Grobmyer
Division of Surgical Oncology
Department of Surgery
University of Florida
Gainesville, Florida

E.W. Scott
Department of Molecular Genetics
  & Microbiology and The McKnight
  Brain Institute
College of Medicine
University of Florida
Gainesville, Florida

1 Particle Technology—Then and Now—A Perspective

Reg Davies

CONTENTS

INTRODUCTION

In the fall of 2001, the Institution of Chemical Engineers (IChemE) in the United Kingdom initiated a new Particle Technology initiative entitled the U.K. Particle Technology Forum. This included the Leslie J. Ford lecture, in honor of Leslie J. Ford, a prominent advocate for particle technology both in ICI, and the then Science and Engineering Research Council, U.K. I was invited as the first L.J. Ford lecturer, and delivered a perspective of particle technology at the millennium. Five years later I was invited to the 50th Anniversary of the Society for Powder Technology in Tokyo to present a keynote lecture on the growth of particle technology in the Americas over the past 50 years, and a perspective of the future.

In the fall of 2005, the Particle Engineering Research Center at the University of Florida organized the International Symposium on the Role of Adsorbed Films and Particulate Systems in Nano and Biotechnologies, with speakers from a number of universities and research centers mentioned in my two previous talks, some of whom were invited to submit topical review articles that constitute the present volume.

This introductory chapter is primarily based on my perspectives and contains excerpts from both talks with an updated summary of the situation as it stands today. It also contains material from reviews presented at the Fifth World Congress in Particle Technology (1,2,3,4), and some of those presented at the 2005 symposium in Gainesville, FL. The perspective of history is always a product of the historian. Some of the dates I have quoted may be inaccurate, and there will definitely be some omissions. My goal was to endeavor to show how the field began, how it has grown, and where it stands today. I hope this achieves this goal.

SELECTED HISTORICAL MILESTONES

In 1945, Dallavalle predicted that the science and technology of fine particles would be of major importance to future consumer products. It is significant to note that America hardly listened, whereas Japan and Northern Europe listened well, and developed the field. However, there was a thrust for particle science and technology in the U.S.A., but it was concentrated in mining, where particle processes were critical to national and economic resource development. It has stayed strong in mining over the years. In Japan, the Research Association of Powder Technology was established in 1957, but renamed The Society of Powder Technology Japan in 1976, and Japan began to focus strongly on the field through the leadership of Iinoya. The Institüt für Mechanische Verfahrenstechnik was established at Karlsruhe, West Germany, in the mid-fifties by Hans Rumpf. This group studied particle technology fundamentals, not so much in the chemistry aspects, but in mechanical engineering supported by physics and mathematics. Graduates from this environment gave Germany a competitive advantage in the particle technology field. It suffered a setback in 1976 by the premature death of Rumpf but the institute regrouped and survived. The Particle Technology Group at Loughborough University in England was specializing in fluidization but expanded to broader particle technology research in 1963. This was one year after the School of Powder Technology had been formed at the University of Bradford. Interest in particle technology at the academic level was high but it was only in the late sixties that diverse industries began to seek active participation in the subject in the United States. Here industry-developed technology disconnected from the mining fraternity. Dallavalle had by now coined the term "micromeritics" and had set in motion advances in chemical engineering, which would impact the field during the next 30 years. It was not that the United States did not have individual leaders who were recognized and respected by the world community, but rather that it lacked organization in the particle technology field. There were exceptions. There were none better than the aerosol community who led the world in instrumentation development and light-scattering research sponsored by DOE and DOD in the post-war years.

As Chicago was prominent in the 1930s in the Depression, so Chicago was the birthplace of some of the early-organized particle technology groups in the United States. Illinois Institute of Technology Research Institute (IITRI) began to develop fine particle technology in 1964 and held the first fully international conference in particle science and technology in 1973. IITRI also organized the first multi-industrial consortium of 32 major corporations to categorize and evaluate particle characterization techniques. This program ran from 1967 to 1973. The Fine Particle Society was conceived in Chicago in 1967 and incorporated the following year. Similarly, the Bulk Solids Handling Conference was begun in Chicago by Abraham Goldberg in 1975. Both continue to the present day.

About this time, government support for fundamental research in particle science began to wane, except for environmental interests. Industry/academic partnerships began to flourish. Organizations such as the Particulate Solids Research Institute (PSRI) and the International Fine Particle Research Institute (IFPRI) were formed in the 1970s. The first review of particle technology organizations in Japan was completed by Iinoya in 1983 on an IFPRI grant, showing the widespread activity in the field in Japanese academic and government laboratories. The Association of Powder Process Industry and Engineering, Japan (APPIE) was formed in 1971 as an informal group of 70 members. Their intent was to provide a means of exchanging technical and business information. It was approved as a corporation by MITI in 1981.

The IChemE Particle Technology Subject Group was formed in England in 1980 under the chairmanship of Professor Don Freshwater. It continues today and has a membership of around 300. Les Ford was its first vice-chair and assumed the position of chair in 1991.

The SERC Specialty Promoted Program (SPP) in Particle Technology was started in late 1982 in England under the chairmanship of Professor John Bridgwater and the coordinator, Les Ford. In total, it funded approximately 100 programs at a funding of 6.5 million pounds sterling for a decade.

Germany held the first World Congress in Particle Technology in Nürnberg in 1986, spearheaded by Professor Kurt Leschonski. Clearly in Europe and Japan, particle science and technology were becoming well-organized and widespread. The Fine Particle Society was the main U.S. organization and was cosponsor of the World Congress.

U.S. industry was becoming more aware of the potential benefits of particle applications. E. I. du Pont de Nemours & Company particle technology group assessed the importance of particle technology to its line of products in 1984. This assessment showed that a surprising 62% of its products were in particle form, and a further 18% of its products contained dispersed particles in its portfolio of shaped products, for example, film, fiber, composites, etc. These percentages were reduced in later years by movement away from bulk chemicals, but they were widely quoted in proposals for particle science and technology funding around the world, and were the basis for the development of Particle Science & Technology Group (PARSAT) within DuPont. In 1985, Ed Merrow reported on the importance of solids processing in the chemical industry, particularly the effects of solids feed on start-up potential of 39 U.S. and Canadian plants. Particle problems such as pluggage, attrition, uneven flow, stickiness/adhesion, and cohesion were the principal causes of delays in plant start-up. Merrow later showed that the introduction and use of new technology when all heat and mass balances around the equipment were not known had similar effects. Merrow's appeal to U.S. industry was "pay more attention to solids processing and do your chemical engineering." With this impetus, U.S. academia in addition to industry began to awaken to the potential of improved solids processing to process optimization. The Center for Advanced Materials Processing was initiated at Clarkson as was the Particulate Materials Center at Penn State University. And so, the final decade of the second millennium opened with the Second World Congress in Particle Technology in Kyoto in 1990. In the United States, an explosion of interest in the 1990s replaced the apathy of the 1980s.

The AIChE Particle Technology Forum was formed in 1993 in an attempt to better link particle science and technology with engineering. The first International Particle Technology Forum took place in Denver in 1994, followed by the second in San Diego in 1996, and the third in Miami in 1998. The Denver meeting should have been the 3rd World Congress in Particle Technology, but legal issues rendered this impossible.

The National Science Foundation (NSF) Engineering Research Center in Particle Science & Technology was initiated at the University of Florida in 1993 to provide a focus for U.S. research and education in the field through the millennium.

The state of New Jersey supported a Center in Particle Coating Technology at New Jersey Institute of Technology (NJIT) in 1997. Centers thrived in solids transport at Pittsburgh, fluidization at Ohio State, agglomeration at CCNY, particulate systems at Purdue, aerosol technology at UCLA, Caltech, Cincinnati, Minnesota, and others.

The "official" 3rd World Congress in Particle Technology was hosted by IChemE in Brighton, England, in 1998.

So how did the world view particle technology at the end of the millennium? Particle technology in Japan still thrived. APPIE in Japan had close to 300 industrial member corporations and over 70 academic, government, and supporting individual members. However, Japan had just lost two of its particle technology leaders with the deaths of professors Iinoya and Jimbo in 1998 and 1999, respectively. Before his premature death in May 1999, Professor Genji Jimbo was active in promoting the Asian Professors' Particle Technology Workshops, linking Japan with Korea, Taiwan, China, Thailand, Singapore, Malaysia, and Vietnam. Although Japan had been the best-organized and major Asian force in particle technology over the past 40 years, the other countries were themselves organizing thriving particle technology institutions. Thus, progress was perhaps best illustrated in the 1997 published report on the Second East Asian Professors' Meeting at the Tokyo Garden Palace Hotel in 1996.

China, for example, was shown to have 15,000 active members in the various academic societies that related to powder technology. These focused mainly on the process of coal energy production. Russia was active in the field, with powder preparation; powder compaction and coatings highlighted as major thrusts. Four of eight Malaysian universities had active particle science and technology programs. The Standard and Industrial Research Institute of Malaysia, incorporated in 1996, offered technology developments in ceramics and metals. Particle technology was shown to be active and expanding in Indonesia, the Philippines, and Vietnam.

Susan A. Roces summarized the Asian scene by delegating development into four stages. Japan had the initial development in the 1960s with South Korea, Taiwan, Hong Kong, and Singapore in the second stage. Malaysia, Thailand, and Indonesia were in the third batch with the Philippines and Vietnam the remaining countries. Although omitted, China probably falls into the second stage.

In the Pacific Rim, Australia too had focused its particle technology via the Chemica conferences, and the new Australian Research Council Center in Particle and Multi-Phase Flow has been created at New Castle. Australia hosted the 4th World Congress in Particle Technology in 2002.

The European Federation of Engineers' Working Parties was still strong— particle technology continued to be featured at the Nürnberg conferences—many particle technology chairs still existed in Germany, where the next generation of professors was establishing itself. Although Loughborough and Bradford were less dominant in U.K. particle technology, universities at Leeds, Surrey, Birmingham, University of Manchester Institute of Science and Technology, U.K. (UMIST), UCL, Imperial, and Herriot-Watt were in ascendancy. In other parts of Europe, centers at Delft, ETH Zurich, Albi, Porsgrunn, and others confirmed the widespread interest in the field. IChemE continued to thrive as indicated by several new initiatives such as the successful 1998 Brighton World Conference, and the annual U.K. Particle Technology Forum with the L.J. Ford Lecture. A soft solids initiative that focused on foams, pastes, gels, microemulsions, and general rheology, was supported by grants of 7.0 million pounds sterling and other support from 30 companies.

Industrial consortia continued to expand. PSRI and IFPRI celebrate 30 years of existence in 2008.

Overall, Europe and Japan have successfully changed the "old guard" and the future is bright. This transformation was also occurring in the United States where young vigorous leadership is in place for the future. Of significance is the emergence of Leeds University as the pre-eminent university in Particle Technology in England. Activity has increased in Birmingham and in Sheffield where young leadership is pushing forward new initiatives.

So on four continents there are strong interest, vibrant organizations, and diverse networks building in the particle science and technology field. A New World Congress Council was formed by which selections could be made for the future hosting of world congresses in the field every four years. The sixth will be in 2010 in Germany, and the seventh in Shanghai, China, in 2014.

One notable omission in all the previous reviews on the technology had been South and Central America. It had been observed that a large participation of scientists and engineers was present at world congresses in particle technology in the past, yet no one knew much about their activities. In order to remedy this, IFPRI supported a grant, with Professor Sorrentino in Venezuela, to review activities in this part of the world and align it with other world activities.

In 2006, the United States hosted the Fifth World Congress on Particle Technology in Orlando, Florida. This meeting, WCPT5, set all-time attendance records for world congress particle technology. Since its initiation in Germany in 1986, the congress had become a regularly featured international event in the subject.

Following these developments, other universities expanded their graduate programs to include particle-related research. In aerosol technology, the United States had long been a world leader, but now other aspects of particle-related systems were added. Today, it is recognized by industry and government alike, but conventional particle science and technology still remains under-funded. It is the emergence of nanobio- and nanomedico-technologies that have caused a resurgence of interest in particle-related research.

In industry, expertise has been lost through early retirements, group reorganizations, and staff reductions. In academia, champions have been lost due to retirement and death. In the past five years, many champions of the subject have died, and new leadership is required to carry the subject forward. Let us now take a more detailed look at academic centers and active university groups in the United States and add some perspective by briefly looking at industrial and consulting group activity.

PARTICLE ENGINEERING RESEARCH CENTER (PERC), UNIVERSITY OF FLORIDA

Formed in 1994, PERC received 11 years of funding from the National Science Foundation. In September 2005, this funding ceased and PERC is reorganizing to become an Industry/University Research Cooperative Center (I/UCRC).

Core Team: B. Moudgil, H. el-Shall, B. Koopman, W. Sigmund, Chang Won-Park, R. Singh, J. Curtis, K. Johanson, S. Svornos, K. Powers, D. Dennis, D. Shah, V. Jackson.

Completed Research: Achievements in PERC have included novel nanoparticles for drug detoxification, smart nanotubes for drug delivery, bacteria detection using dye-doped nanoparticle-antibody conjugates, carbon nanotube-tipped AFM research, selective flocculation process for solid-solid separations, computational code for granulation and mixing, atomic flux coating process, new instrumentation including an on-line slurry densitometer, Ewald method for bacterial adhesion measurements, multiwavelength-multiangle spectrometers for joint particle property measurement, laser-induced breakdown spectrometer for on-line phosphate analysis, cohesive powder rheometer, and the process modification of filter surfaces for the removal of microorganisms. A fully equipped laboratory is available not only for member use, but also for other industrial and academic services.

Current Drive: Through a possible I/UCRC in collaboration with Columbia University, conventional particle technology research will be conducted to meet the needs of the 45 industrial member companies. Federally supported research will be organized around nano-, bio-, medico-, and agro-technologies. This will be in collaboration with Shands Hospital, the Brain Imaging Institute on campus, and the Departments of Medicine, Anesthesiology, Pharmacy, and Agriculture. Florida is well qualified for this work as a recent poll by *The Scientist* magazine ranked Florida number eight in the nation's top ten best places in the United States to conduct life sciences research. The criteria cited for the ranking were excellent facilities, good peer relations, complementary faculty expertise, institutional management, commensurate salary, and tenure positions.

RESEARCH CENTER FOR STRUCTURED ORGANIC COMPOSITES—RUTGERS UNIVERSITY, PURDUE, NJIT, AND THE UNIVERSITY OF PUERTO RICO

Core Team: F. Muzzio, A. Cuitino, K. Morris, C. Velazquez, B. Glasser

Vision: This center is to be the national focal point for science-based development of structured organic composite products and their manufacturing processes in the pharmaceutical, nutraceutical, and agrochemical industries.

Mission: Develop a scientific foundation for the optimal design of structured organic composites. Develop science and engineering methods for designing, scaling, optimizing, and controlling relevant manufacturing processes. Establish effective educational and technological transfer vehicles. Produce faster, more reliable, less expensive drug products with less-expensive manufacturing processes through new technologies. This center has just begun and will be funded through 2017.

PARTICULATE RESEARCH CENTER, I/UCRC, PENN STATE UNIVERSITY

Core Team: J. H. Adair, D. Agrawal, D. Green, G. Messing, V. Puri, C. Randall, D. Velegol, and W. White (Penn State); R. Haber (Rutgers University), and W. Kronke (University of New Mexico)

Research Goals: This center aims to support industrial and government member research and manufacturing interests by developing engineering and scientific foundations for the manufacturing of advanced particulate materials. Focus has now shifted from granular particles to nanoparticulate materials. Research projects span synthesis, processing, and characterization. A fully equipped characterization facility is available for member use. Studies are also in progress on powder compaction including time-dependent elastic-viscoplastic modeling. Other programs include studies on segregation in granular systems and the mechanics of die filling processes. Some emphasis is placed on the development of new instrumentation and models to improve understanding of powder mechanics and design. The Particle Technology and Crystallization Center, led by D. Hatziavramides, supports research in nucleation and crystallization that will result in faster development of new pharmaceutical products. This is in collaboration with Purdue, Massachusetts Institute of Technology (MIT), and Argonne National Laboratory.

CENTER FOR ADVANCED MATERIALS PROCESSING (CAMP), CLARKSON UNIVERSITY

Core Team: E. Matijevic, G. Campbell, R. Partch, D.H. Rasmussen

Research Areas:

Nanosystems: Nanoparticle synthesis, nanocomposites, self-assembly, biomaterials, and biological systems.

Colloidal Dispersions and Processing: Polymer blends, foams, surfactants, gels, colloids, catalysts, and sols.

Particle Transport, Deposition, and Removal: Modeling of fluid flows, flow visualization, wet particulate cleaning systems.

Chemical Mechanical Polishing: Metal and dielectric film polishing, abrasives, post-CMP cleaning, modeling of fluid flow, heat and mass transfer.

Particle Synthesis and Properties: Micro- and nanoparticle synthesis, inorganic and organic composites, optical, magnetic and electrical systems adhesion and coagulation.

Thin Films and Coatings: Coated particles and fibers, chemical vapor deposition, adhesion.

CENTER FOR ENGINEERED PARTICULATES, NJIT

Core Team: R. Dave, E. Dreizin, M. Huang, B. Khusid, R. Pfeffer, M. Howley, S. Watano

Research Topics:

Coated Particles: Coating nano- and submicron particles onto micron-sized particles or polymeric film coating on particles.

Designer Particles: Synthesis of particles with tailored properties, for example, size, shape, surface, surface morphology.

Composite Particles: Nano-structured composites, for example, mechano-alloying, and microgranulation.

Process Research: Mechanical processing, supercritical fluid processing, hydrothermal processing, and microarc processing.[3]

PARTICLE TECHNOLOGY & CRYSTALLIZATION CENTER, ENERGY & SUSTAINABILITY INSTITUTE, ILLINOIS INSTITUTE OF TECHNOLOGY, CHICAGO

Core Team: D. T. Wasan, H. Arastoopour, D. Gidaspow

Research Interests:

Colloid and Interface Science: Inter-particle force measurement and modeling and colloid stability.

Simulation of Flow: In circulating, bubbling, and rotating fluidized beds.

Mathematical Modeling: Of multi-type particle flow and cohesive particle flow.

Simulation and Modeling: Of nanoparticle flow and of solid/solid flow in food processing systems. Simulation of particle/crystal growth, and particle agglomeration linking population balance and CFD models.

Size Reduction: Pulverization of polymeric and elastomeric materials using a solid-state shear extrusion process.

SOME OTHER UNIVERSITY ACTIVITY

There are many universities in the United States doing particle technology, mainly as a result of the nanotechnology funding. Most want a piece of that funding. The following list of universities is mentioned because these are the ones that support PTF in its endeavors and they employ people who run for office. In consequence, these, along with many mentioned above, represent the new leadership of U.S. particle technology.

Caltech: M. Hunt and R. Flagan have research interests in particle-particle wall collisions, effect of vibration on powder flows, energy dissipation in shearing flows, DEM modeling, heat transfer in particle-laden flows, submicron aerosol measurements in the environment, and nanoaerosols.

City College of New York, CCNY: G. Tardos has research interests in binder granulation of fine powders, strength and morphology of solidifying bridges in dry granules, X-ray tomography to study porosity and morphology of tablets and granules, measurement of forces in flowing powders.

Cincinnati: This university began extensive studies in coagulation and growth of submicron aerosols under the guidance of Professor S. Pratsinis. Some of this work continues with Professor Gregory Beaucage along with Pratsinis as a consultant. Also Professor K. Bauckhage, originally at Bremen (now retired), is active in Cincinnati working on droplets and sprays.

Colorado: A. Weimer and C. Hreyna research modeling and scale-up of gas-fluidized beds, DEM studies on hydrodynamics and segregation in granular flows, cohesion, modeling of aerosol flows, nanoparticles, and nanoparticle synthesis.

Ohio State: Dominated by the extensive work of Professor L. S. Fan—bubbling and turbulent gas-solid fluidized beds, electrical capacitance tomography, electrostatic phenomena in gas-solids flows, Oscar process development and demonstration, Carbonox development and demonstration.

Pittsburgh: Professors G. Klinzing and J. McCarthy do theoretical and experimental measurements in lean and dense phase pneumatic conveying, design guidelines in solids handling, heat transfer in granular media, computational and experimental aspects of mixing and blending.

Princeton: Professor S. Sundaresan studies meso-scale structures in gas-solid flows, the role of cohesion and wall friction on fluidization/defluidization behavior, simulations of particle flows, constitutive models for the rheology of cohesive powders. Professor W. Russell conducts theoretical studies on dense suspension flow and rheology.

UCLA: Although no longer active, Professor Sheldon Friedlander is recognized as the "father" of aerosol technology in the United States. Work continues to emerge from UCLA but mention is made here to honor one of the "old guard" of U.S. particle technology.

Utah: The University of Utah was one of the pioneering universities leading mineral processing and highlighting particle technology. Led for some time by Professors John Herbst, and Rajamani, Utah now collaborates widely with other universities in studies using X-ray microtomography. Professor J. D. Miller has linked with Particle Engineering Research Center (PERC), Florida, to investigate segregation and homogeneity in powder shear testers. The instrumentation has also been used to study colloid deposition on surfaces.

West Virginia: Professor R. Turton studies tablet coating in rotating pan coaters, DEM of coating processes and the development of novel video-imaging techniques using tracer particles.

In addition to these centers and professors who have provided leadership in the field of particle technology, there are many more that are new to the subject or on the edge of providing leadership to the field in the United States. In the final program of WCPT5, the following universities submitted papers and are worthy of more research into their vision and objectives for future reviews: Purdue, Missouri-Columbia, Central Connecticut State, Tulane, Iowa State, Wisconsin, Washington-St Louis, MIT, Missouri-Rolla, Akron, Rowan, New York, Illinois, Mississippi State, Michigan, Duke, Maryland, Lehigh, Rensselaer Polytechnic, Kentucky, East Carolina, Auburn, Texas Tech, Carnegie Mellon, Georgia, Delaware, Minnesota, Houston, New Mexico, and Xavier, Louisiana.

INDUSTRIAL CONSORTIA

INTERNATIONAL FINE PARTICLES RESEARCH INSTITUTE (IFPRI)

Although a truly international organization, IFPRI was incorporated in the state of Delaware in 1978, and, hence, is an American organization. IFPRI is an industrially sponsored consortium supporting fundamental research in particle technology at universities worldwide. Currently it is supported by 27 companies. IFPRI research continues to be focused in five areas: suspensions of particles in liquids, particle synthesis, particle breakage, dry powder flow, and characterization.

PARTICULATE SOLIDS RESEARCH INSTITUTE (PSRI)

Like IFPRI, PSRI was incorporated in Delaware about the same time and is an American company. Unlike IFPRI, PSRI supported its research at one location. It was directed by Professor Fred Zenz and work was conducted at Manhattan College. Then it was transferred to the Institute of Gas Technology under the direction of Dr. Ted Knowlton. Research is done on hydrodynamics of circulating fluidized beds and riser reactors. Both experimental and theoretical standpipe and cyclone design and operation are conducted.

OTHER INDUSTRIAL ACTIVITY

Many companies have been strong supporters of particle science and technology through the past 40 years. Among those are DuPont, Dow, ExxonMobil, Proctor & Gamble, Merck, Millennium, Pfizer, Eastman Chemical, Kodak, PPG, and others.

The subject would not be as quantitative without the support and development of characterization equipment by instrument companies. Some of those U.S. companies who helped initiate and continue to support the particle technology business are Coulter Electronics (the Coulter Counter celebrated 50 years in the medical and industrial characterization field in 2006), Thermo Systems Incorporated, Micromeritics, Particle Measuring Systems (with the acquired Royco and HIAC businesses), Quantachrome Corporation, and Microtrac.

Consulting activities continue through Jenike and Johanson, J. Johanson Consulting, California; K. Johanson Consulting, Florida; McCrone Associates and Particle Data Laboratories-Chicago.

This section would not be complete without mentioning the role of AIChE in U.S. particle technology. AIChE took a fledgling particle technology program under its wing in 1993, called it PTF, and encouraged it to grow to world status. PTF thrives today as a result of AIChE foresight. However, complacency is dangerous, and perhaps the engineering particle technology perspective of PTF under-emphasized new technologies. John Texter with the American Chemical Society (ACS) developed highly pertinent conferences through the Particles 2000–2006 series. These attracted strong support and rivaled PTF in its influence on the particle technology community. Particle technology in the United States is well represented by these two organizations. Unlike previous world congresses that focused on conventional technology, the WCPT5 in the United States broadened its perspective to include nanotechnology in its various applications.

NANOTECHNOLOGY

The field of nanotechnology involves the manufacture and manipulation of materials at the molecular level. It is forecasted to change the way companies make products ranging from fibers to water sensors. Experts estimate that nanotechnology will be incorporated into 15% of global manufacturing output by 2014, a $2.6 billion industry.

Currently the United States leads the world in organized research, and, in fact, has one quarter of the investment from all nations of the world. When nanotechnology was introduced in 2000, the 2001 budget was $464 million in the United States. The 2007 budget has projected $1.3 billion. Twenty-five federal agencies will dispense this research funding.

In 2003, the United States issued the National Nanotechnology Initiative (NNI) in which the outline of U.S. nanotechnology research was defined. In this initiative, the particle range of nanotechnology was specified to be 1–100 nanometers. Since 2000, this has been the fastest growing particle research area in the world. At the current time, 7–8% of all U.S. publications have a nano-link, and 30–40% of all publications come from the United States. However, due to world competition, this is slowly decreasing. There are 10,000 more patents per year in the field now than five years ago. Six hundred companies are engaged in nanotechnology R&D in manufacturing, sale, and use in the United States. Of these companies, 57.6% have a product on the market. Of these, the largest percentage of products is in the biomedical/life sciences field. Companies with the most patents are IBM, Intel, and L'Oreal.

The U.S. Government currently funds 3,000 research projects in nanotechnology. Over 40 nanotechnology centers, networks, and user facilities are now constructed, and many more are scheduled in the year ahead. The 2007 budget defines seven areas of technology where funding is projected. These are as follows.

FUNDAMENTAL NANOSCALE PHENOMENA AND PROCESSES

- Nanomaterials
- Nanoscale Devices and Systems
- Instrumentation Research, Metrology, and Standards

- Nanomanufacturing
- Major R&D Centers, Facilities and Instrument Acquisitions
- Societal Dimensions

Societal dimensions include the overriding question of nanotoxicology. Should this prove to be damaging to the society, then the whole scope of nanotechnology could be severely curtailed.

In the Americas, government investment is being made in Argentina, Brazil, Canada, Mexico, and the United States.

CHALLENGES

Despite the widespread acceptance and growth of particle technology, many problems and challenges remain. Some of the challenges facing industry include

- The resolution of major conventional particle technology related design, operation and manufacturing problems despite the shrinking workforce.
- The solving of conventional unit operation problems that will not be aided by federal resources. Government expects industry to fund these programs.
- How to replace expertise that has been lost via retirement and staff reductions
- How to help managers realize the importance of the field knowing that their interest is minimal in technology.
- How to foster noncompetitive collaborative work on common problems.
- How to continue to support academic research when money is tight.
- How to target and hire graduates with a working knowledge of the subject.
- How to support basic training and continuing education in the subject when funding is cut for these enterprises.

Some of the challenges to academia include

- How to provide basic knowledge training in Particle Science and Technology (PS&T) in college and continuing education despite competitive program interests.
- How to prepare for the change in diversity expected in 2014 when 50% of the U.S. population is projected to be Hispanic. Should we foster better communications with Latin America and Spanish particle technology organizations now and look for opportunities?
- How can we improve industry/university relations and help meet real needs while experiencing workforce reductions?
- How do we persuade academia to continue with experimental facilities, though expensive, instead of rewarding modeling papers of dubious relevance?
- How do we persuade academia that business is for industry and innovative research is for universities?

- How will we overcome the tremendous overload in information that soaks up time and availability?

SUMMARY

Particle Science and Technology in the world is alive and kicking in 2008. In many parts of the Americas, mining is still the focus. In the United States, the subject has moved more into the mainstream of science and engineering and is no longer a stand-alone subject. Funding will always be a problem for conventional technology.

Fifty years ago mining and unit operations were the driving forces. They still are in Central and South America and Canada; whereas today, in the United States, pharmacy, biology, life sciences, health care, microelectronics, food, and agriculture drive government research and, along with nanotechnology, are driving particle technology innovations in the United States and other parts of the world. Nanotechnology has burst onto the front pages of the newspapers and even a president of the United States mentioned it in a State of the Union address. It is hailed as the future new technology. Experts say that nanotechnology will be incorporated into 15% of global manufacturing output by 2014, a 2.6 trillion market. It is only the possibility of unique properties that fuel the nano-engine, but the current worldwide concern for nanotoxicology could rapidly slow down the progress. A number of review articles that follow this chapter present a glimpse of exciting new applications of engineered particles in the nano- and biotechnologies. They range from advances in measurement of fundamental forces between particles to synthesizing multifunctional particulate systems for novel noninvasive bioimaging, diagnosis, and therapies. It is possible that some of the new particulate systems may exhibit unwarranted ecological and biomedical toxicological effects, but at the same time they can be envisioned also to provide some of the needed environmental remediation solutions. The thought of the new nano-bio opportunities in medicine and health care is truly exciting. Are they possible? In his youth, my father could never imagine that men would fly or go to the moon, so these new frontiers of science and medicine should not be beyond imagination.

I do believe that more people are working in particle science and technology in the United States today than ever before. There is certainly more academic activity and it is widely spread. These academics may not recognize the subject as we do, so to them it might remain fundamental but obscure.

ACKNOWLEDGMENTS

I thank the IChemE and the Society of Powder Technology—Japan for honoring me with invitations to present my view of particle technology in various parts of the world. These made it possible for me to prepare this chapter. I also thank the University of Florida and Professor Brij Moudgil for supporting the preparation of both this chapter and the earlier presentations. I am grateful to my countless colleagues and mentors who have guided and supported me in this field for the past 50 years.

REFERENCES

1. Sorrentino, J.A. Particle Technology in Latin America, *IFPRI Research Review*, 2003.
2. Davies, R. Particle Technology—A View at the Millennium, L.J. Ford Lecture, IChemE Meeting, 2001.
3. Dhodapkar, S. A View from the Americas, Proc. WCPT5, Orlando, Florida, 2006.
4. Davies, R. Particle Technology in the New World, 50th Anniversary Symposium of the Society of Powder Technology, Japan, November 2006.

2 The Production of Nanoparticles: The "Top-Down" Approach, with Emphasis on Aerosol Routes, Especially Electrohydrodynamic Atomization

J.C.M. Marijnissen, J. van Erven, and K.-J. Jeon

CONTENTS

INTRODUCTION

For the production of (nano)particles, two fundamentally different main routes can be distinguished. The first one is by building them from molecules, such as in gas phase aerosol reactors. The second one is by disintegration of bigger structures into (nano)fractions. Here the second one, the "top-down" route, will be considered. Different top-down techniques exist such as grinding, liquid atomization, lithography

15

and etching, and others where both disintegration and building-up play a role, as in furnace evaporation/condensation. Attention will be only given here to liquid atomization with the consequent droplet-to-particle conversion. From the several atomization methods, we are only interested in methods that break up in rather uniform droplets, so we limit ourselves to jet breakup in the laminar flow region (Lefebvre, 1989). Most emphasis is paid to a very promising technique, Electrohydrodynamic Atomization (EHDA) or Electrospraying. EHDA is a method to produce very fine droplets from a liquid (atomization) by using an electric field. By applying the right conditions, droplets can be monodispersed from nanometers until several micrometers can be produced. By means of an example, that is, the production of nanoplatinum particles, a generic way to produce nanoparticles from a multitude of different precursors is given.

THEORETICAL BACKGROUND

EHDA refers to a process where a liquid jet breaks up into droplets under influence of electrical forces. Depending on the strength of the electric stresses in the liquid surface relative to the surface tension stress, and depending on the kinetic energy of the liquid jet leaving the nozzle, different spraying modes will be obtained (Cloupeau and Prunet-Foch, 1994; Grace and Marijnissen, 1994). For the production of nanoparticles in our case, the so-called Cone-Jet mode is the relevant one. In this mode, a liquid is pumped through a nozzle at low flow rate (μl/hr to ml/hr). An electric field is applied between the nozzle and some counter electrode. This electric field induces a surface charge in the growing droplet at the nozzle. Due to this surface charge, and due to the electric field, an electric stress is created in the liquid surface. If the electric field and the liquid flow rate are in the appropriate range, then this electric stress will overcome the surface tension stress and transform the droplet into a conical shape, the Taylor cone (Taylor, 1964). The tangential component of the electric field accelerates the charge carriers (mainly ions) at the liquid surface toward the cone apex. These ions collide with liquid molecules, accelerating the surrounding liquid. As a result, a thin liquid jet emerges at the cone apex. Depending on the ratio of the normal electric stress over the surface tension stress in the jet surface, the jet will break up due to axisymmetric instabilities, also called varicose instabilities, or because of varicose instabilities and also lateral instabilities, called kink instabilities (Hartman et al., 2000). At a low stress ratio in the varicose break-up mode, the desired monodisperse droplets are produced.

The droplets produced by EHDA carry a high electric charge close to the Rayleigh charge limit (Hartman et al., 2000). To avoid Rayleigh disintegration of the droplets (Davis and Bridges, 1994; Smith et al., 2002), which happens when the mutual repulsion of electric charges exceeds the confining force of surface tension, a result here is the evaporation of the droplets. To make the droplets manageable, they have to be completely or partially neutralized. A possible method of discharging, which is used in this study, is with ions of opposite charge created by corona discharge.

To estimate the right conditions and operational parameters to produce nanodroplets of a certain size, scaling laws can be used. Fernández de la Mora and

Loscertales (de la Mora and Loscertales, 1994) and Gañán-Calvo et al. (Gañán-Calvo et al., 1997) developed scaling laws that estimate the produced droplet size (or jet diameter) and the electric current required for a liquid sprayed in the Cone-Jet mode as function of liquid flow rate and liquid properties. Hartman refined the scaling laws for EHDA in the Cone-Jet mode using his theoretically derived models for the cone, jet, and droplet size (Hartman et al., 1999; Hartman et al., 2000). For the current scaling for liquids with a flat radial velocity profile in the jet, which is appropriate here because of the high conductivity of the solution, Hartman derived the following relation

$$I = b(\gamma K Q)^{\frac{1}{2}} \qquad (2.1)$$

where Q is the flow rate (m³/s), I is the current through the liquid cone (A), γ is the surface tension (N/m), K is conductivity (S/m), and b is a constant, which is approximately 2.

The droplet diameter for the varicose break-up mode is given by equation (2.2).

$$d_{d,v} = c\left(\frac{\rho \varepsilon_0 Q^4}{I^2}\right)^{\frac{1}{6}} \qquad (2.2)$$

where $d_{d,v}$ is the droplet diameter for varicose break-up and c is a constant, which is approximately 2. Substituting equation (2.1) into equation (2.2) yields:

$$d_{d,v} = \left(\frac{16 \rho \varepsilon_0 Q^3}{\gamma K}\right)^{\frac{1}{6}} \qquad (2.3)$$

For a spherical particle, the diameter of the (final) platinum particle (d_p) is related to the droplet diameter (equation [2.3]) by equation (2.4):

$$d_p = \sqrt[3]{f \frac{\rho_{droplet}}{\rho_{particle}} d_{droplet}^3} \qquad (2.4)$$

where f is the mass fraction of platinum in the solution (–), $\rho_{droplet}$ is the density of the solution, and $\rho_{particle}$ is the density of the platinum particle (kg/m³).

This paper describes the production of platinum nanoparticles by EHDA. Other authors report already on the production of nanoparticles by EHDA (Rulison and Flagan, 1994; Hull et al., 1997; Ciach, Geerse, and Marijnissen, 2002; Lenggoro et al., 2000), but besides presenting two methods to produce platinum nanoparticles by EHDA, which is new, our methods can, according to us, be seen as generic ways to produce well-defined nanoparticles of many different compositions on demand.

The two different EHDA configurations, which have been used, relate to the two different routes of the decomposition step of the platinum precursor into platinum. In the first one, the precursor droplets are collected on a support and heat treated afterward. In the second route, the produced precursor droplets are kept in airborne state, neutralized, and heat treated before collection. Platinum nanoparticles produced in this way have already been used in microscale catalytic soot oxidation experiments.

The results of these experiments have been published in a paper by Seipenbusch and others (Seipenbusch et al., 2005).

EXPERIMENTAL

The two production routes of platinum nanoparticles using EHDA are described later. In both routes, the droplets are produced from a solution of chloroplatinic acid ($H_2PtCl_6.6H_2O$ Alfa-Aesar 99.9%) in ethanol. When heated above 500°C, the platinum precursor will decompose into platinum, gaseous hydrochloric acid, and chlorine (Hernandez and Choren, 1983). In the first route, the EHDA-produced chloroplatinic acid particles are deposited on a carrier support. After deposition, the support is placed in a tubular furnace and the particles are decomposed forming platinum nanoparticles. In the second route, the produced droplets are neutralized and ducted in an airborne state through a tubular furnace where they decompose. After ducting into the tubular furnace, the particles are deposited on a substrate, such as a TEM grid.

The two different routes have different setups. The first one, with "off-line heating" is referred throughout the text as the *capillary plate setup* and the second one, with "in-flight heating" as the *aerosol reactor setup*.

CAPILLARY PLATE SETUP

The capillary plate setup is shown in Figure 2.1. Droplets are produced by pumping (Harvard PHD2000) a 1-wt% solution of chloroplatinic acid in ethanol ($K = 4 \cdot 10^{-2}$ S/m, $\gamma = 0.022$ N/m) through a capillary (B). The flowrate of the solution was 13 μl/hr. The required electrical field is created by applying a voltage between the capillary (B) (inner diameter 60 μm, outer diameter 160 μm) and a grounded counter electrode (D) using a high voltage power supply (C) (FUG HCL 14-12500). For the experiments conducted in this study, the potential difference between B and D was 1.26 kV and the distance between the tip of the capillary (B) and the carrier support (E) was 1 mm. The droplets are deposited on the carrier support (E), which in principle can be any material that is heat resistant at the decomposition temperature of chloroplatinic acid and is conductive to discharge the droplets. In this study, thin plates of silicon, with a 0.4-μm oxidized top layer, of about 20 by 20 mm were used as carrier support. The setup was operated at room temperature. After evaporation of the solvent, the support with the chloroplatinic acid nanoparticles was placed in a tubular furnace for 10 min at T = 700°C to decompose the deposited chloroplatinic acid particles into platinum particles. The particles were examined before and after decomposition by an SEM (Hitachi Model S-4700).

AEROSOL REACTOR SETUP

The aerosol reactor setup is shown in Figure 2.2. The setup can be divided in two sections, A and B. Section A is the production part, which is based on the Delft Aerosol Generator (Meesters et al., 1992). In section B, the chloroplatinic acid particles are decomposed, in the airborne state, during their transport through the tubular furnace. A blowup of the production area, section A, is shown in the upper part of Figure 2.2.

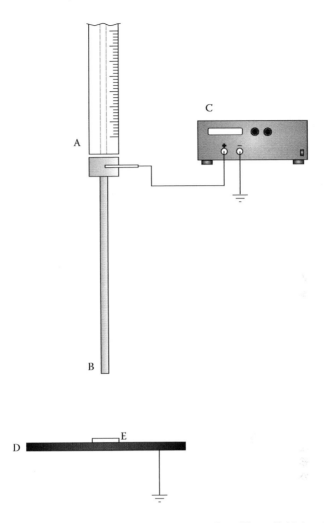

FIGURE 2.1 Capillary plate setup. A. Syringe; B. metal capillary; C. high-voltage power supply; D. grounded plate; E. Si/SiO$_2$ support.

A 0.2-wt% solution of chloroplatinic acid in ethanol (K = 1 · 10^{-2} S/m, γ = 0.022 N/m) was pumped (Harvard PHD2000) through a metal capillary (I.D. 60 µm, O.D. 160 µm) with a flowrate of 8 µl/hr. In this setup, a ring is used as counter electrode. The ring is connected to a high voltage power supply (FUG HCL 14 12500), but at a lower voltage than the capillary, respectively, 5.57 kV and 8.8 kV. The distance between the ring and capillary is approximately 15 mm. The potential difference between the nozzle and the ring creates the field to produce the droplets, which will pass through the ring. In this way, the droplets are not deposited as in the capillary plate setup, but are kept in airborne state.

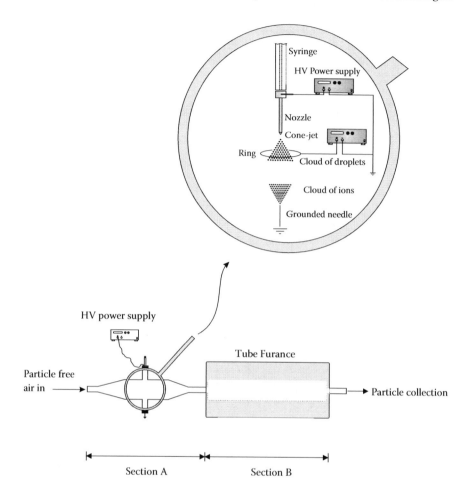

FIGURE 2.2 Aerosol reactor setup. In section A, the particles are generated and dried. In section B, the dried chloroplatinic particles are decomposed to form platinum particles.

To discharge the highly charged droplets, a grounded needle is used in this setup. The needle has a sharp tip and the high electric field strength there creates a corona discharge, supplying ions of opposite charge for the neutralization. The distance between the tip of the needle and the ring is 60 mm.

The chloroplatinic acid particles are then ducted into a tubular furnace ($T = 700°C$) with filtered air ($\phi_v = 1.5$ l/min). The residence time is estimated to be 2 minutes. After ducting into the furnace, the platinum nanoparticles are deposited on a TEM grid. The deposition takes place by two phenomena: thermophoresis and diffusion. In the beginning, thermophoresis is important because the TEM grid is cold compared to the gas. When the grid has been heated up, diffusion will be the dominant process of deposition. After deposition, the nanoparticles are examined by an HR-TEM (Philips CM30UT).

RESULTS AND DISCUSSION

A small area of Si/SiO$_2$ substrate with chloroplatinic acid particles, produced by the capillary-plate setup, is shown in Figure 2.3a. The surface concentration was obtained by spraying for 5 seconds. The spot sizes, as seen in Figure 2.3a, vary between 80 nm–120 nm. Substituting the values of the different variables as described in the experimental section in the scaling laws (equation [2.3]) and using equation (2.4), yields a particle size of 63 nm (here in equation [2.4], f is the mass fraction of the chloroplatinic acid in ethanol, $\rho_{droplet}$ is the density of ethanol, and $\rho_{particle}$ is the density of chloroplatinic acid). Realizing that some deformation might occur during deposition of still wet particles, the measured and calculated values correspond well.

Figure 2.3b shows the particles after the decomposition of the chloroplatinic acid in a tubular furnace for 10 min at 700°C. It can be seen that the original chloroplatinic acid particles are formed into clusters of supposedly platinum particles of 5 to 15 nm. This is caused by the fact that platinum does not evaporate at 700°C, while the other decomposition products are gaseous.

Platinum particles produced by the aerosol reactor setup with the settings mentioned in the previous section are shown in Figure 2.4. In Figure 2.4a, a TEM micrograph of a single particle of approximately 8 nm is shown. The produced particles are not charged and can therefore form agglomerates. An example of such an agglomerate is shown in Figure 2.4b. Elemental analysis using EDX showed that the particles only contain platinum (see Figure 2.5). The TEM pictures also prove that the platinum particles are crystalline. Using the values of the variables as described in the experimental section, the scaling laws (equations [2.3] and [2.4]) predict a particle size of 13 nm. By observing different areas of the TEM grid, we noticed that the particle size of nonagglomerated particles was very similar. To get an estimation of the size, a limited number of particles was measured giving an average size on the order of 10 nm.

Since the aim of this chapter is to show the ability of EHDA to produce (metal) nanoparticles of specified size, in our case, platinum nanoparticles, which are used, for example, for microscale catalytic experiments, we did not try to measure the production rate. Yet we will give an estimation of realizable production rates. For the capillary plate setup, this is straightforward. Most droplets produced will reach the plate. So dividing the flow rate by the volume of the initial droplet gives the number of precursor particles per second. With a flowrate of 13 μl/hr and an initial calculated droplet size of 426 nm, this is $8.9 \cdot 10^7$ droplets per second, yielding through the decomposition step about an order of magnitude more platinum particles of about 10 nm. For the aerosol reactor setup, it is a bit more complicated. Again, the droplet production rate can be estimated by dividing the flowrate of 8 μl/hr by the initial calculated droplet volume (d = 421 nm) giving $5.7 \cdot 10^7$ droplets per second. However, between the droplet production and the collection of platinum particles, different forms of particle losses will take place. The first one occurs because with the configuration used here, the discharging efficiency of the highly charged droplets is not known. Geerse (2003) suggests that it might be low; however, no quantification is given. The nonneutralized fraction may undergo Rayleigh disintegration and/ or deposit on the walls of the setup.

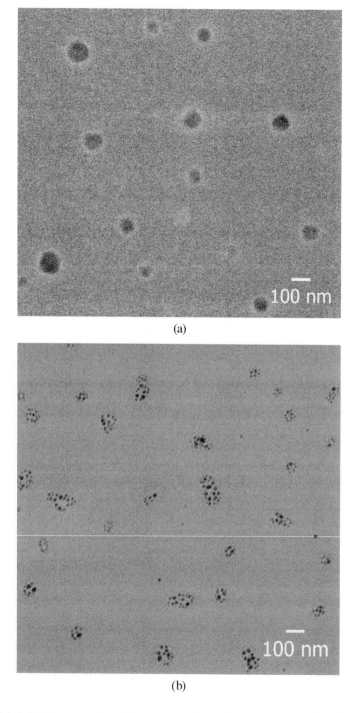

(a)

(b)

FIGURE 2.3 SEM images of particles produced by capillary plate setup; (a) before and (b) after 10 minutes decomposition at 700°C.

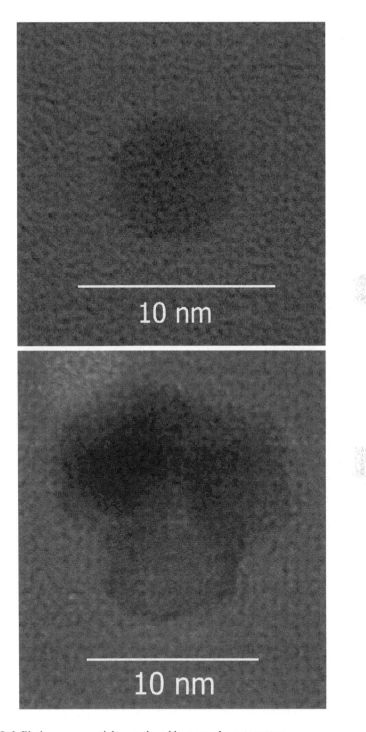

FIGURE 2.4 Platinum nanoarticles produced by aerosol reactor setup.

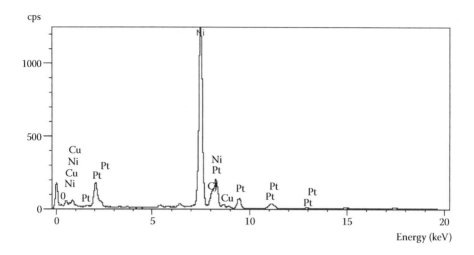

FIGURE 2.5 TEM-EDX spectrum of a platinum particle (Cu & Ni peaks are from TEM grid).

The loss of the neutral particles in the setup is dominated by diffusion and, to a lesser extent, by thermophoresis. Several researchers derived relations for particle penetration P (loss $= 1 - P$) as a result of diffusion to the walls. In these relations the dimensionless deposition parameter μ plays an important role (Hinds, 1999).

$$\mu = \frac{D \cdot L}{Q} \tag{2.5}$$

where D is the diffusion coefficient, L is the distance that the particles travel in a tube, and Q is the volumetric flow rate of the carrier gas. For calculating the deposition parameter, we split the setup in two different sections. The first section is the aerosol generation part with a temperature of 20°C and assumed particle size of 63 nm and the second section is the furnace with a temperature of 700°C and particle size of 10 nm. For the first part $\mu = 2.0 \cdot 10^{-5}$ and for the second part $\mu = 8.6 \cdot 10^{-3}$.

The formula derived by Gormley and Kennedy (1949) can be used for calculating the penetration of both sections

$$P = 1 - 5.50\mu^{\frac{2}{3}} + 3.77\mu \qquad \text{for } \mu < 9 \cdot 10^{-3} \tag{2.6}$$

This gives for our particles a total penetration P of 80%, which means a loss of 20%. So the efficiency of neutralization of the initial droplets is probably the limiting factor in the total production efficiency.

CONCLUSIONS

It has been shown that platinum nanoparticles can be produced by EHDA using two methods. The first method uses the capillary plate setup to deposit precursor particles on a plate and subsequent decomposition will form platinum particles. The second method uses the aerosol reactor setup, which keeps the produced precursor particle aerosolized after discharging the droplets. These particles are ducted through a furnace where they are decomposed, forming platinum particles.

Using the capillary plate method, clusters of platinum nanoparticles were made, with particle size of 5–15 nm. The cluster size depends on the original precursor droplet size and precursor concentration. This method can, for example, be used to deposit very efficiently metal nanoparticles on supports. The production rate for one nozzle is small, but the system can easily be scaled-up by using a number of parallel spraying nozzles. Single and agglomerated nanoparticles of platinum were made using the aerosol reactor setup with a primary particle size of 8–12 nm. From TEM observations, it was found that the latter particles are crystalline and EDX showed that the particles were platinum. In the latter method, different mechanisms causes particle losses between the initial droplet production phase and the final collection of the particles. The main contribution seems to come from the incomplete neutralization of the initial droplets with the configuration used. So it is important to develop a more efficient way of neutralization.

As a final remark, we would like to state that EHDA offers a unique possibility to produce monosized nanoparticles on a small scale. Here we demonstrate the production of platinum nanoparticles, but it is evident that other precursors can be used to produce a wide variety of nanoparticles, if we realize that we can have the reduction step take place under different temperatures and atmospheric conditions. Increasing the production rate can be done by using a number of parallel spraying nozzles. It should also be mentioned that EHDA not only can be employed to produce nanoparticles, but can also be used to make fibers with nanodiameter dimensions (Li and Xia, 2004) and nano (structured) layers (e.g., Chen et al., 1999).

ACKNOWLEDGMENTS

Dr. P. J. Kooyman of DCT/NCHREM, Delft University of Technology, Delft, the Netherlands, is acknowledged for performing the electron microscopy investigations.

REFERENCES

Chen, C.H., Emond, M.H.J., Kelder, E.M., Meester, B., and Schoonman, J., 1999, Electrostatic sol-spray deposition of nanostructured ceramic thin films. *Journal of Aerosol Science* **30**, 959–967.

Ciach, T., Geerse, K.B., and Marijnissen, J.C.M, 2002, EHDA in particle production, in *Nanostructured Materials*, Kanuth, P. and Schoonman, J., eds. Kluwer Academic, Dordrecht, the Netherlands.

Cloupeau, M. and Prunet-Foch, B., 1994, Electrohydrodynamic spraying functioning modes: A critical review. *Journal of Aerosol Science* **25**, 1021–1036.

Davis, E.J. and Bridges, M.A., 1994, The Rayleigh Limit of Charge revisited—Light-Scattering from exploding droplets. *Journal of Aerosol Science* **25**, 6, 1179–1199.

De la Mora, J.F. and Loscertales, I.G. 1994, The current emitted by highly conducting Taylor cones. *Journal of Fluid Mechanics* **260**, 155–184.

Gañán-Calvo, A.M., Davila J., and Barreo, A. 1997, Current and droplet size in the electrospraying of liquids. Scaling laws. *Journal of Aerosol Science* **28**, 249–275.

Geerse, K.B., 2003, Applications of Electrospray: From people to plants. PhD thesis, Delft University of Technology.

Gormley, P.G. and Kennedy, M. 1949. Diffusion from a stream flowing through a cylindrical tube. *Proceedings of the Royal Irish Academy* **52A**, 163–169.

Grace, J.M., and Marijnissen, J.C.M. 1994, A review of liquid atomization by electrical means. *Journal of Aerosol Science* **25**, 6, 1005–1019.

Hartman, R.P.A., Brunner, D.J., Camelot, D.M.A., Marijnissen, J.C.M., and Scarlett, B. 1999, Electrohydrodynamic atomization in the cone-jet mode physical modeling of the liquid cone and jet. *Journal of Aerosol Science* **30**, 7, 823–849.

Hartman, R.P.A., Brunner, D.J., Camelot, D.M.A., Marijnissen, J.C.M., and Scarlett, B. 2000, Jet break-up in electrohydrodynamic atomization in the cone-jet mode. *Journal of Aerosol Science* **31**, 1, 65–95.

Hernandez, J.O. and Choren, E.A., 1983, Thermal stability of some platinum complexes. *Thermochimica Acta* **71**, 3, 265–272.

Hinds, W.C., 1999, *Aerosol Technology,* 2nd ed. John Wiley & Sons, Inc.

Hull, P., Hutchison, J., Salata, O., and Dobson, P., 1997, Synthesis of nanometerscale silver crystallites via a room-temperature electrostatic spraying process. *Advanced Materials* **9**, 5, 413–417.

Lefebvre, A.H., 1989, *Atomization and Sprays.* Hemisphere Publishing, New York.

Lenggoro, I., Okuyama, K., de la Mora, J., and Tohge, N., 2000, Preparation of ZnS nanoparticles by electrospray pyrolysis. *Journal of Aerosol Science* **31**, 1, 121–136.

Li, D. and Xia, Y., 2004, Electrospinning of nanofibers: Reinventing the wheel? *Advanced Materials* **16**, 1151–1170.

Meesters, G., Vercoulen, P.H.W., Marijnissen, J.C.M., and Scarlett, B., 1992, Generation of micron-sized droplets from the Taylor cone. *Journal of Aerosol Science* **23**, 1, 37–49.

Rulison, A.J. and Flagan, R.C., 1994, Synthesis of yttria powders by electrospray pyrolysis. *The Journal of the American Ceramic Society* **77**, 3244–3250.

Seipenbusch, M., van Erven, J., Schalow, T., Weber, A.P., van Langeveld, A.D., Marijnissen, J.C.M., and Friedlander, S.K., 2005, Catalytic soot oxidation in microscale experiments. *Applied Catalysis B: Environmental* **55**, 31–37.

Smith, J.N., Flagan, R.C., and Beauchamp, J.L., 2002, Droplet evaporation and discharge dynamics in electrospray ionization. *Journal of Physical Chemistry A* **106**, 42, 9957–9967.

Taylor, G.I., 1964, Disintegration of water drops in an electric field. *Proceedings of the Royal Society* **A280**, 383–397.

3 Use of Self-Assembled Surfactants for Nanomaterials Synthesis

M. Andersson, A.E.C. Palmqvist, and K. Holmberg

CONTENTS

INTRODUCTION

Synthesis of inorganic materials with nanosized dimensions can take advantage of the ability of surfactants to self-assemble into well-defined structures. These structures are used as a kind of template for the synthesis. This approach of nanomaterials preparation has triggered substantial interest both in the surface chemistry and the materials chemistry community. The accuracy and reproducibility of the self-assembly process has been seen as means of achieving control of materials architecture on the nanometer scale. This is a biomimetic approach; a wide variety of biological structural materials are made by a templating process with the use of surface active compounds—primarily surface active polymers but also low molecular weight polar lipids. The number of papers dealing with surfactant-templated synthesis of inorganic materials has increased dramatically in recent years and several review papers deal with various aspects of the technique (1–12). Existing and potential applications for the synthesized materials range from biomaterials (e.g., artificial bone), to technical products such as catalysts, magnetic particles, and pigments.

This chapter summarizes and discusses the development in the three most important areas of surfactant-assisted synthesis of nanomaterials: microemulsion-based synthesis of nanoparticles, preparation of mesoporous materials from surfactant liquid crystals, and surfactant-mediated crystallization. The latter is often connected to one of the former topics. The primary nanoparticles formed may, in a subsequent step, form superstructures and the surfactant may induce specific morphologies of these. The entire topic is vast and focus is being put on the more recent development.

MICROEMULSION-BASED NANOPARTICLE SYNTHESIS

Boutonnet and others were the first to systematically study the preparation of metal nanoparticles from water-in-oil microemulsions (13,14). The synthesis procedure is simple. Two microemulsions are formulated, one with a metal salt or a metal complex dissolved in the water pools and one with a reducing agent, such as sodium borohydride or hydrazine, present in the pools. Alternatively, if the objective is to prepare nanoparticles of a metal salt AB (e.g., silver chloride), one microemulsion containing a water-soluble salt AX (silver nitrate), and one containing a water-soluble salt YB (sodium chloride) are mixed. Here it is important that the salt YX (sodium nitrate) has a high water solubility to avoid parallel formation of two solid compounds. Due to their small size, the droplets are subject to Brownian motion. They collide continuously and in doing so dimers and other aggregates will form. These aggregates are short-lived and rapidly disintegrate into droplets of the original size. There is also transport of ions in the form of ion pairs through the hydrocarbon domain. As a result of the exchange processes occurring in the system, the content of the water pools of the two microemulsions will become distributed evenly over the entire droplet population. The reaction leading to the solid material, such as a reduction of a metal salt to the metal or formation of an insoluble salt, will occur in the droplets. The overall kinetics of the system are usually such that the particles cease to grow when they have reached a size comparable with the size of the starting microemulsion. Thus, for systems involving fast reaction steps within the droplets, the overall reaction kinetics are mainly governed by transport of species through the hydrocarbon domain and by droplet fusion, which governs the size and the size distribution of the nanoparticles (12). This means that the term "templating and casting" should not be used in a strict sense. The water droplets of the starting water-in-oil microemulsion should not be seen as a mold that is being filled with product during the course of the reaction. It should rather be viewed as a compartmentalization of the reaction solution, which affects the local reagent concentrations and particle nucleation process, the mixing and diffusion rates of reagents, and the interactions between formed nanoparticles in solution.

After completed reaction, a fine suspension of nanoparticles can be obtained and this suspension coexists with a microemulsion consisting of water droplets in oil. Assuming that the water droplets of the two microemulsions used in the preparation of the nanoparticles AB consist of 10% aqueous solutions of the salts AX and YB, that the reaction goes to completion, and that the size of the final particles are the same as that of the droplets of the two starting microemulsions (which is reasonable as a rough approximation), the product mixture will contain suspended particles and salt-free water droplets in a ratio of roughly 10 to 90. The inorganic nanoparticles are most likely surrounded by a water film, and both the particles and the water droplets will be stabilized by a monolayer of surfactant. The aligned surfactants will direct the polar head group into the water and the hydrophobic tails will be oriented into the continuous hydrocarbon domain. Thus, in colloidal chemistry nomenclature, two water-in-oil microemulsions containing different salts in the water pools are transformed into a complex system in which a fine, solid suspension coexists with a water-in-oil microemulsion in which the water pools are more or less salt-free. The process is illustrated in Figure 3.1a.

The particles can be maintained as a fine suspension for a long time. They are sometimes amorphous (15,16) but Pileni, who pioneered the area, has shown that crystalline nanoparticles can be obtained by proper choice of conditions, and, in particular, by using an anionic surfactant with a counterion, Me^+ or Me^{2+}, that is the cation of the metal, Me, to be generated in the reaction (16,17).

Heterogeneous catalysis is one of the most important applications for metal nanoparticles and the catalytically most active sites on the particles are often the metal atoms at edges and boundaries. Reaction in microemulsions is attractive as a method to make catalytically active particles because not only can the particles be made small, but the method also offers some control of both size and shape, which is important for furthering the understanding of activity and selectivity of catalyst nanoparticles. The microemulsion-based synthesis has also been evaluated for preparation of alloys and bimetallic (nonalloy) particles with extremely small dimensions of the individual domains. To prepare an alloy or a mixture of two metals, one may simply combine the two metal salts at the desired molar ratio into one microemulsion and mix this microemulsion with one containing the reducing agent. Alternatively, one may mix three microemulsions, one containing a salt of the first metal, one containing a salt of the second metal, and one containing the reducing agent. As an example, mixed platinum-palladium nanoparticles in the size range of 5–40 nm can be prepared by mixing a nonionic surfactant-based microemulsion containing both platinum and palladium salts in the water pools with a hydrazine-containing microemulsion (18). TEM, in combination with elemental analysis (X-ray energy dispersive spectroscopy), showed that the nanoparticles consisted of a crystalline platinum core onto which very small amorphous palladium particles were bound. A TEM micrograph of such a bimetallic particle is shown in Figure 3.2. The likely reason why platinum formed the core of the particle is that the platinum salt is somewhat easier reduced than the palladium salt. Figure 3.1b illustrates the process for the case where three microemulsions are combined. The resulting platinum-palladium nanoparticles showed high activity for carbon monoxide oxidation. Interestingly, Wu, Chen, and Huang found, in contrast, strong indications that their synthesis, which was based on sodium bis(2-ethylhexyl)sulfosuccinate (AOT) as surfactant and hydrazine as reducing agent, resulted in Pt-Pd alloy particles (19). The reason for this difference is not clear, but the results indicate the importance of suitable surfactant-metal ion interactions for the formation of crystalline alloy nanoparticles in microemulsions. In addition, unpublished work from our group indicates that such suitable interactions, as well as controlled formation kinetics, are important also for the generation of binary

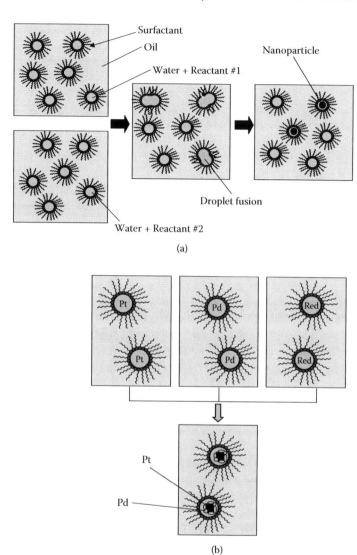

FIGURE 3.1 (a) Preparation of suspended nanoparticles by mixing two water-in-oil micro-emulsions. (b) Preparation of mixed platinum-palladium nanoparticles.

compounds, rather than the kinetically favored single-component nanoparticles, in the Co-Sb and Bi-Te systems (20).

SIZE OF THE NANOPARTICLES

The microemulsions of interest for the synthesis of inorganic nanoparticles are of water-in-oil type; that is, they consist of small water droplets, surrounded by a mono-layer of surfactants and dispersed in a continuous oil domain. (A microemulsion is a

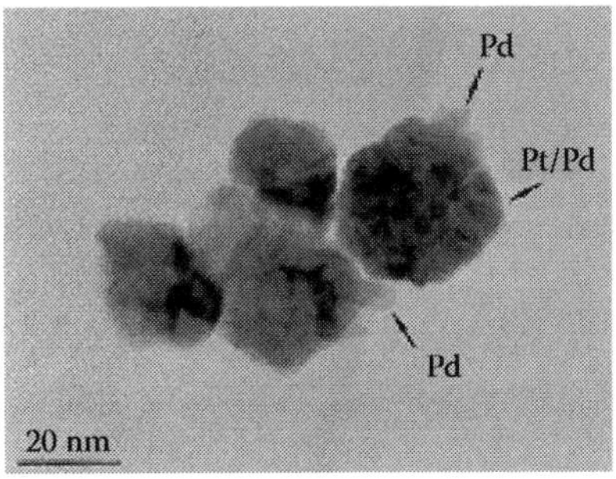

FIGURE 3.2 TEM micrograph of platinum-palladium nanoparticles prepared from a water-in-oil microemulsion.

macroscopic one-phase system; therefore, it is more stringent to call the continuous medium a domain rather than a phase.) The size of the water pools of the microemulsions is a function of the water to surfactant ratio used in the formulation, often denoted W_0. It has been demonstrated that for constant surfactant concentration there is a linear correlation between water concentration and size of the water droplet. Thus, the droplet size is proportional to W_0. For the commonly used surfactant AOT, the droplet radius, Rd, can be directly obtained from the W_0 value by the relationship Rd (nm) = 0.175 W_0 (21). The W_0 value is particularly important in the use of microemulsions as media for bioorganic synthesis because it has been found that a maximum in enzymatic activity often occurs around a value of W_0 at which the size of the droplet is equivalent to, or slightly larger than, that of the entrapped enzyme (22).

As mentioned in the Introduction, the inorganic nanoparticles formed by reaction in water-in-oil microemulsions are often of the same order as the size of the starting microemulsion. In some cases, such as for silver sulfide (23), and sulfated zirconia (24), both prepared in AOT-based microemulsions, there is an almost linear correlation between the W_0 value and the size of the resulting nanoparticles. However, in the majority of cases, there is no straightforward correlation between the droplet and the particle sizes. In several cases, such as various cadmium salts (CdS, CdTe, and CdMnS) (25), platinum (26), silver (23), and copper (23,27), there seems to be some relation between the two at lower water content, that is, at low values of W_0; but as the droplet size increases the particle diameter levels out and stays almost constant above a certain value of W_0 (16).

Particle size control in microemulsion-based synthesis is an intriguing and important issue. Härelind Ingelsten and others have found that, at least for platinum particle formation, the choice of surfactant is not decisive. Approximately the same particle diameters were obtained when an anionic surfactant (AOT) was used when nonionic surfactants (ethoxylated dodecanol with slight variations in the number of oxyethylene groups) were employed (26). A range of experimental findings has been summarized in a review by López-Quintela (12) and extended in a recent paper by Barroso (28):

Particle size increases with reactant concentration.
Particle size decreases if the concentration of one of the reactants is increased far beyond the concentration of the other reactant.

> Particle size increases with an increase in surfactant film flexibility. The film
> flexibility can be increased by several means: incorporating an alcohol as
> cosurfactant, decreasing the oil molecular weight, approaching the insta-
> bility border of the microemulsion, etc.
> Particle size may increase with an increase in microemulsion droplet size (see
> discussion earlier).
> Particle size becomes larger as the critical nucleus increases (28).

The groups of López-Quintela and Tojo have performed Monte Carlo simulations and these seem to correlate well with the experimental results (28–32). The
Monte Carlo simulations were made under the assumption that the chemical reaction
occurring in the microemulsion droplets is much faster than the exchange of materials within the droplets. If the chemical reaction is slow compared to the exchange
rate, a pseudophase model can instead be used to explain the growth kinetics (33).
The Monte Carlo simulations took the following parameters into account (12): the
film flexibility through a parameter f, which is proportional to $\kappa^{-3/2}$ with κ being the
curvature elastic modulus; the interdroplet exchange rate constant; the concentration
of reactants; the reactant excess ratio; the droplet size (which is controlled by the
water-to-surfactant ratio, see discussion before); the volume fraction of the droplets;
and the critical nucleus. In addition, the possibility of autocatalysis of the reaction
by the subnanoparticles formed and the possibility of ripening were included in the
simulations. One important general observation that came out of these and other (34)
Monte Carlo simulations is that particle formation in compartmentalized media is
very different from that in bulk.

If the particle formation is slower than the microemulsion droplet interchange
dynamics, information regarding the reaction kinetics and the influence of the
microemulsion droplet size and shape are obtained when studying the particle formation. Many fewer studies have been performed according to this case and they
often concern more complicated reactions, such as hydrolysis and condensation reactions of alkoxides during the formation of silica particles (33). In a recent study
performed by Andersson, Pedersen, and Palmqvist, a novel microemulsion-based
method for the formation of silver nanoparticles, where the dual functionality of a
nonionic surfactant to act both as template and as reducing agent was utilized (35).
As the reducing agent used is the surfactant in the microemulsion, the silver particle
formation is not influenced by the dynamics of the microemulsion, which made it
possible to study; specifically, the templating effect of the droplets and the kinetics
of particle formation. Using different ratios of the nonionic surfactant Brij 30 (a
fatty alcohol ethoxylate) and the ionic surfactant AOT (which does not give rise to
silver reduction), the influence of the droplet size could be examined. Small angle
X-ray scattering (SAXS) was used for determining the size, size distribution, and
shape of the droplets, as well as the influence of the ratio of the two surfactants. In
addition, the silver nanoparticle size was determined. It was found that the droplet
size decreased with increasing amount of AOT present and that at 100% AOT the
droplets had a cylindrical rather than a spherical shape. Also from the SAXS analysis made on the formed Ag nanoparticles, a general trend toward larger particle size

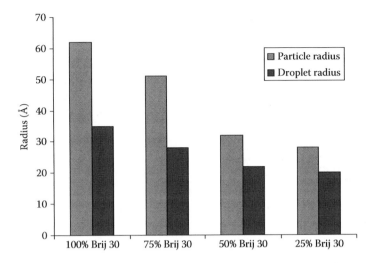

FIGURE 3.3 Microemulsion droplet radius and silver nanoparticle radius obtained by reaction in microemulsions using different combinations (molar ratio indicated) of the alcohol ethoxylate Brij 30 and the anionic surfactant AOT.

was found when the relative amount of Brij 30 was increased. These results support the conclusion that larger microemulsion droplets have a tendency to give larger particles (see Figure 3.3). The study shows that the starting droplet size influences the size of the formed nanoparticles even though the actual sizes are different. The shape of the microemulsion droplets, however, did not appear to influence the shape of the formed silver nanoparticles (35).

SHAPE OF THE NANOPARTICLES

The metal nanoparticles formed in the microemulsion-based synthesis are not always spherical. They can have very specific shapes and the factors that govern the shape have been much discussed in the literature. It has been postulated that the structure of the particles obtained is governed by the structure of the template, that is, spherical water droplets, elongated droplets, rodlike water channel, etc. In some cases, this seems to be the case. A particularly striking example is Pileni's work on preparation of copper nanoparticles from microemulsions with varying internal structures (36,37). Reaction in a microemulsion consisting of spherical water droplets gave spherical particles, the diameter of which increased with increasing value of W_0. Reaction in a microemulsion consisting of interconnected water cylinders yielded cylindrical copper nanocrystals (together with some spherical particles). Finally, reaction in a complex system consisting of a water-in-oil microemulsion coexisting with a lamellar phase gave a mixture of particles with different shapes: spheres, cylinders, flat objects, etc. Another example of how the templated structure governs the structure of the nanocrystals is the synthesis of cadmium sulfide particles in microemulsions where the composition has been changed in a systematic way (38). The structure of the CdS nanoparticles formed nicely matched the structure of the water domains of the starting microemulsion.

The situation is far more complex than that, however. As was discussed earlier, for silver there seems not to be any correlation between the shapes for starting droplets and formed particles. There are many other examples of a poor, or nonexistent, such correlation. There must be other factors influencing the process. One such parameter is the presence of salt, and the anion, in particular, has been found to strongly influence the crystallization behavior in some cases. An excellent example of this is again the formation of copper nanocrystals made from copper bis(2-ethylhexyl)sulfosuccinate (AOT) with sodium as counterion exchanged by the divalent cuprous ion (16,39). Reduction with hydrazine was made in the presence of a constant concentration of NaF, NaCl, NaBr, or NaNO$_3$. The presence of F$^-$ gave small cubes, Cl$^-$ gave long rods, Br$^-$ resulted in cubes, and NO$_3^-$ gave a variety of shapes. The addition of salt did not much affect the internal structure of the microemulsion so the effect is not due to direct templating. The control of the morphology must be related to selective adsorption of the ions on different facets during the crystal growth. There are other examples of how the presence of Br$^-$ induces a cubic morphology of nanocrystals grown in a microemulsion (16,40).

SYNTHESIS OF NANOPARTICLES FROM MICELLAR SOLUTIONS OF SURFACTANTS

Nanoparticles can also be obtained from micellar solutions of anionic surfactants using as counterion the ion of the metal that is to be formed as nanoparticles. The most commonly used surfactant for the purpose is dodecyl sulfate and a wide variety of metal salts have been used as counterion. Copper nanoparticles have been synthesized from a micellar solution of copper dodecyl sulfate using sodium borohydride as reducing agent (41). It has been found that the particle size can be varied by using a mixture of copper dodecyl sulfate and sodium dodecyl sulfate as surfactant. Increasing the amount of sodium dodecyl sulfate while keeping the concentration of copper dodecyl sulfate constant, gave smaller particles, most likely as a result of fewer Cu(II) ions per micelle.

Various types of magnetic particles have also been prepared by this route and the results have been summarized in a review (23). Silver nanoparticles prepared according to this procedure have been commercialized. Such nanoparticles have found use as effective disinfectants.

THE PHASE TRANSFER METHOD OF MAKING NANOPARTICLES

Metal nanoparticles can also be synthesized in a water-hydrocarbon two-phase system using a phase transfer agent to transport the metal ion into the organic phase, a method developed for gold particles by Brust and colleagues and later refined also for silver particles (42–45). The phase transfer agent is usually a lipophilic cationic species, such as a tetraalkylammonium salt, which gives an ion pair that is readily

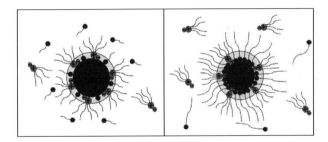

FIGURE 3.4 Schematic model of the palisade layer surrounding the growing platinum nanoparticle. Short-chained alkylamines (*left picture*) have higher solubility in toluene than long-chained alkylamines (*right picture*). The palisade layer obtained with the long-chained alkylamines will be more static and access of new platinum complexes will be restricted as compared to the situation with the short-chained alkylamines. The ion pairs shown in the pictures are composed of negatively charged platinum ions and quaternary ammonium compounds containing three long alkyl chains and one methyl group.

soluble in the organic phase. A reducing agent, such as sodium borohydride, is added to the water phase. The phase transfer agent also transfers water into the nonpolar phase and small metal particles are generated in these aqueous droplets. A stabilizing agent is needed in order to control the growth of the nanoparticles. Most work has been directed toward synthesis of noble metal nanoparticles, of interest for catalysis and several other applications, and alkyl thiols have been the most commonly used stabilizing agent. Sulfur is a poison for many catalysts, however, and alkyl amines of different chain lengths have also been explored as alternative stabilizing agents (46). An inverse relationship was found between the length of the alkyl chain of the alkyl amine and the size of the nanoparticles obtained. It has been postulated that this effect is related to the packing of the amphiphilic molecules in the monolayer surrounding the aqueous droplets. Whereas a long-chain alkyl amine is likely to form a tightly packed palisade layer together with the phase transfer agent, shorter alkyl amines, which are more soluble in the hydrocarbon phase (which is often toluene), will not give such an ordered interface. Transport of metal ions in the form of an ion pair will occur more readily through the less ordered palisade layer, leading to larger particles, as is schematically shown in Figure 3.4 (46).

DEPOSITION OF THE PARTICLES ON A SOLID SUPPORT

Heterogeneous catalysis has from the beginning (13,14) been seen as the most prominent potential application for the synthesized nanoparticles but many other promising uses exist in areas such as electronics, information technology, and biotechnology (47). Almost invariably, the nanoparticles formed need to be deposited on some kind of solid support. For catalyst applications, the particles, usually noble metals such as platinum or palladium, are deposited on a carrier material, such as γ-alumina, silica, or titania. There are various ways in which this can be done. One procedure is to add a solvent that dissolves the surfactant and is miscible with both oil and

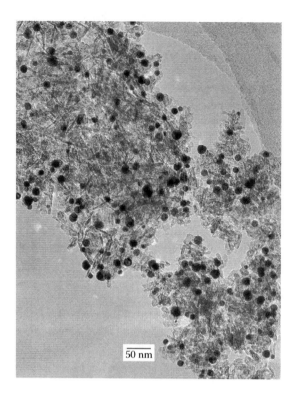

FIGURE 3.5 Platinum nanoparticles (black spots) deposited on alumina.

water. The surfactant will then be removed from the surface of the small particles and precipitation will occur due to gravitational forces. Tetrahydrofuran (THF) is a suitable solvent for the purpose and the procedure has been described in several papers (13,14,48). The support material, in the form of a powder, is usually added first and the solvent is subsequently poured into the mixture, which is vigorously stirred. The ready-made catalyst, that is, the support onto which the noble metal particles are deposited, can be filtered off and the surfactant can subsequently be removed by rinsing with more solvent. It has been found that the rate at which THF is added is crucial: Too fast a rate will lead to extensive particle agglomeration (48). Under optimized conditions, a relatively even distribution of noble metal particles across the support surface can be obtained although some agglomeration into larger particles inevitably occurs. Figure 3.5 shows a typical TEM micrograph of a calcined Pt/Al$_2$O$_3$ sample.

Kim and others (49,50) and Ikeda and colleagues (51) have described an alternative deposition method. The support is first synthesized using a microemulsion as reaction medium. A metal alkoxide precursor of the metal oxide support is added to a microemulsion that contains ammonium hydroxide. The alkoxy group will be removed by hydrolysis and the metal oxide will form. A microemulsion containing the noble metal salt and another microemulsion containing a reducing agent are added during the course of formation of the support material. The noble metal particles formed will spontaneously adhere to the metal oxide surface. The technique has the drawback that some of the metal particles become embedded in the support material, which reduces the available surface area of the active phase (50).

A third deposition method involves centrifugation of the nanosized particles followed by redispersion in water with the help of a surfactant (48). The dispersion is then added to an alumina slurry. By selection of surfactant and by control of pH, the charge on both the noble metal particle and the alumina support can be controlled. (Alumina has a point of zero charge of 8.5–9.0.) It seems that a weak interaction between the alumina and the dispersed nanoparticle, as obtained with a nonionic surfactant as dispersing agent, gives the best result. In such a system, surface-induced agglomeration of the primary particles is suppressed and the catalytic activity (carbon monoxide oxidation over Pt/Al_2O_3) becomes high.

MESOPOROUS MATERIALS

Mesoporous materials are defined as materials that have at least one dimension in the size range 2–50 nm (by mesoporous materials we mean highly ordered mesoporous materials). Biological systems exhibit numerous such structures that may be organic or inorganic or, which is very common, organic-inorganic hybrid materials. Potential technical applications of mesoporous materials are separation, adsorbents, catalyst supports, electrodes for solar cells and fuel cells, etc.

Synthesis of mesoporous materials by the use of surfactants as structure-directing agents started around 1990 when it was demonstrated that mesoporous silica could be obtained by the templating action of cationic surfactants (52–54). A tetraalkylorthosilicate, such as tetraethylorthosilicate (TEOS), was used as silica precursor and silica with pores in the mesoporous regime and with relatively narrow size distribution could be obtained by proper choice of surfactant, surfactant concentration, and pH. Either basic or acidic conditions can be used for hydrolysis of the precursor molecule, which initiates the polymerization. The number of research papers dealing with the topic has grown tremendously during the last decade, as is shown in Figure 3.6, which is based on data compiled through a search in the database SciFinder on the topic mesoporous silica. The majority of papers deal with mesoporous silica but a substantial number of other oxides, such as alumina, titania, and zirconia, and mixed oxides, have also been synthesized using the same general method. There are a number of recent reviews that cover the topic (8,10,11,55). Important early contributions have been made by the groups of Stucky and Chmelka at the University of California–Santa Barbara (56–58). The surfactant-assisted route of making mesoporous materials has proven versatile and simple to use. The mechanism involved in the formation of the materials has been subject to considerable debate, however.

FORMATION MECHANISM

A relatively recent review by Patarin, Lebeau, and Zana summarizes the current view on the formation mechanisms (10) and an earlier review on the same topic was published in 1999 (59). Already in the first publication by Beck and others (52) it was stated

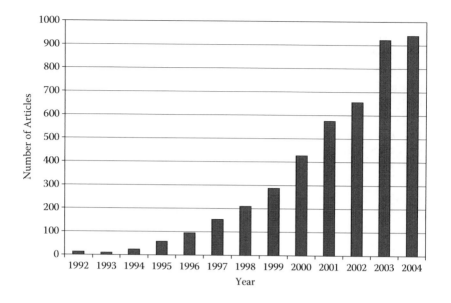

FIGURE 3.6 The development of the research field of mesoporous silica presented as number of articles published per year in journals covered by SciFinder.

that there are two alternative routes to arrive at the mesoporous material, as illustrated in Figure 3.7. In the first of these (Figure 3.7, Route 1), a relatively low concentration of a cationic surfactant is used. Using a spin probe (5-doxyl stearic acid) to sense the motional changes of the surfactant molecules by Electroparamagnetic Resonance (EPR) during the course of reaction, the process has been shown to proceed in two distinct stages (60,61). The first stage, which occurs over 10–15 minutes after addition of TEOS to the micellar solution of a quaternary ammonium surfactant, is the formation of silicate oligomers, which are negatively charged and which interact with the positively charged surfactant micelles. Thus, the water-soluble oligomeric silica replaces the surfactant counterion and becomes a polymeric counterion for the surfactant micelle and in doing so it changes the spontaneous curvature of the self-assembled surfactant (62). The change in curvature induces a change from micellar solution into a hexagonal packing of the surfactant with an aqueous solution of oligomeric and polymeric silica constituting the surrounding water domain. Thus, silicate condensation into higher molecular weight polymers and orientational ordering of the surfactant assemblies occur simultaneously.

After the rapid first stage a slower process follows, which involves further silica condensation. This leads to stiffening of the water domains that surround the hexagonal arrays of self-assembled surfactant, a process that can be monitored by EPR as a slow increase of the order parameter of the spin probe while the rotational parameter stays constant (60). Approximately the same picture of the formation mechanism emerged from a study where in situ EPR was combined with X-ray diffraction (63). It was shown that there is a hydrophobicity gradient in the materials: The sides of the pores, which are rich in silanol groups, are hydrophilic whereas the siloxane-rich pore corners are hydrophobic.

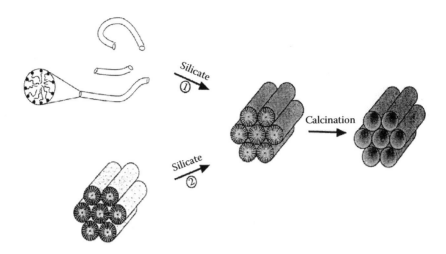

FIGURE 3.7 Two routes of preparing mesoporous materials. In Route 1, a solution of rodlike micelles is transformed into a surfactant liquid crystalline phase of hexagonal geometry by the influence of silicate oligomers and polymers generated from a precursor such as tetraethylorthosilicate. In Route 2, a silicate precursor is added to a preformed hexagonal surfactant liquid crystalline phase.

In another work, Zana and coworkers proposed a slightly different mechanism of formation of the mesoporous material (64). Based on fluorescence probing techniques, they showed that addition of oligomeric silicate did not much affect the size and shape of cationic surfactant micelles. They also showed that the exchange of surfactant counterion from bromide to silicate only occurred to a small extent. In light of this, they concluded that the exchange process is not the decisive event. They proposed instead that the silicate pre-polymers formed interact with surfactant molecules in a cooperative manner, forming mixed micellar aggregates. This picture is analogous to the generally accepted view of how organic polyelectrolytes interact with oppositely charged surfactants. Thus, the mechanistic hypothesis put forward by Zana and colleagues differs from that discussed earlier in that the silicate oligomers (or pre-polymers) interact with individual surfactant molecules, not with entire micelles. The micelles simply act as reservoirs for the free surfactant species. The situation is complex, however, and the mechanism may differ from case to case. The same group has studied formation of mesostructured, hexagonal alumina using the anionic surfactant sodium dodecyl sulfate as templating agent (65). Again using fluorescence probing techniques, it was found that the positively charged inorganic pre-polymer remained bound to the micellar surface via electrostatic interactions. Just prior to precipitation there is a change from a micellar structure to a hexagonal array.

Route 2 of Figure 3.7 shows the second path to mesoporous materials. This involves formation of a surfactant liquid crystalline phase as a ready-made template followed by polymerization within this template. The principle of starting with a surfactant liquid crystal, solidifying the aqueous domain and producing the porous material by removal of the surfactant is illustrated in Figure 3.8. This approach has

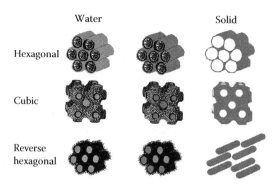

FIGURE 3.8 Schematic illustration of how a surfactant liquid crystal is first solidified and then transformed into a mesoporous material by removal of the surfactant. Surfactant removal can be made either by washing or by calcination.

mainly been used for the preparation of mesostructured hexagonal materials but materials with a bicontinuous cubic structure have also been synthesized starting from surfactant liquid crystals with cubic geometry. Nonionic surfactants, in particular block copolymers made up of hydrophilic polyoxyethylene segments and hydrophobic polyoxypropylene segments, have been popular. Recent development in the field has been compiled in a comprehensive review (11).

The latter method, sometimes referred to as the direct templating method, appears more straightforward than the route starting from a micellar solution. Once the aqueous phase behavior of the surfactant, in the presence of the polymer precursor, has been established, a composition is chosen such that the mixture forms the desired phase; for example, a hexagonal phase and polymerization is initiated by a change in pH. The phase transition from micellar solution to liquid crystalline phase, which might be difficult to predict and which is not always fully understood (see discussion before), is avoided. The situation is not always without complications, however. Sometimes the polymerization process affects the phase behavior such that the composition gradually moves away from the liquid crystalline phase, which is the intended template structure. For example, Su and coworkers prepared silica materials in a preformed hexagonal liquid crystalline phase made from nonionic surfactants of the fatty alcohol ethoxylate type (66,67). The reaction gave a disordered wormlike mesoporous material. When the surfactant concentration was reduced to a value below that required for the hexagonal liquid crystal to form, the reaction led to formation of a well-ordered mesoporous material.

A somewhat different approach for the formation of mesoporous materials is the so-called evaporation-induced surfactant self-assembly (EISA) technique developed by Brinker and colleagues. Here a mixture of surfactant, acid, and a metal alkoxide precursor, which have the compositions corresponding to the wanted liquid crystalline phase, is prepared. The mixture is dissolved in excess of ethanol, which results in a low viscous solution. Then, upon evaporation of ethanol, the liquid crystalline phase is reformed. It can then act as a template according to the direct templating method, as described above. Utilizing this technique, thin films of highly ordered mesoporous

BF   DF

FIGURE 3.9 TEM micrographs of mesoporous titania with incorporated silver. The *left picture* shows a bright field (BF) micrograph of the highly ordered mesoporous cubic structure of titania. The dark spots are silver particles situated within the titania. The *right picture* shows a dark field (DF) micrograph taken on mesoporous cubic titania. Here the bright spots are titania crystallites and the very dark spots are silver.

material can be formed. For example, partly crystalline mesoporous titania thin films, with silver nanoparticles incorporated within the pores, supported onto glass slides have been produced (68). Figure 3.9 shows TEM micrographs of thin silver-containing mesoporous titania films. This material was found to be photocatalytically active as evidenced by the photodegradation of stearic acid as a model reaction.

The two general routes to arrive at mesoporous materials have been compared by Coleman and Attard (69). They prepared mesoporous silica in aqueous solutions of a nonionic surfactant of fatty alcohol ethoxylate type with a systematic variation of the surfactant concentration. Starting with a dilute micellar solution and increasing the concentration while remaining in the micellar region gave a progression from a low-ordered to a high-ordered hexagonal structure up to a certain surfactant concentration above which there was a decrease in ordering (still at a concentration below the limit of the micellar region in the binary surfactant-water system). Increasing the concentration further so that the composition reached the lower limit of the liquid crystalline region of the binary system again gave a material with high order. A further increase of surfactant concentration reduced the order. This set of experiments is a nice piece of evidence for the existence of the two separate routes to synthesize the mesoporous materials, one starting with a micellar solution and the other one starting with a surfactant liquid crystalline phase. The work is also a good illustration of how delicate the process is with respect to surfactant concentration. Obviously, each synthesis requires a trial-and-error exercise in order to optimize the reaction conditions. Recent progress has been made in predicting syntheses recipes for mesoporous metal oxides using surfactant phase diagrams (70,71).

STRUCTURE OF THE MESOPOROUS MATERIAL

Since the mesoporous materials are made from templates of surfactants, either formed *in situ* or preformed (Routes 1 and 2, respectively, of Figure 3.7), and since the size of these templates are known, the pore dimension of the mesoporous material can, at least in principle, be tailor-made. The surfactant orients itself in a liquid crystal such that its hydrophilic part extends into the aqueous domain, that is, the domain that solidifies during the course of the reaction. The hydrophilic blocks seem to be firmly anchored in the inorganic domains, as evidenced by substantial restriction of the molecular mobility, while the hydrophobic blocks remain flexible (72,73). The surfactant hydrophobic tail will make up the core of the structure and will remain unaffected by the polymerization/solidification. After completed reaction, the surfactant is either washed away or burned off to give the porous material. When the surfactant is removed there will be a void where the surfactant hydrophobic tails had been. Since a surfactant in lyotropic liquid crystals, of both hexagonal and cubic geometry, aligns in bilayers with the hydrophobic tails pointing toward each other, the pore dimension should equal twice the length of the surfactant tail. For a block copolymer of polyoxyethylene-polyoxypropylene-polyoxyethylene (EO-PO-EO) type, the pore dimension should equal the length of the polyoxypropylene block. This is often the case but it has been shown that the pore dimension is not always independent of the length of the polyoxyethylene segment of EO-PO-EO block copolymers (and probably for nonionic surfactants in general). Increasing the polyoxyethylene block length while keeping the polyoxypropylene block constant or adding EO homopolymer, that is, poly(ethylene glycol), leads to an increase in the mesopore diameter (74). In addition to the mesopores with dimensions in the size range 2–50 nm, the material produced from templates of nonionic surfactants and block copolymers usually contains pores of much smaller dimension, so-called micropores. These are believed to originate from the protrusion of the polyoxyethylene blocks into the inorganic domains.

A number of parameters influence the structure formed: choice of precursor, choice of surfactant, presence of specific ions, condensation rate (which, in turn, is mainly governed by the pH and the temperature used). It seems that a low reaction rate favors formation of a large, well-ordered crystalline material. For instance, faceted single crystals of mesoporous materials with cubic geometry were obtained by running the reaction for one week at 0°C, instead of shorter time at room temperature, in which case a less-ordered product was obtained (75). It is likely that the control of crystal morphology that can be obtained by low temperature is due to the reaction then proceeding under thermodynamically controlled conditions, which is not always the case otherwise (76).

It is well known that the solution behavior of ethoxylated nonionic surfactants is very temperature sensitive. This also holds true for EO-PO-EO block copolymers. An increase in temperature leads to a dehydration of the polyoxyethylene chain or chains, leading to a reduction in size of this block. This effect on the surfactant packing parameter leads to a change in the spontaneous curvature of a self-assembled surfactant film. Increased temperature will lead to a transition from less elongated to more elongated micelles and to a transition from hexagonal, via micellar cubic and lamellar, to bicontinuous cubic liquid crystals. Using the direct templating method (Figure 3.7, Route 2), one would expect that the temperature-controlled mode of self-assembly in solution would be transferred to the corresponding structures in the mesoporous material. Using an EO-PO-EO block copolymer as template this has been found to be the case (77). Several recent studies, from Stucky's group (78) and from others (11), have shown that the rules known from the solution chemistry of surfactants can be used to give good predictions of the structure of the mesoporous material.

Inorganic ions may drastically affect the crystallization pattern of nanoparticles and this phenomenon was briefly discussed earlier. Similar effects have been seen in the formation of mesoporous materials and a correlation between effect on curvature and position in the so-called Hofmeister series has been demonstrated (79). A striking example of the effect of inorganic anions on the structure obtained is found in a recent paper by Che and others (80). Synthesizing silica from triethylorthosilicate with the cationic surfactant cetyltrimethylammonium bromide as templating species gave different structures depending on the acid that was used for hydrolysis of the precursor. HBr gave a 2D-hexagonal *p6mm* structure, H_2SO_4 resulted in formation of facetted single crystals of 3D-hexagonal *P6$_3$/mmc* geometry, HCl gave the cubic *Pm-3n* mesostructure, and HNO_3 yielded bicontinuous mesoporous silica with *Ia3d* symmetry. Obviously, the difference in interaction of the different anions with the surfactant headgroup layer influences the critical packing parameter, and, thus, the spontaneous curvature, such that mesoporous structures with distinctly different geometries are obtained.

For applications that involve transport in or out of the material, the bicontinuous cubic phase is of special interest because of its symmetrical three-dimensional structure. In an attempt to prepare mesoporous alumina with bicontinuous, cubic symmetry, of interest as catalyst support material, a monoolein-based bicontinuous phase was used as template (81). Monoolein is a particularly suitable amphiphile for the purpose because in the binary monoolein-water system, a large body-centered phase is formed with the surfactant bilayer forming an infinite periodic minimal surface of the gyroid type (space group *Im3m*) (82). Aluminum nitrate was used as starting material and polymeric alumina was formed by careful adjustment of the pH to around 10. Removal of the surfactant gave a mesoporous material with a narrow pore size distribution centered at 38 Å, which fitted well with the length of two extended hydrophobic tails. Mesoporous silica with a similar bicontinuous cubic structure was later synthesized using an EO-PO-EO block copolymer as template (83). Under normal conditions this procedure leads to a two-dimensional hexagonal structure but addition of NaI directed the synthesis into the material with cubic symmetry, another example of the effect of salt addition. The mesostructured material obtained had unusually large unit cell dimensions, which is interesting from a practical point of view.

The vast majority of work on mesoporous materials deal with silica and the silica source is usually an alkoxide, such as tetraethylorthosilicate (TEOS) or tetramethylorthosilicate (TMOS). These alkoxides are convenient as starting material but they are expensive and the products made from them can only find use in applications where price is not a major issue. Thus, mesoporous silica produced from an alkoxide as precursor has a price level that limits large-scale applications.

The economic considerations have recently triggered an interest in the use of inexpensive inorganic silicate as starting material. Aqueous sodium silicate, water glass, is a commodity chemical with a price level an order of magnitude lower than that of the alkoxides. Use of such raw materials instead of the organosilicates will radically change the production price of the products and may open up completely new fields of application. Preparation of mesoporous silica from inorganic silica sources has been the subject of a recent review (84). Recent results from our laboratory indicate

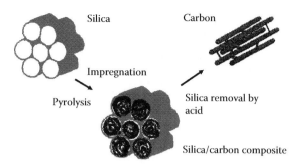

FIGURE 3.10 Preparation of mesoporous carbon from mesoporous silica.

that mesoporous materials with a high degree of order can be obtained by both of the routes that are illustrated in Figure 3.7.

MESOPOROUS CARBON

Hydrophobic mesoporous material can also be made in the form of carbon through a double templating method. Hexagonal mesoporous silica, obtained in the normal way, is impregnated with a polymerizable, carbon-rich material. Furfuryl alcohol has been used for the purpose (85). The organic material is polymerized, yielding a continuous organic polymer in the formerly empty channels of the mesoporous material. Calcination of the material transforms the polymer into graphite. There is some shrinkage during the polymerization and calcination procedures, and best results are obtained if the cycle of addition of organic material, polymerization, and calcination is repeated a few times (86,87). The silica is subsequently removed by treatment with hydrofluoric acid, leaving mesoporous carbon with cubic geometry. This multistep procedure, which is illustrated in Figure 3.10, results in a surprisingly narrow pore size distribution, as can be seen from Figure 3.11. Such mesoporous carbon is of interest as electrode material in fuel cells, etc.

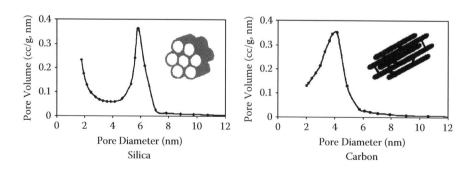

FIGURE 3.11 Pore size distribution curve of mesoporous carbon and of the silica used as template in the synthesis.

SURFACTANT-MEDIATED CRYSTALLIZATION

Self-assembled surfactants can affect crystallization of metal salts and oxides, often in a way that is difficult to predict. Not only the surfactant but also the counterion of an ionic surfactant, or the presence of extra salt, can be used for the purpose, as discussed earlier. In bioceramic products such as bone, teeth, shellfish shells, etc., nature provides a very high degree of control of the growth of the crystals and this is believed to be the key to the excellent mechanical properties of these products. There is no doubt that important properties of man-made ceramic product can be improved if the crystals that make up the products were more homogeneous with respect to morphology, size, and shape.

The formation of titanium dioxide crystals by hydrothermal treatment of a micro-emulsion is a striking example of how a surfactant can affect the crystallization pattern (88,89). TiO_2 exists either as anatase, brookite, or rutile. Using titanium(IV) butoxide as precursor and a nonionic surfactant as templating agent, it was found that anatase was formed exclusively when HNO_3 was used to catalyze the hydrolysis of the precursor while rutile was obtained when HCl was employed as acid. The anatase consisted of spherical particles with narrow size distribution and the rutile consisted of elongated needles. The reaction occurred at a temperature of 120°C, which was much above the existence region of the starting microemulsion (89). Hence, there is no direct templating in this case. Both the rutile and the anatase were found to be photochemically active but they catalyzed the breakdown of a test substance, phenol, by different mechanisms (89).

The wide applicability of surfactant self-assembly for inorganic structure direction has been further studied for various forms of calcium carbonate structures that have been synthesized using different types of liquid crystalline phases resulting in the formation of different morphologies and polymorphs (90). Here the approach was to form the desired liquid crystalline phase from an aqueous solution of calcium nitrate using ammonia to adjust the pH. The mixture was placed in an autoclave and exposed to carbon dioxide gas. The gas dissolves in the water, generating CO_3^{2-}, which reacts with calcium ions and forms $CaCO_3$. At low concentrations of $CaCl_2$, the $CaCO_3$ product formed showed high surface areas, predominantly of the vaterite structure with crystallites of fibrillar shape (Figure 3.12a and Figure 3.12c). At high concentrations, the $CaCO_3$ formed was instead mainly calcite with crystallites having flake morphologies (Figure 3.12b and Figure 3.12d) and lower specific surface areas. Also X-ray diffraction showed that the samples of low surface areas consisted predominantly of calcite while samples with high surface areas consisted of vaterite. It was concluded that the vaterite form of calcium carbonate is amenable for templated growth, while the calcite polymorph proved less susceptible to templating. The fibrils formed in the reverse hexagonal phase were found to have a diameter of approximately 8–10 nm, comparable to the water channels (8 nm) in this phase. The diameter of these fibrils remained the same, regardless of the specific surface area. The $CaCl_2$ concentration seems to be determining the ratio of calcite/vaterite in the product and thus the specific surface area of the calcium carbonate samples formed.

It has been shown that the surfactants do not only act as templates in the formation of nanosized particles but they also influence the self-assembly of these into

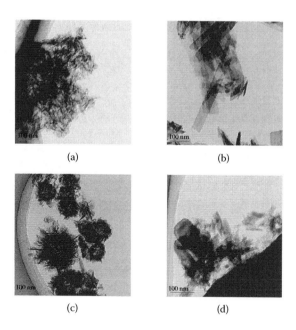

(a)   (b)

(c)   (d)

FIGURE 3.12 TEM micrographs of the products from the syntheses of $CaCO_3$ in hexagonal liquid crystals using a $CaCl_2$ concentration of (a) 0.5 wt.% and (b) 20 wt.%, and in reversed hexagonal liquid crystals using a $CaCl_2$ concentration of (c) 0.5 wt.%, and (d) 20 wt%.

superstructures at different hierarchical levels (91). Crystallization of barium sulfate, barite, in a microemulsion based on AOT as surfactant gave long crystalline nanofibers, which self-assembled into superstructures (92). When another barium salt, barium chromate, was used instead, primary cuboids that organized in well-defined morphologies were obtained (91). The control of the ordering of the nanocrystals is believed to be due to a soft epitaxial match between crystal structure and surfactant packing, resulting in an enthalpic driving force for formation of the specific superstructure.

Similar ordering of primary particles has been demonstrated by Andersson and coworkers for the case of silver (93). A silver nitrate salt was slowly reduced in a reverse hexagonal phase made from a surface-active polymer consisting of polyoxyethylene and polyoxypropylene blocks ($EO_{20}PO_{70}EO_{20}$). Silver nanoparticles, 3 nm in diameter, were formed and these self-assembled into fibers, several millimeters long. TEM micrographs clearly showed the spontaneously aligned particles, Figure 3.13. An interesting observation made in this work is that the silver particles could be formed without any added reducing agent. The surfactant acted both as template and as reducing agent (93,94). It was later confirmed that surfactants containing polyoxyethylene chains have the ability to reduce a noble metal salt to the metal while at the same time becoming oxidized. The oxidation occurs via hydroperoxide formation and results in the formation of aldehydes and other breakdown products (95).

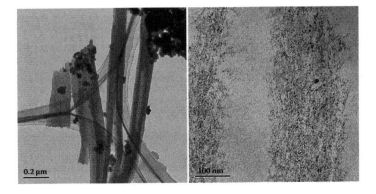

FIGURE 3.13 Alignment of silver nanoparticles into fibers. The silver particles were generated in a reverse hexagonal surfactant liquid crystal. The *right picture* shows that the fibers consist of aligned nanoparticles.

Surfactants made up by two hydrophilic blocks, so-called double-hydrophilic surfactants, have been found to possess interesting qualities with respect to inducing crystallization. These molecules are not surfactants in the true meaning of the word since they are not very surface-active and they do not much reduce the oil-water or air-water interfacial tensions. One of the hydrophilic blocks contains functional groups that interact strongly with a specific mineral surface. Thus, a double-hydrophilic surfactant adsorbs at the surface of such minerals with one block while the other block extends out in solution and may assist in structure formation. The reduction in interfacial energy obtained by soft epitaxy can drastically affect the crystallization process. The concept, explored by Qi, Cölfen, and Antonietti (96,97), is versatile since one can obtain a high specificity by the choice of functional group in the polymer segment. Barium sulfate, barite, has been crystallized in the presence of polymers with varying functional groups and a range of different superstructures was obtained. The double-hydrophilic surfactants contained poly(ethylene glycol) (PEG) as "inert" block and blocks with carboxylate groups, amino acid residues, sulfonate groups, or phosphonate groups as functionalized block. Typical examples are

- Poly(ethylene glycol)-*block*-poly(ethylene imine)-poly(acetic acid)
- Poly(ethylene glycol)-*block*-poly(ethylene imine)-poly(sulfonic acid)
- Poly(ethylene glycol)-*block*-poly(methacrylic acid)
- Poly(ethylene glycol)-*block*-poly(aspartic acid)
- Partially phosphonated poly(ethylene glycol)-*block*-poly(methacrylic acid)

The control of the morphology when the crystals are grown under the influence of the specific surfactant is remarkable. For instance, the phosphonated copolymer, at pH 5, gave large bundles of barite nanofilaments ranging from 20 to 30 nm in diameter, each of which is a single crystal elongated along the crystallographic direction. The ordering is probably due to preferential adsorption of the phosphonated block copolymer on specific crystalline faces.

Several studies show that the surfactant not only affects the crystallization into single particles, but also influences the ordering of these into complex superstructures. This can lead to the most amazing structures and the phenomenon has been referred to as "brick and wall control" (8). Only small variations of the structure-ordering surfactant can have large effects. As an example, crystallization of barite in the presence of a double-hydrophilic surfactant-containing carboxylate groups on the functionalized block can give very different final morphologies depending on the pH, that is, on the degree of deprotonization of the carboxylic acid groups, see Figure 3.14 (97). It is believed that the difference in the packing of the subcrystals is caused by differences in probability of nucleation on the side-surfaces of the rodlike primary particles. This, in turn, depends on the strength of the binding of the functionalized block to these surfaces.

The use of surfactants with specific functional groups to control the crystallization process bears resemblance with the use of surface-active polymers to inhibit crystallization of calcium carbonate and other metal salts in process water. This procedure, usually referred to as antiscaling, is routinely used in many industrial processes, such as wood pulping. Kjellin and colleagues have demonstrated that polymers containing one polyoxyethylene block and one polyoxypropylene block (EO_7PO_{31} and $EO_{41}PO_{31}$) and containing a diphosphate group at one end reduce the amount of calcium carbonate scaling (98,99). The effect is most pronounced near the cloud point, where the polymer is most surface-active. Most likely, the surface-active diphosphate functionalized polymer binds to the growing $CaCO_3$ crystal by a combination

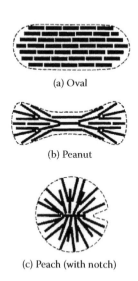

(a) Oval

(b) Peanut

(c) Peach (with notch)

FIGURE 3.14 Schematic illustration of how barite subcrystals can grow into particles of different shape: (a) oval, through growth without side-surface nucleation; (b) peanut, through side-surface nucleation and growth at limited tilt angles; (c) peach, with side-surface nucleation and growth at random tilt angles. (From Qi, L.M., Cölfen, H., and Antonietti, M., *Chem. Mater.* 12, 2392, 2000. With permission.)

of complexation with lattice calcium and adsorption due to the hydrophobic effect, thus affecting the crystallization pattern of the calcium carbonate. The functionalized block copolymer also influences the ratio of calcite to vaterite formed. The more hydrophobic polymer, that is, EO_7PO_{31}, completely inhibited vaterite growth.

ACKNOWLEDGMENTS

We thank Per Kjellin, Kjell Wikander, Andreas Berggren, and Tomasz Witula for putting results from current research at our disposal.

REFERENCES

[1] M.-P. Pileni, J. Phys. Chem. 97 (1993) 6961.
[2] V. Pillai, P. Kumar, M.J. Hou, P. Ayyub, D.O. Shah, Adv. Colloid Interface Sci. 55 (1995) 241.
[3] J. Sjöblom, R. Lindberg, S.E. Friberg, Adv. Colloid Interface Sci. 95 (1996) 125.
[4] A. Corma, Chem. Rev. 97 (1997) 2373.
[5] M.-P. Pileni, I. Lisiecki, L. Motte, C. Petit, J. Tanori, N. Moumen, in: D.O. Shah (Ed.), Micelles, Microemulsions and Monolayers, Marcel Dekker, New York (1998) 289.
[6] D.M. Dabbs, I.A. Aksay, Annu. Rev. Phys. Chem. 51 (2000) 601.
[7] A. Stein, B.J. Melde, R.C. Schroden, Adv. Mater. 12 (2000) 1403.
[8] M. Antonietti, Curr. Opin. Colloid Interface Sci. 6 (2001) 244.
[9] K. Holmberg, B. Jönsson, B. Kronberg, B. Lindman, Surfactants and Polymers in Aqueous Solution, 2nd ed., Wiley, Chichester (2002).
[10] J. Patarin, B. Lebeau, R. Zana, Curr. Opin. Colloid Interface Sci. 7 (2002) 107.
[11] A.E.C. Palmqvist, Curr. Opin. Colloid Interface Sci. 8 (2003) 145.
[12] M.A. López-Quintela, Curr. Opin. Colloid Interface Sci. 8 (2003) 137.
[13] M. Boutonnet, J. Kizling, R. Touroude, G. Marie, P. Stenius, P., Catal. Lett. 9 (1991) 347.
[14] M. Boutonnet, J. Kizling, P. Stenius, Colloids Surfaces 5 (1982) 209.
[15] H. Weller, Angew. Chem. Int. Ed. 32 (1993) 41.
[16] M.-P. Pileni, Nature Mater. 2 (2003) 145.
[17] A. Courty, I. Lisiecki, M.-P. Pileni, J. Chem. Phys. 116 (2002) 8074.
[18] M. Yashima, L.K.L. Falk, A.E.C. Palmqvist, K. Holmberg, J. Colloid Interface Sci. 268 (2003) 348.
[19] M.-L. Wu, D.-H. Chen, T.-C. Huang, J. Colloid Interface Sci. 243 (2001) 102.
[20] A. Saramat, A.E.C. Palmqvist, unpublished results.
[21] P.D.I. Fletcher, B.H. Robinson, R.B. Freedman, C. Oldfield, J. Chem. Soc. Faraday Trans. 1 81 (1985) 2667.
[22] K. Holmberg, in: P. Kumar, K.L. Mittal (Eds), Handbook of Microemulsion Science and Technology, Marcel Dekker, New York (1999) 713.
[23] M.-P. Pileni, Langmuir 13 (1997) 3266.
[24] H. Althues, Kaskel, Langmuir 18 (2002) 7428.
[25] M.-P. Pileni, Catal. Today 58 (2000) 151.
[26] H. Härelind Ingelsten, R. Bagwe, A.E.C. Palmqvist, M. Skoglundh, C. Svanberg, K. Holmberg, D.O. Shah, J. Colloid Interface Sci. 241 (2001) 104.
[27] S. Qiu, J. Dong, G. Chen, J. Colloid Interface Sci. 216 (1999) 230.
[28] F. Barroso, M. de Dios, C. Tojo, M.C. Blanco, M.A. López-Quintela, Coll. Surf. A (2005) 270–271.

[29] C. Tojo, M.C. Blanco, M.A. López-Quintela, Langmuir 13 (1997) 4527.

[30] C. Tojo, M.C. Blanco, M.A. López-Quintela, Curr. Topics Colloid Interface Sci. 4 (2001) 103.

[31] S. Quintillán, C. Tojo, M.C. Blanco, M.A. López-Quintela, Langmuir 17 (2001) 7251.

[32] M.A. López-Quintela, J. Rivas, M.C. Blancko, C. Tojo, C., in: L.M. Liz Marzán, P.V. Mamat (Eds.), Nanoscale Materials, Kluwer Academic Plenum Publ., New York (2002).

[33] K. Osseo-Asare, F.J. Arriagada, J. Colloid Interface Sci. 218 (1999) 68.

[34] Y. Li, C.-W. Park, Langmuir 15 (1999) 952.

[35] M. Andersson, J.S. Pedersen, A.E.C. Palmqvist, Langmuir 21 (2005) 11387.

[36] I. Lisiecki, A. Filankembo, H. Sack-Kongehl, K. Weiss, M.-P. Pileni, J. Urban, Phys. Rev. B 61 (2000) 4968.

[37] M.-P. Pileni, Langmuir 17 (2001) 7476.

[38] B.A. Simmons, et al. Nano Lett. 2 (2002) 263.

[39] A. Filankembo, M.-P. Pileni, J. Phys. Chem. B 104 (2000) 5865.

[40] C.C. Chen, J.J. Lin, Adv. Mater. 13 (2001) 136.

[41] I. Lisiecki, F. Billoudet, M.P. Pileni, J. Phys. Chem. 100 (1996) 4160.

[42] M. Brust, M. Walker, D. Bethell, D. J. Schiffrin, R. Whymanet, Chem. Comm. (1994) 801.

[43] M. Brust, J. Fink, D. Bethell, D. J. Schiffrin, C. Kielyet, Chem. Comm. (1995) 1655.

[44] J.R. Heath, C. M. Knobler, D. V. Leff, J. Phys. Chem. B. 101 (1997) 189.

[45] B.A. Korgel, S. Fullam, S. Connolly, D. Fitzmaurice, J. Phys. Chem. B. 102 (1998) 8379.

[46] K. Wikander, K. Holmberg, C. Petit, M.-P. Pileni, Langmuir 22 (2006) 4863.

[47] J.R. Heath, Acc. Chem. Res. 32 (1999) 388.

[48] H. Härelind Ingelsten, J.-C. Béziat, K. Bergkvist, A.E.C. Palmqvist, M. Skoglundh, H. Qiuhong, L.K.L. Falk, K. Holmberg, Langmuir 18 (2002) 1811.

[49] W.Y. Kim, T. Hanaoka, M. Kishida, K. Wakabayashi, K., Appl. Catal. A 155 (1997) 283.

[50] W.Y. Kim, H. Hayashi, M. Kishida, H. Nagata, K. Wakabayashi, Appl. Catal. A 169 (1998) 157.

[51] M. Ikeda, S. Takeshima, T. Tago, M. Kishida, K. Wakabayashi, Catal. Lett. 58 (1999) 195.

[52] J.S. Beck, J.C. Vartuli, W.J. Roth, M.E. Leonowicz, C.T. Kresge, K.D. Schmitt, C.T.W. Chu, D.H. Olson, E.W. Sheppard, S.B. McCullen, J.B. Higgins, J.L. Schlenker, J. Amer. Chem. Soc. 114 (1992) 10834.

[53] C.T. Kresge, M.E. Leonowicz, W.J. Roth, J.C. Vartuli, J.S. Beck, Nature 359 (1992) 710.

[54] T. Yanagisawa, T. Shimizu, K. Kuroda, C. Kato, Bull. Soc. Chem. Japan 63 (1990) 988.

[55] K. Holmberg. J. Colloid Interface Sci. 274 (2004) 355.

[56] Q. Huo, D.I. Margolese, U. Ciesla, D.G. Demuth, P. Feng, T. E. Gier, P. Sieger, A. Firouzi, B.F. Chmelka, F. Schüth, and G.D. Stucky, Chem. Mater. 6 (1994) 1176.

[57] A. Firouzi, D. Kumar, L.M. Bull, T. Besier, P. Sieger, Q. Huo, S.J. Walker, J.A. Zasadzinski, C. Glinka, J. Nicol, D. Margolese, G.D. Stucky, B.F. Chmelka, Science 267 (1995) 1138.

[58] A. Firouzi, F. Atef, A. Oertli, G.D. Stucky, B.F. Chmelka, J. Am. Chem. Soc. 119 (1997) 3596.

[59] J.Y. Ying, C.P. Mehnert, M.S. Wong, Angew. Chem. Int. Ed. 38 (1999) 56.

[60] J. Zhang, Z. Luz, H. Zimmermann, D. Goldfarb, J. Phys. Chem. B 104 (2000) 279.

[61] J. Zhang, D. Goldfarb, Microporous Mesoporous Mater. 48 (2001) 143.

[62] A. Galarneau, F. Di Renzo, F. Fajula, L. Mollo, B. Fubini, M.F. Ottaviani, J. Colloid Interface Sci. 201 (1998) 105.

[63] M.F. Ottaviano, A. Galarnneau, D. Desplantier-Giscard, F. Di Renzo, F. Fajula, Microporous Mesoporous Mater. 44–45 (2001) 1.

[64] J. Frasch, B. Lebeau, M. Soulard, L. Patarin, R. Zana, Langmuir 16 (2000) 9049.

[65] L. Sicard, B. Lebeau, J. Patarin, R. Zana, Langmuir 18 (2002) 74.

[66] J.L. Blin, A. Léonard, B.L. Su, Chem. Mater. 13 (2001) 3542.

[67] G. Herrier, J.L. Blin, B.L. Su, Langmuir 17 (2001) 4422.

[68] M. Andersson, H. Birkedal, N. Franklin, T. Ostomel, S. Buecher, A.E.C. Palmqvist, G.D. Stucky, Chem. Mater. 17 (2005) 1409.

[69] N.R.B. Coleman, G.S. Attard, Microporous Mesoporous Mater. 44–45 (2001) 73.

[70] M. Klotz, A. Ayral, C. Guizard, L. Cot, J. Mater. Chem. 10 (2000) 663.

[71] P.C.A. Alberius, K.L. Frindell, R.C. Hayward, E.J. Kramer, G.D. Stucky, B.F. Chmelka, Chem. Mater. 14 (2002) 3284.

[72] S.M. De Paul, J.W. Zwanziger, R. Ulrich, U. Wiesner, H.W. Spiess, J. Amer. Chem. Soc. 121 (1999) 5727.

[73] R. Ulrich, A. Du Chesne, M. Templin, U. Wiesner, Adv. Mater. 11 (1999) 141.

[74] C. Gölter-Spickermann, Curr. Opin. Colloid Interface Sci. 7 (2002) 173.

[75] Y. Sakamoto, M. Kaneda, O. Terasaki, D. Zhao, J.M. Kim, G.D. Stucky, H.J. Shin, R. Ryoo, Nature 408 (2000) 449.

[76] S. Che, Y. Sakamoto, O. Terasaki, T. Tatsumi, Chem. Mater. 13 (2001) 2237.

[77] P. Kipkemboi, A. Fogden, V. Alfredsson, K. Flodström, Langmuir 17 (2001) 5398.

[78] J.M. Kim, Y. Sakamoto, Y.K. Hwang, Y.-U. Kwon, O. Terasaki, S.-E. Park, G.D. Stucky, J. Phys. Chem. B 106 (2002) 2552.

[79] E. Leontidis, Curr. Opin. Colloid Interface Sci. 7 (2002) 81.

[80] S. Che, S. Lim, M. Kaneda, H. Yoshitake, O. Terasaki, T. Tatsumi, J. Amer. Chem. Soc. 124 (2002) 13962.

[81] N. Cruise, K. Jansson, K. Holmberg, J. Colloid Interface Sci. 241 (2001) 527.

[82] K. Larsson, K., Lipids—Molecular Organization, Physical Functions and Technical Applications, Oily Press, Scotland (1994).

[83] K. Flodström, et al. J. Amer. Chem. Soc. 125 (2003) 4402.

[84] A. Berggren, A. Palmqvist, K. Holmberg. Soft Matter 1 (2005) 219.

[85] F. Kleitz, S.H. Choi, R. Ryoo, Chem. Commun. 17 (2003) 2136.

[86] K. Wikander, H. Ekström, A.E.C. Palmqvist, A. Lundblad, K. Holmberg, G. Lindbergh, Fuel Cells: From Fundamental to Systems, Fuel Cells 6 (2006) 21–25.

[87] K. Wikander, A.E.C. Palmqvist, Micropor. Mesopor. Mater.

[88] M. Wu, J. Long, A. Huang, Y. Luo, Langmuir 15 (1999) 8822.

[89] M. Andersson, L. Österlund, S. Ljungström, A.E.C. Palmqvist, J. Phys. Chem. B 106 (2002) 10674.

[90] P. Kjellin, M. Andersson, A.E.C. Palmqvist, Langmuir 19 (2003) 9196.

[91] M. Li, S. Mann, Langmuir 16 (2000) 7088.

[92] M. Li, H. Schnablegger, S. Mann, Nature 402 (2000) 393.

[93] M. Andersson, V. Alfredsson, P. Kjellin, A.E.C. Palmqvist, Nano Lett. 2 (2002) 1403.

[94] M. Andersson, H. Härelind Ingelsten, A.E.C. Palmqvist, M. Skoglundh, K. Holmberg, in: B.H. Robinson (Ed.), Self-Assembled Systems, IOS, UK (2003) 105.

[95] F. Currie, M. Andersson, K. Holmberg, Langmuir 20 (2004) 3835.

[96] L.M. Qi, H. Cölfen, M. Antonietti, Angew. Chem. Int. Ed. 39 (2000) 604.

[97] L.M. Qi, H. Cölfen, M. Antonietti, Chem. Mater. 12 (2000) 2392.

[98] P. Kjellin, K. Holmberg, M. Nydén, Colloids Surfaces A 194 (2001) 49.

[99] P. Kjellin, Colloids Surfaces A 212 (2003) 19.

4 Synthesis and Engineering of Polymeric Latex Particles for Hemodialysis
Part I—A Review

S. Kim, H. El-Shall, R. Partch, and B. Koopman

CONTENTS

INTRODUCTION

End stage renal disease (ESRD) is a chronic condition in which kidney function is impaired to the extent that the patient's survival requires removal of toxins from the blood by dialysis therapy or kidney transplantation. The National Kidney Foundation estimates that over 20 million Americans had chronic kidney disease in 2002 (National Kidney Foundation, 2002). The number of people with ESRD is rapidly increasing in the United States with approximately 96,295 incident and 406,081 prevalent patients, including 292,215 on dialysis, and 113,866 with a functioning graft in 2001. It is projected that there will be more than 2.2 million ESRD patients by 2030 (USRDS, 2003). The expenditure for the ESRD treatment program had reached $22.8 billion, 6.4% of the Medicare budget in 2001. Due in part to the limited availability of kidneys for transplantation, hemodialysis is the primary clinical treatment for patients with ESRD.

The central element of a hemodialysis machine is the semi-permeable membrane that allows for selective transport of low molecular weight biological metabolites from the blood. One limitation of current dialysis technologies is the inability to efficiently remove middle molecular weight toxins such as β_2-microglobulin (β_2M) and interleukin 6 (IL-6). β_2M is a causative protein of Dialysis-Related Amyloidosis (DRA), a disease arising in patients with chronic kidney failure as a serious complication of long-term hemodialysis treatment (Gejyo et al., 1985). β_2M deposition in tissue is the primary cause of destructive arthritis and carpal tunnel syndrome (Drueke, 2000; Vincent et al., 1992).

Although attempts have been made to increase the efficiency of middle molecular weight toxin removal by changes in the membrane pore size and the use of innovative materials to adsorb these toxins (Ronco et al., 2001a; Samtleben et al., 1996a), removal efficiency is not as high as those achieved by a normal healthy kidney. Traditional membranes have a number of processing and performance limitations (Leypoldt, Cheung, and Deeter, 1998; Westhuyzen et al., 1992), such as a restricted choice of surface chemistries and limited control of porosity. The development of novel engineering membrane technology is needed to remove middle molecule toxins.

Polymeric latex particles have received increasing attention in areas such as solid phase supports in biological applications (Piskin et al., 1994). Examples include immunoassay (Chern et al., 2003; Radomska-Galant and Basinska, 2003), DNA diagnostic (Elaïssari et al., 1998), cell separation, and drug delivery carriers (Kurisawa, Terano, and Yui, 1995; Luck et al., 1998; Yang et al., 2000), etc. This is because of the well-defined colloidal and surface characteristics of the particles. By using a seed emulsion polymerization method it is possible to synthesize monodisperse latex particles with various particle size ranges and surface chemistries. Functionalized core-shell latex particles can be introduced by multistep emulsion polymerization. Core particles are synthesized in the first stage of the polymerization and the functional monomer is added in the second stage. This is done without any emulsifier addition to prevent the production of new homopolymer particles (Keusch and Williams, 1973). The core-shell particles are useful in a broad range of applications because of their improved physical and chemical properties over their single-component counterparts (Lu, Keskkula, and Paul, 1996; Nelliappan et al., 1997). Through the development of a hemodialysis membrane using monodisperse latex particles, improvements in advanced separation treatment for patients with ESRD may be realized. This requires the maximum removal of middle molecular weight proteins with minimal removal of other beneficial proteins such as albumin. Thus, an understanding of the fundamental interactions between the particles and biopolymers is vital to maximize the performance of this membrane technology. This chapter reviews the state of the art related to synthesis and engineering of polymeric latex particles for hemodialysis.

ADVANCES IN MEMBRANE TECHNOLOGY

Dialysis for blood purification is widely used in the treatment of ESRD. Hemodialysis (HD) techniques use a semipermeable membrane to replace the filtration role of the kidney. The membranes used in HD can be broadly classified into those based on cellulose and those manufactured from synthetic copolymers. These membranes come in various shapes such as sheets, tubular structures, or hollow fiber assemblies (Ronco et al., 2001a). The first attempt at blood dialysis using a cellulose-based membrane occurred in 1913. Abel, Rowntree, and Turner (1990) from the Johns Hopkins Medical School described a method whereby the blood of a living animal may be submitted to dialysis outside the body using a membrane based on cellulose and returned to the natural circulation without exposure to air, infection by microorganisms, or any alteration that would necessarily be prejudicial to life. This same technique is still used today; however, the device has been modified over the years as better membranes were developed and the anti-coagulant, heparin, has become available.

Cellulose membranes have been widely used for the treatment of renal failure from 1928, when the first human dialysis was performed, until the mid-1960s. The basic molecular structure of cellulose is a polymer with all repeat units containing hydroxyl (OH) groups. The realization that such groups imparted undesirable qualities on the material with respect to blood contact behavior was discovered in the early 1970s and since then has been the focus of development of modified cellulose membranes having some of the hydroxyl groups converted to benzyl ether groups. The result is a molecular mosaic of hydrophobic (benzyl) and hydrophilic (hydroxyl and cellulose) regions.

Kolff (Van Noordwijk, 2001) studied the rotating drum artificial kidney for patients with acute renal failure in 1943. Cellophane tubing was used for the membrane with heparin as the anticoagulant. For the next 17 years, hemodialysis therapy was performed by this method but only for patients with acute reversible renal failure. Vascular access required repeated surgical insertions of cannulas (slender tubes) into an artery and vein, and limited the number of treatments that a patient could receive in order to minimize the amount of vascular damage.

Initially, the need for dialysis in patients with acute renal failure was determined mainly by the development of signs and symptoms of uremia. After dialysis, some time might elapse before uremic manifestations returned to warrant a sequential treatment of dialysis. Many patients with acute renal failure, secondary to accidental or surgical trauma were hypercatabolic, but the interdialytic interval might be prolonged because of anorexia or use of a low-protein diet. However, Obrien and his coworkers (Obrien et al., 1959) showed that patient well-being and survival were improved by what they termed prophylactic daily hemodiaysis, or administration of the treatment before the patient again became sick with uremia. Their report in 1959 was the first description of daily hemodialysis.

Development of membrane accessories such as a shunt has also been an area of focus for treatment improvement. In 1960, the development of a shunt (Quinton, Dillard, and Scribner, 1960), a flexible polytetrafluoroethylene (PTFE or Teflon) tubing, made many more hemodialysis treatments possible for chronic kidney failure patients. PTFE has a nonstick surface and is relatively biocompatible, leading to minimized blood clotting in the shunt.

Synthetic membranes are prepared from engineered thermoplastics such as polysulfone (PSf), polyamide (PA), and polyacrylonitrile (PAN) by phase inversion or precipitation of a blended mixture resulting in the formation of asymmetric and anisotropic structures. Figure 4.1 shows a fiber type of the Polyflux S membrane consisting of polyamide (PA), polyacrylethersulfone (PAES), and polyvinylpyrrolidone (PVP) with the integral three-layer structure. The skin layer on the inside fiber-type membrane contacts blood and has a very high surface porosity and a narrow pore size distribution. This layer constitutes the discriminating barrier deciding the permeability and solute retention properties of the membrane. The skin layer is supported by a thick sponge-type structure with larger pores providing mechanical strength and very low hydrodynamic resistance.

PSf is a widely used membrane material for the hemodialysis application (Malchesky, 2004) because of its thermal stability, mechanical strength, and chemical inertness.

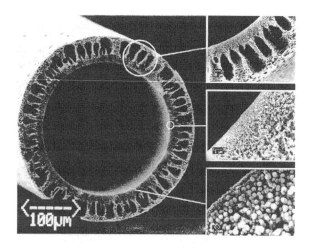

FIGURE 4.1 Cross-sectional SEM image view of the Polyflux S (polyamide + polyacrylethersulfone + polyvinylpyrolidone) dialysis membranes (Deppisch et al., 1998).

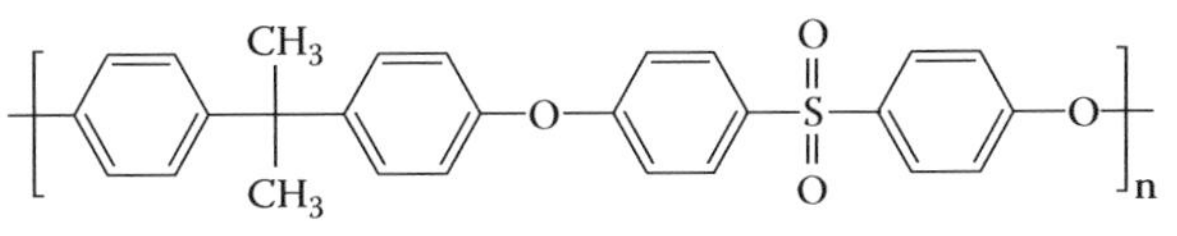

FIGURE 4.2 The chemical structure of polysulfone (PSf).

According to a report from the National Surveillance of Dialysis-Associated Disease (NSDAD) in the United States, over 70% of hemodialysis membranes were PSf-based (Bowry, 2002). This is most likely because PSf has many advantages over other materials. This synthetic polymer is one of few materials that can withstand sterilization by steam, ethylene oxide, and γ-radiation. PSf membranes can be prepared by conventional immersion precipitation methods into many different shapes including porous hollow fiber or flat sheet hemodialysis membranes. The material also has a high permeability to low molecular weight proteins and solute, and high endotoxin retention. The chemical structure of PSf is shown in Figure 4.2.

There is one major disadvantage to PSf. The hydrophobic nature of the PSf causes serious complications through the activation of the complement alternative pathway leading to the adsorption of serum proteins onto the membranes (Singh et al., 2003). Anticoagulants are added during dialysis therapy to avoid blood clotting, but this does not completely eliminate the problem. In order to overcome this disadvantage of the PSf membrane, various studies have been performed to change the material's surface properties. These investigations include hydrophilic polymer coating (Brink et al., 1993; Higuchi et al., 2003; Kim, Fane, and Fell, 1988); layer grafting onto PSf membrane (Mok et al., 1994; Pieracci, Crivello, and Belfort, 2002; Song et al., 2000; Wavhal and Fisher, 2002); and chemical reaction of hydrophilic components onto the membrane surface (Blanco, Nguyen, and Schaetzel, 2001; Guiver et al., 1993; Higuchi and Nakagawa, 1990; Higuchi, Mishima, and Nakagawa, 1991, 1993; Nabe, Staude, and Belfort, 1997). Hydrophilic monomers, 2-hydroxy-ethylmethacrylate (HEMA), acrylic acid (AA), and methacrylic acid (MMA), have also been grafted onto PSf membrane to increase flux and Bovine Serum Albumin (BSA) retention. Hancock, Fagan, and Ziolo (2000) synthesized polysulfone/poly(ethylene oxide) (PEO) block copolymers to improve the resistance to platelet adhesion. Kim, Kim, and Kim (2003) also studied blending a sulfonated PEO acrylate diblock copolymer into PSf in order to reduce platelet adhesion and enhanced biocompatibility. PEO is a commonly used biomaterial due to its excellent resistance to protein adsorption and inherent biocompatibility (Hariss, 1992). Kim et al. (2005) studied a self-transformable copolymer to enhance the hydrophilicity of an asymmetric PSf membrane with an ultra-thin skin layer. The polymer had an entrapped diblock copolymer containing a hydrophilic block of poly(ethylene glycol) (PEG)-SO3 acrylate and a hydrophobic block of octadecylacrylate (OA). Molecular dynamic (MD) simulations were performed as a function of copolymer density to optimize interfacial structure information. McMurry (2004) developed a strategy using an amphiphilic graft copolymer added to PSf membranes by introducing polysulfone-g-poly(ethylene glycol). When compared to unmodified PSf, these graft copolymer and resulting blend membranes hold promise for biomedical device applications.

Polyamide (PA) membranes have also been used for hemodialysis because of their mechanical strength in both wet and dry conditions. Polyamide consists of aromatic or/and aliphatic monomers with amide bonding (-CONH-), also known as a peptide bond. The basic amide bond in polyamide is shown in Figure 4.3. R_1 and R_2 can be either aromatic or aliphatic linkage groups.

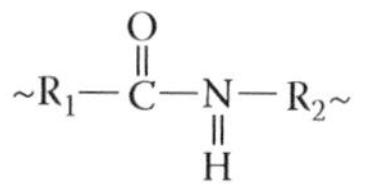

$$\sim R_1 - \overset{\overset{\displaystyle O}{\|}}{C} - \overset{}{N} - R_2 \sim$$
$$\underset{H}{\overset{\|}{}}$$

FIGURE 4.3 The chemical structure of polyamide (PA).

Panichi and co-workers (1998) evaluated the biocompatibility of the PA membrane and concluded that PA hemofiltration was a highly biocompatible technique due to the use of a synthetic membrane with a sterile re-infusion fluid and the convective removal of the activated anaphylatoxins and β_2-microglobulin (β_2M). The PA-based membrane Polyflux (manufactured by Gambro GmbH, Germany), blended with polyamide, polyarylethersulfone, and polyvinylpyrrolidone (PVP), was able to clean small molecules such as urea, creatinine, and phosphate, as well as decrease β_2M amounts by 50.2% (Hoenich and Katopodis, 2002). Due to the nonselectivity of the membrane, removal of these unwanted materials also led to the undesirable loss of beneficial proteins during therapy. Meier et al. (2000) evaluated different immune parameters using a modified cellulose low-flux hemophan and synthetic high-flux PA membrane during a 1-year period in chronic hemodialysis patients. They found that the 1-year immunological evaluation of hemodiaysis membrane biocompatibility was associated with changes in the pattern of chronic T-cell actiovation.

Polyacrylonitrile (PAN) is another commonly used membrane material because it is inherently hydrophilic and has been commercialized for ultrafiltration and microfiltration (Scharnagl and Buschatz, 2001). PAN is a semicrystalline polymer and the mechanical properties strongly depend on the crystalline structures. The chemical structure of PAN is shown in Figure 4.4.

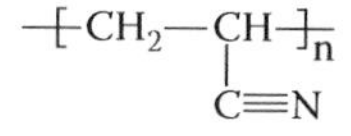

$$-\!\!\left[CH_2-\underset{\underset{C\equiv N}{|}}{CH}\right]_n$$

FIGURE 4.4 Chemical structure of polyacrylonitrile (PAN).

The addition of additives such as polyvinylpyrrolidone (PVP) as a pore-forming agent gives PAN membranes more flexible processing parameters and increased performance (Jung et al., 2005). PAN membrane performance has been optimized through copolymerization with many other vinyl monomers including glycidyl methacrylate (Godjevargova, Konsulov, and Dimov, 1999; Hicke et al., 2002); N-vinylimidazol (Godjevargova et al., 2000); hydroxyl ethyl methacrylate (Bhat and Pangarkar, 2000; Ray et al., 1999); methacrylic acid (Ray et al., 1999); vinyl pyrrolidone (Ray et al., 1999); acrylic acid (Trotta et al., 2002); acrylamide (Musale and Kulkarni, 1997); and vinylchloride (Broadhead and Tresco, 1998). These monomers provide a reactive group for enzyme immobilization, improved mechanical strength, solvent resistance, pervaporation, permeation flux, antifouling, and biocompatibility. Because of this, PAN-based copolymer membranes have great potential for the treatment of hemodialysis in an artificial kidney. This material can also be used for other applications like the treatment of wastewater, the production of ultrapure water, biocatalysis together with separation, and methanol separation by pervaporation.

Lin, Liu, and Yang (2004) studied the modification of PAN-based dialyzer membranes to increase the hemocompatibility by the immobilization of chitosan and heparin conjugate on the surface of the PAN membrane. When a foreign material is exposed to blood, plasma proteins are adsorbed, clotting factors are activated, and a nonsoluble fibrin network, or thrombus, is formatted (Goosen, Sefton, and Hatton, 1980). The result of this research was that the biocompatible chitosan polymer and a blood anticoagulant heparin prevented blood clotting. They showed prolonged coagulation time, reduced platelet adsorption, thrombus formation, and protein adsorption. Nie et al. (2004) studied PAN-based ultrafiltration hollow-fiber membranes (UHFMs). In order to improve the membrane performance, acrylonitrile (AN) was copolymerized with other functional monomers such as maleic anhydride and α-allyl glucoside. They found that the number and size of macrovoid underneath the inner surface of membrane decreased by increasing the amount of solvent DMSA in the internal coagulant. The water flux of the UHFMs also decreased while the bovine serum albumin rejection increased minutely. Godjevargova, Dimov, and Petrov (1992) modified the PAN-based membrane with hydroxylamine and diethylaminoethylmethacrylate to improve membrane dialysis properties. Formed functional groups like primary amine, oxime, and tertiary amine groups provided the membrane with more hydrophilic properties and a substantial increase in the permeability of the membranes.

The wide use of filtration in practice is limited by membrane fouling. Solute molecules deposit on and in the membrane in the process of filtration, causing dramatic reduction in flux through the membrane. Fouling occurs mostly in the filtration of proteins. Three kinetic steps are involved in the fouling of UF membranes according to Nisson (1990). The first step is the transfer of solute to the surface. The second step is the transfer of solute into the membrane until it either finally adsorbs or passes through after a set of adsorption-desorption events. The third step includes surface binding accompanied by structural rearrangement in the adsorbed state (Ko et al., 1993). Bryjak, Hodge, and Dach (1998) studied the surface modification of a commercially available PAN membrane to develop superior filtration properties with less fouling by proteins. The PAN membrane was immersed in excess NaOH solution to convert some of the surface nitrile groups into carboxylic groups by the hydrolysis process. This modified PAN membrane was not so severely fouled in the Bovine Serum Albumin (BSA) filtration test. The pore size, however, decreased during the hydrolysis process, leading to a significant reduction in flux and made the membrane less productive in the ultrafiltration (UF) mode.

SORBENT TECHNOLOGY

Over the last three decades, sorbent technology (Castino et al., 1976; Korshak et al., 1978; Malchesky et al., 1978) has been further developed to increase the efficiency of dialysis, or replace it, for the treatment of ESRD. Sorbents remove solutes from solution through specific or nonspecific adsorption, depending on both the nature of the solute and the sorbent. Specific adsorption contains tailored ligands, or antibodies, with high selectivity for target molecules. Specific adsorbents have been used in autoimmune disorders such as idiopathic thrombocytopenic purpura (Snyder et al., 1992) and for the removal of lipids in familial hyper cholesterolemia (Bosch et al., 1999). Nonspecific adsorbents, such as charcoal and resins, attract target molecules through various forces including hydrophobic interactions, ionic (or electrostatic) attraction, hydrogen bonding, and van der Waals interactions.

New dialysate with sorbents has become an accepted modification of dialysis, and sorbent hemoperfusion is gaining ground as a valuable addition to dialysis, especially as new sorbents are developed (Winchester et al., 2001). Hemoperfusion is defined as the removal of toxins or metabolites from circulation by the passing of blood, within a suitable extracorpoteal circuit, over semipermeable microcapsules containing adsorbents such as activated charcoal (Samtleben et al., 1996b), various resins (Ronco et al., 2001b), albumin-conjugated agarose etc. Novel adsorptive carbons with larger pore diameters have been synthesized for potential clinical use (Mikhalovsky, 1989). Newly recognized uremic toxins (Dhondt et al., 2000; Haag-Weber, Cohen, and Horl, 2000) have resulted in several investigations on alternatives to standard, or high-flux, hemodialysis to remove these molecules. These methods include hemodiafilteration with (de Francisco et al., 2000) or without (Takenaka et al., 2001; Ward et al., 2000) dialysate regeneration using sorbents, as well as hemoperfusion using such adsorbents as charcoal and resins.

Kolarz and Jermakowiczbartkowiak (1995) and Kolarz et al. (1989) studied the hyper-crosslinked sorbent prepared from styrene and divinylbenzene (DVB) for a hemoperfusion application. They found that the pore structure of a swollen sorbent was changed by additional crosslinking with α,α'-dichloro-p-xylene in the presence of a tin chloride catalyst and in a dichloroethane solution. They also realized that the hemocompatibility was useful for hemoperfusion and could be imparted to the sorbents by introducing sulfonyl groups at a concentration of about 0.2 mmol/g.

A special polymeric adsorbing material (BM-010 from Kaneka, Japan) has been investigated by another group (Furuyoshi et al., 1991) for the selective removal of β2M from the blood of dialysis patients. The adsorbent consists of porous cellulose beads modified with hexadecyl groups that attract β2M through a hydrophobic interaction. The adsorption capacity of this material is 1 mg of β2M per 1 ml of adsorbent. Several small clinical trials were performed using a hemoperfusion cartridge containing 350 ml of these cellulose beads in sequence with a high-flux hemodialyzer. During 4–5 hours of treatment, about 210 mg of β2M were removed, thus reducing the concentration in the blood by 60–70% of the initial level (Gejyo et al., 1993; 1995; Nakazawa et al., 1993).

RenalTech developed a hemoperfusion device, BetaSorb, containing the hydrated cross-linked polystyrene (PS) divinylbenzene (DVB) resin sorbents with a pore structure designed to remove molecules between 4 and 30 kDa (Winchester et al., 2002). In this case, solute molecules are separated according to their size based on their ability to penetrate the porous network of the beaded sorbents. The resin beads were prepared with a blood-compatible coating, and confirmed to be biocompatible *in vivo* in animals (Cowgill and Francey, 2001).

LIMITATION OF CURRENT HEMODIALYSIS TREATMENT

Hemodialysis is a widely used life-sustaining treatment for patients with ESRD. However, it does not replace all of the complex functions of a normal healthy kidney. As a result, patients on dialysis still suffer from a range of problems, including infection, accelerated cardiovascular disease, high blood pressure, chronic malnutrition, anemia, chronic joint and back pain, and a considerably shortened life span. One significant limitation of the current dialysis technology is the inability to efficiently remove larger toxic molecules. This is mainly because of the broad pore size distribution reducing the selective removal of toxins, and unsatisfied biocompatibility, causing complications such as inflammation, blood clotting, calcification, infection, etc.

Dialysis purifies the patient's blood by efficiently removing small molecules such as salts, urea, and excess water. However, as toxic molecules increase in size, their removal rate by hemodialysis substantially declines. Typically, only 10–40% of these middle molecular weight toxins (300–15,000 Da) are removed from the blood during a dialysis session (Vanholder et al., 1995). These toxins then reach an abnormally high level and begin to damage the body. One such toxin, $\beta_2 M$, causes destructive arthritis and carpal tunnel syndrome, by joining together like the links of chain to form a few very large molecules and deposit damaging the surrounding tissues (Lonnemann and Koch, 2002). This is also a main cause of mortality for long-term dialysis patients. Other middle molecule toxins appear to inhibit the immune system and may play a significant role in the high susceptibility to infections in dialysis patients. Still others are believed to impair the functioning of several other body systems, such as the hematopoietic and other endocrine systems. This may contribute to accelerated cardiovascular disease, the leading cause of death among dialysis patients, as well as clinical malnutrition, which affects up to 50% of this patient population.

Over the last decade, polymeric dialysis membranes have been developed to increase the capacity for removing middle molecular weight toxins by changing the pore size of dialyzer membranes and using new materials that adsorb these toxins for improved removal characteristics. However, removal efficiency is not as high as those achieved by a normal healthy kidney.

LATEX PARTICLES

The first polymer synthesized using emulsion polymerization was a rubber composed of 1,3-butadiene and styrene made during World War II in the United States. The Dow Chemical Company has been a major manufacturer of polystyrene, including latex used in paint formulations. The theory of emulsion polymerization, in which a surfactant is used, was established by Harkins (1947) and by Smith and Ewart (1947). By 1956, the technology was complete, including the method of building larger diameter particles from smaller ones. The product by this emulsion polymerization is referred to as latex, a colloidal dispersion of polymer particles in water medium (Odian, 1991). Latexes are currently undergoing extensive research and development for a broad range of areas, including adhesives, inks, paints, coatings, drug delivery systems, cosmetics, medical assay kits, gloves, paper coatings, floor polish, films, carpet backing, and foam mattresses. The relatively well-known and easily controlled emulsion process is one of the main advantages for these applications. Therefore, polymeric latex particles prepared by emulsion polymerization can be a candidate for the medical applications because of the easy control of the particle size and morphology as well as flexible surface chemistry to be required.

THE COMPONENTS FOR EMULSION POLYMERIZATION

The main components of the emulsion polymerization process are the monomer, a dispersing medium, a surfactant, and an initiator. Available monomers are styrene, butadiene, methylmethacrylate, acrylic acid, etc. The dispersing medium is usually water, which will maintain a low solution viscosity, provide good heat transfer, and allow transfer of the monomers from the monomer droplets into micelles and growing particles consisted of and surrounded by surfactants, respectively. The surfactant (or emulsifier) has both hydrophilic and hydrophobic segments. Their main functions are to provide the nucleation sites for particles and aid in the colloidal stability of the growing particles. Initiators are water-soluble inorganic salts, which dissociate into free radicals to initiate vinyl polymerization. A chain transfer agent such as mercaptan may be added to control molecular weight.

PARTICLE NUCLEATION

Free radicals are produced by dissociation of initiators at the rate on the order of 10^{13} radicals per milliliter per second in the water phase. The location of the polymerization is not in the monomer droplets but in micelles because the initiators are insoluble in the organic monomer droplets. Such initiators are referred to as oil-insoluble initiators. This is one of the big differences between emulsion polymerization and suspension polymerization where initiators are oil-soluble and the reaction occurs in the monomer droplets. Because the monomer droplets have a much smaller total surface area, they do not compete effectively with micelles to capture the radicals produced in solution. It is in the micelles that the oil-soluble monomer and water-soluble initiator meet, and is favored as the reaction site because of the high monomer concentration compared to that in the monomer droplets. As polymerization proceeds, the micelles grow by the addition of monomer from the aqueous solution whose concentration is refilled by dissolution of monomer from the monomer droplets. There are three types of particles in the emulsion system: monomer droplets; inactive micelles in which polymerization is not occurring; and active micelles in which polymerization is occurring, referred to as growing polymer particles.

The mechanism of particle nucleation goes on by two simultaneous processes: micellar nucleation and homogeneous nucleation. Micellar nucleation is the entry of radicals, either primary or oligomeric radicals formed by solution polymerization, from the aqueous phase into the micelles. In homogeneous nucleation, solution-polymerized oligomeric radicals are becoming insoluble and precipitating onto themselves or onto dead oligomer. The relative levels of micellar and homogeneous nucleation are variable with the water solubility of the monomer and the surfactant concentration. Homogeneous nucleation is favored for monomers with higher water solubility and low surfactant concentration, and micellar nucleation is favored for monomers with low water solubility and high surfactant concentration. It has also been shown that homogeneous nucleation occurs in systems where the surfactant-insoluble monomer, styrene (Hansen and Ugelstad, 1979; Ugelstad et al., 1980), is created by micellar nucleation and a water-insoluble monomer, such as vinyl acetate (Zollars, 1979), is formed by homogeneous nucleation.

A third reaction has been proposed, referred to as coagulative nucleation. In this reaction, the major growth process for the first-formed polymer particles (precursor particles) is coagulation with other particles rather than polymerization of monomer. The driving force for coagulation of precursor particles, several nanometers in size, is their relative instability compared to larger-sized particles. The small size of a precursor particle with its high curvature of the electrical double layer permits low surface charge density and high colloidal instability. Once the particles become large enough in size to maintain high colloidal stability, there is no longer a driving force for coagulation and further growth of particles takes place only by the polymerization process.

TYPES OF PROCESSES FOR EMULSION POLYMERIZATION

There are three types of production processes used in emulsion polymerization: batch, semicontinuous (or semibatch), and continuous. In the batch type process, all components are added at the beginning of the polymerization. As soon as the initiator is added and the temperature is increased, polymerization begins with the formation and growth of latex particles at the same time. There is no further process control possible once the polymerization is started. In the semi-continuous emulsion polymerization process, one or more components can be added continuously. Various profiles of particle nucleation and growth can be generated from different orders of component addition during polymerization. There are advantages to this process such as control of the polymerization rate, the particle number, colloidal stability, copolymer composition, and particle morphology. In the continuous process, the emulsion polymerization components are fed continuously into a product while the product is simultaneously removed at the same rate. High production rate, steady heat removal, and uniform quality of latexes are advantages of the continuous polymerization processes.

Other available methods include the intermittent addition and shot addition of one or more of the components. In the shot addition process, the additional components are added at one time, during the later stages of the polymerization, prior to complete conversion of the main monomer. This method has been used successfully to develop water-soluble functional monomers such as sodium styrene sulfonate (Kim et al., 1989).

CHEMISTRY OF EMULSION POLYMERIZATION

Emulsion polymerization is one type of free radical polymerization and can be divided into three distinct stages: initiation, propagation, and termination. The emulsion polymerization system is shown in Figure 4.5.

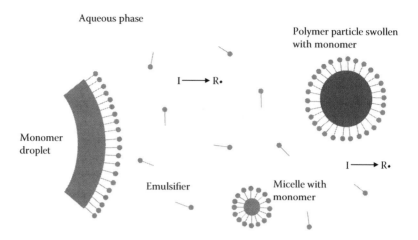

FIGURE 4.5 Emulsion polymerization system. (Radicals [R·] are created from initiators [I]. Monomer is transferred from larger monomer droplet into micelles by emulsifier. Initiated polymer particle by radicals keeps growing until monomers are all consumed. The reaction is performed in aqueous media.)

In the initiation stage, free radicals are created from an initiator by heat or an ultraviolet radiation source. The initiator with either peroxide groups (-O-O-) such as sodium persulfate, or azo groups (-N=N-) such as azobisisobutyronitrile, is commonly used for emulsion polymerization. The primary free radicals created from initiator react with the monomer for initiation of polymerization. In the propagation stage, the polymer chain grows by monomer addition to the active center, a free radical reactive site. There are two possible modes of propagation: head-to-head addition and head-to-tail addition. The head-to-tail mode is the predominant configuration of the polymer chain, a result of steric hindrance of the substituted bulky group. In the termination stage, polymer chain growth is terminated by either coupling of two growing chains forming one larger polymer molecule or by disproportionation, which involves moving a hydrogen atom from one growing chain to another, forming two polymer molecules, one having a saturated end-group and the other with an unsaturated end-group.

SEED EMULSION POLYMERIZATION

Polymer latex particles have received increasing interest because of the versatility of the many applications heterophase polymerization processes, such as emulsion, dispersion, microemulsion, seeded emulsion, precipitation, etc., offer. Especially, to prepare well-defined microspheres having monodisperse and various particle sizes as well as surface group functionalities, it is necessary to use seed particles prepared by emulsion polymerization and enlarge them to a desired size in the further stages of reactions.

Polymeric particles are required to have a uniform particle size in many applications, such as chromatography, where they are used as a packing material.

Morphological control of latex particles is also important for many practical applications (Schmidt, 1972). Seed emulsion polymerization (or two-stage emulsion polymerization) is a useful method to achieve both monodisperse particle size and morphological design. In seeded emulsion polymerization (Gilbert, 1995), preformed "seed" latex is used to control the number of particles present in the final latex. The advantage of seeded emulsion polymerization is that the poorly reproducible process of particle nucleation can be bypassed, so that the number concentration of particles is constant and known. Various mechanisms have been proposed for the growth of latex particles (Okubo and Nakagawa, 1992; Ugelstad et al., 1980) using this polymerization technique. The initial seed particle preparation step is well known and relatively easy to perform because, at the relatively small particle sizes (0.2 to 0.5 micron), the particle growth process can be readily controlled by the use of an emulsifier. Enough emulsifier is used to prevent coagulation but not enough to cause the formation of new particles. As the particles are grown to larger sizes in successive seeding steps, it becomes increasingly difficult to maintain a stable, uncoagulated emulsion without forming new particles and thereby destroying monodisperisty of the latex. Recently, there have been several reports concerning the reaction kinetics of seed emulsion polymerization and the development of latex morphologies over the course of the reaction (Chern and Poehlein, 1990; Delacal et al., 1990; Lee, Chiu, and Chern, 1995; Lee, 2000, 2002).

POLYSTYRENE LATEX PARTICLES

A number of papers have described the synthesis of polystyrene latex particles bearing various functional surface groups such as carboxyl (Lee et al., 2000; Reb et al., 2000; Tuncel et al., 2002); hydroxyl (Tamai, Hasegawa, and Suzawa, 1989); mercapto (Nilsson, 1989); epoxy (Luo et al., 2004; Shimizu et al., 2000); acetal (Izquierdo et al., 2004; Santos and Forcada, 1997); thymine (Dahman et al., 2003); chloromethyl (Izquierdo et al., 2004; Park, Kim, and Suh, 2001; Sarobe et al., 1998); amine (Cousin et al., 1994; Ganachaud et al., 1995; Miraballes-Martinez and Forcada, 2000; Miraballes-Martinez et al., 2001); ester (Nagai et al., 1999), etc. In order to produce these functionalized particles, different methods for particle preparation must be used. The polymer emulsion with core-shell morphology of latex particles is one of them. This is a multistep emulsion polymerization process in which the polystyrene "core" particle is synthesized in the first stage and the functional monomer is added in the second stage of the polymerization without any emulsifier postfeeding to prevent the production of new homopolymer particles, thus forming the functionalized polymer "shell" on the "core" particle (Keusch and Williams, 1973). There are requirements to limit secondary nucleation and encourage core-shell formation in seeded emulsion polymerization, including the addition of smaller seed particles at high solid content to increase particle surface area, low surfactant concentration to prevent formation of micelles, and the semi-continuous addition of monomer to create a starved-feed condition and keep the monomer concentration low. There are some advantages (Hergeth et al., 1989) of dispersions with polymeric core-shell particles. First, it is possible to modify the interfacial properties of polymer particles in the aqueous phase by the addition of only very small amounts of a modifying agent during the last period of the reaction. Thus, these core-shell particles are useful in a broad range of applications since they always exhibit improved physical and chemical properties over their single-component counterparts (Lu, Keskkula, and Paul, 1996; Nelliappan et al., 1997). In this way, the improvement of film properties of such dispersions is straightforward and inexpensive. The other is that polymers with a core-shell structure are perfect model systems for investigating the material properties of polymer blends and composites because of their regular distribution of one polymer inside a matrix polymer and because of the simple spherical geometry of the system. The core-shell properties usually depend on the structures of latex particles. Chen and his co-workers (Chen et al., 1991b, 1992b, 1993) reported the morphological development of core shell latex particles of polystyrene/ poly (methyl methacrylate) during polymerization. Before the research by Chen and co-workers, Min et al. (1983) reported the morphological development of core shell latex of polystyrene (PS)/polybutylacrylate (PBA) by seeded emulsion polymerization as a function of the addition method of PS. They found that the percentage of grafting PS to the PBA was greatest for the batch reaction, and the PBA-PS core-shell particles with a high degree of grafting remained spherical upon aging test because of the emulsifying ability of graft copolymer.

VARIOUS APPLICATIONS OF LATEX PARTICLES

Latex particles are applicable to a wide range of areas such as biomedical applications (Piskin et al., 1994), especially as the solid phase such as in immunoassays (Chen et al., 2003; Radomska-Galant and Basinska, 2003), DNA diagnostic, drug delivery carriers (Luck et al., 1998; Kurisawa et al., 1995; Yang et al., 2000), blood cell separations, and column packing reagents. Thus, protein adsorption on polymeric solid surfaces has become a center of attention. Of particular interest are microspheres, defined as "fine polymer particles having diameters in the range of 0.1 to several microns," which can be used as functional tools by themselves or by coupling with biocompounds. Singer and Plotz (1956) first studied microspheres, or latex agglutination tests (LATs), by using monodisperse polystyrene (PS) and polyvinyltoluene polymer particles as the support on which the biomolecules were going to adsorb. The biomolecule adsorption, however, was limited by possible desorption of the adsorbed species or loss of specific activity of the complex formed. Since this work was published, the latex particle applications for immunoassay have been rapidly and widely studied and developed.

Latex agglutination tests, or latex immunoassays, start with tiny, spherical latex particles with a uniform diameter and similar surface properties. The particles are coated with antibodies (sensitized) through the hydrophobic interaction of portions of the protein with the PS surface of the particles. If sensitized particles are mixed with a sample containing antigen, urine, serum, etc., they will become agglutinated and visibly agglomerate. Latex tests are inexpensive as compared with the other techniques (Bangs, 1988).

Unipath (Percival, 1996) has manufactured a range of immunoassays based on a chromatographic principle and deeply colored latex particles for use in the home and clinical environments. The latex particles, which are already sensitized with a monoclonal antibody, can detect any antigens bound to the surface of the latex particles. Some detectable pollutants include estrogen mimics, which induce abnormalities in the reproductive system of male fishes and lead to a total or partial male feminization. Rheumatoid factor (RF) in different age subpopulations has also been evaluated according to a patient's clinical status by using a rapid slide latex agglutination test for qualitative and semiquantitative measurement in human serum and the latex immunoassay method (Onen et al., 1998). Magalhaes, Pihan, and Falla (2004) studied a diagnostic method of contamination of male fishes by estrogen mimics, using the production of vitellogenin (VTG) as a biomarker. This was based on a reverse latex agglutination test, developed with monoclonal antibodies specific to this biomarker. Premstaller, Oberacher, and Huber (2000; Premstaller et al., 2001) have prepared a porous poly(styrene-divinylbenzene) (PS-DVB) polymer monolith to use for highly efficient chromatographic separation of biomelecules such as proteins and nucleic acids. They used a progeny, a mixture of tetrahydrofuran and decanol, to fabricate a micropellicular PS-DVB backbone. Legido-Quigley, Marlin, and Smith (2004) have developed the monolith column to obtain further chromatographic functionality to the column by introducing chloromethylstyrene in place of styrene into the polymer mixture.

Core-shell type monodisperse polymer colloids have been synthesized by Sarobe and Forcada (1998) with chloromethyl functionality in order to improve the biomolecule adsorption through a two-step emulsion polymerization process. They investigated the functionalized particles by optimizing the experimental parameters of the functional monomer, including reaction temperature, the amount and type of redox initiator system used, the type of addition of the initiator system, and the use of washing. They concluded that the relation between the amount of iron sulfate and the persulfate/bisulfite system added should be controlled to obtain monodisperse particles and prevent the premature coagulation of the polymer particles during the polymerization.

A semicontinuous emulsion polymerization technique for latex particle synthesis was performed by Ramakrishnan et al., 2004. They introduced a variety of particle sizes, compositions, morphologies, and surface modifications to fabricate latex composite membranes (LCMs). Arrays of stabilized latex particles on the surface of a microporous substrate form narrowly distributed interstitial pores between the particles, which serve as separation channels. They investigated the membrane performance using gas fluxes, water permeability, and the retention characterization of dextran molecules. From these tests, they concluded that the narrow, discriminating layer made of the latex particles leads to a highly efficient composite membrane.

PROTEINS

Proteins are polyamides comprised of different α-amino acids of varying hydrophobicity (Norde, 1998). Proteins are more or less amphiphilic and usually highly surface active because of the number of amino acid residues having side groups along the polypeptide chain, which contain positive or negative charges (Norde, 1998). The side-chain groups have different chemical properties such as polarity, charge, and size, and influence the chemical properties of proteins as well as determine the overall structure of the protein. For instance, the polar amino acids tend to be on the outside of the protein when they interact with water and the nonpolar amino acids are on the inside forming a hydrophobic core. The covalent linkage between two amino acids is known as a peptide bond. A peptide bond is formed when the amino group of one amino acid reacts with the carboxyl group of another amino acid to form an amide bond through the elimination of water. This arrangement gives the protein chain a polarity such that one end will have a free amino group, called the N-terminus, and the other end will have a free carboxyl group, called the C-terminus (Solomon and Fryhle, 2000). Peptide bonds tend to be planar and give the polypeptide backbone rigidity. Rotation can still occur around both of the α-carbon bonds, resulting in a polypeptide backbone with different potential conformations. Although many conformations are theoretically possible, interactions between the side-chain groups will limit the number of potential conformations and proteins tend to form a single functional conformation. In other words, the conformation, or shape of the protein, is due to the interactions of the chain side groups with one another and with the polypeptide backbone. The interactions can be intermolecular between adjacent or nearby amino acids that are close together, or distant amino acids in one protein molecule; or intermolecular between groups on different polypeptides altogether. These different types of interactions are often discussed in terms of primary, secondary, tertiary, and quaternary protein structure.

The primary amino acid sequence and positions of disulfide bonds strongly influence the overall structure of protein (Norde and Favier, 1992). For example, certain side chains will promote hydrogen bonding between neighboring amino acids of the polypeptide backbone, resulting in secondary structures such as β-sheets or α-helices. In the α-helix conformation, the peptide backbone takes on a "spiral staircase" shape that is stabilized by H-bonds between carbonyl and amide groups of every fourth amino acid residue. This restricts the rotation of the bonds in the peptide backbone, resulting in a rigid structure. Certain amino acids promote the formation of either α-helices or β-sheets due to the nature of the side-chain groups. Some side chain groups may prevent the formation of secondary structures and result in a more flexible polypeptide backbone, which is often called the random coil conformation. These secondary structures can interact with other secondary structures within the same polypeptide to form motifs or domains (i.e., a tertiary structure). A motif is a common combination of secondary structures, and a domain is a portion of a protein that folds independently. Many proteins are composed of multiple subunits and therefore exhibit quaternary structures.

INTERACTION FORCES BETWEEN PROTEINS

Proteins in aqueous solution acquire compact, ordered conformations. The compact structure is possible only if interactions within the protein molecule and interactions between the protein molecule and its environment are sufficiently favorable to compensate for the low conformational entropy (Malmsten, 1998). Protein adsorption studies often focus on structural rearrangements in the protein molecules because of their significance to the biological functioning of the molecules and the important role such rearrangements play in the mechanism of the adsorption process. Knowledge of the major interaction forces that act between protein chains and control the protein structures helps to understand the behavior of proteins at interfaces.

These forces include hydrogen bonding and Coulomb, hydrophobic, and van der Waals interactions.

COULOMB INTERACTION

Most of the amino acid residues carrying electric charge are located at the aqueous boundary of compacted protein molecules. At the isoelectric point (IEP) of the protein, where the positive and negative charges are more or less evenly distributed over the protein molecule, intramolecular electrostatic attractions dominate. Deviation to either more positive or more negative charge, however, leads to intramolecular repulsion and encourages an expanded structure. Tanford (1967) calculated the electrostatic Gibbs energy for both a compact impenetrable spherical molecule (protein) and a loose solvent-permeated spherical molecule (protein) over which the charge is spread out. From the results, he found that the repulsion force was reduced at higher ionic strength due to the screening action of ions.

HYDROGEN BONDING

Most hydrogen bonds in proteins form between amide and carbonyl groups of the polypeptide backbone (Malmsten, 1998). The number of available hydrogen bonds involving peptide units is therefore far greater than that involving side chains. Because α-helices and β-sheets are aligned more or less parallel to each other, the interchain hydrogen bonds reinforce each other. Kresheck and Klotz (1969) examined the role of peptide-peptide hydrogen bonds and concluded that hydrogen bonds alone between peptide units do not stabilize a compact structure of protein. However, because the peptide chain on the inner part of the compacted structure is shielded from water due to other interactions, hydrogen bonding between peptide groups does stabilize α-helical and β-sheet structures.

HYDROPHOBIC INTERACTION

Hydrophobic interaction refers to the spontaneous dehydration and subsequent aggregation of nonpolar components in an aqueous environment. In aqueous solutions of proteins, the various nonpolar amino acid residues will be found in the interior of the compacted molecule, and thus are shielded from water. The intramolecular hydrophobic interaction for the stability of a compact protein structure was first recognized by Kauzmann (1959). If all the hydrophobic residues are buried in the interior and all the hydrophilic residues are at the outermost border of the molecule, intramolecular hydrophobic interaction would cause a compact protein structure. However, other types of interactions, including geometrical restrictions, generally cause a fraction of the hydrophobic residues to be exposed to the aqueous environment.

VAN DER WAALS INTERACTION

The mutual interaction between ionic groups, dipoles, and induced dipoles in a protein molecule cannot be established as long as the protein structure is not known in great detail. Moreover, the surrounding aqueous medium also contains dipoles and ions that compete for participation in the interactions with groups of the protein molecule. Dispersion interactions favor a compact structure. However, because the Hamaker constant for proteins is only a little larger than that of water, van der Waals effects are relatively small (Nir, 1977).

β_2-MICROGLOBULIN (β_2M)

The protein β_2M is of particular interest because it is involved in the human disorder dialysis-related amyloidosis (DRA) (Argiles, 1996; Floege and Ketteler, 2001; Gejyo et al., 1985). DRA is a complication in end stage renal failure patients who have been on dialysis for more than 5 years (Bardin et al., 1986; Drueke, 2000). DRA develops when proteins in the blood deposit on joints and tendons, causing pain, stiffness, and fluid in the joints, as is the case with arthritis. In vivo, β_2M is present as the nonpolymorphic light chain of the class I major histocompatibility complex (MHC-I). As part of its normal catabolic cycle, β_2M dissociates from the MHC-I complex and is transported in the serum to the kidney where the majority (95%) is degraded (Floege and Ketteler, 2001). Renal failure disrupts the clearance of β_2M from the serum, resulting in an increase in β_2M concentration by up to 60-fold (Floege and Ketteler, 2001). By a mechanism that is currently not well understood, β_2M then self-associates into amyloid fibrils and typically accumulates in the musculoskeletal system (Homma et al., 1989). Analysis of *ex vivo* material has shown that the majority of amyloid fibrils in patients with DRA are present as of full-length wild-type β_2M, although significant amounts (~20–30%) of truncated or modified forms of the protein are also present (Bellotti et al., 1998; Floege and Ketteler, 2001). Native β_2M consists of a single chain of 100 amino acid residues and has a seven stranded β-sandwich fold, typical of the immunoglobulin superfamily (Saper, Bjorkman, and Wiley, 1991; Trinh et al., 2002). The normal serum concentration of β_2M is 1.0 to 2.5 mg/L. It is a small globular protein with a molecular weight of 11.8 kDa, a Stokes radius of 16 Å, and a negative charge under physiological conditions (isoelectric point, IEP = 5.7). β_2M contains two β-sheets that are held together by a single disulfide bridge between the cysteines in positions 25 and 81. β_2M cannot be removed completely by current dialysis techniques but through a better understanding of the structure and interaction forces that lead to this structure, it will be possible to more efficiently remove this problematic protein.

ALBUMIN

Albumin is the most abundant protein found in plasma and is typically present in the blood at a concentration of 35–50 g/L. According to extensive studies about its physiological and pharmacological properties, albumin has a high affinity to a very wide range of materials such as electrolytes (Cu^{+2}, Zn^{+2}), fatty acids, amino acids, metabolites, and many drug compounds (Fehske, Muller, and Wollert, 1981; Kragh-hansen, 1981; Peters, 1985). The most important physiological role of the protein is therefore to bring such solutes in the bloodstream to their target organs, as well as to maintain the pH and osmotic pressure of the plasma. Bovine serum albumin (BSA) is an ellipsoidal protein with the dimensions of $140 \times 40 \times 40$ Å (Peters, 1985). The

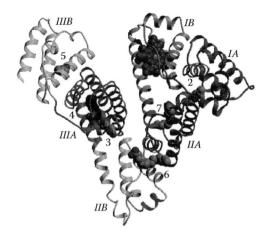

FIGURE 4.6 Secondary structure of human serum albumin (HSA) with sub-domains (Zunszain et al., 2003).

primary structure is a single helical polypeptide of 66 kDa (IEP = 4.7) with 583 residues containing 17 pairs of disulfide bridges and one free cysteine (Dugaiczyk, Law, and Dennison, 1982). BSA has been classified as a soft and flexible protein because it has a great tendency to change its conformation on adsorption to solid surfaces (Carter and Ho, 1994; Kondo, Oku, and Higashitani, 1991; Norde and Favier, 1992; Soderquist and Walton, 1980;) and consists of three homologous domains (I–III) most likely derived through gene multiplication (Brown, 1976). Each domain is composed of A and B subdomains (He and Carter, 1992). The secondary structure of human serum albumin (HSA) is shown in Figure 4.6.

Although all three domains of the HSA molecule have similar three-dimensional structures, their assembly is highly asymmetric. Domains I and II are almost perpendicular to each other to form a T-shaped assembly in which the tail of subdomain IIA is attached to the interface region between subdomains IA and IB by hydrophobic interactions and hydrogen bonds. In contrast, domain III protrudes from subdomain IIB at a 45° angle to form the Y-shaped assembly for domains II and III. Domain III interacts only with subdomain IIB. These features make the HSA molecule heart-shaped.

PROTEIN ADSORPTION

Protein adsorption studies date back to the 1930s. At the beginning, they mainly focused on the determination of the molecular weight, electrophoretic and chromatographic applications. Later, the adsorption mechanism, especially structural rearrangements, was studied. More recently, the relation between protein adsorption and biocompatibility of the sorbent materials has been investigated (Norde, 1998).

There are interaction forces at the interfaces between protein molecules and latex particles. These forces are mainly divided into the following groups: hydrophobic interaction, ionic interaction, hydrogen bonding, and van der Waals interaction (Andrade, 1985).

HYDROPHOBIC INTERACTION

It is known that hydrophobic interaction has a major role in protein adsorption phenomena. The adsorption of proteins on the low-charged latex particles occurs by this interaction force. Generally, monomers such as styrene offer hydrophobic surfaces that protein molecules adsorb to. The amount of adsorbed protein by this interaction force is maximum at the isoelectric point (IEP) of the protein, and the pH at maximum adsorption shifts to a more acidic region with an increase in ionic strength (Kondo and Higashitani, 1992; Shirahama, Suzuki, and Suzawa, 1989; Suzawa et al., 1980, 1982). By the reports (Lee et al., 1988; Suzawa et al., 1980, 1982), protein adsorption was greater on a hydrophobic surface than on a hydrophilic one, implying that hydrophobic interaction is one of most dominant forces in protein adsorption.

IONIC INTERACTION

Negatively charged latex particles have ionic functional groups on their surfaces, such as sulfate and carboxylate anions. Sulfate groups originate from an initiator such as potassium persulfate, and carboxylic groups originate from hydrophilic co-monomers such as acrylic acid (AA) or methacrylic acid (MAA). Ionic bonds are formed between the negative charges of these latex particles and positive surface charges of protein molecules. The conventional low-charged latex particles rarely form these ionic bonds.

HYDROGEN BONDING

A hydrogen bond is a strong secondary interatomic bond that exists between a bound hydrogen atom (its unscreened proton) and the electrons of adjacent atoms (Carllister, 1999). Protein can be adsorbed on hydrophilic polar surfaces through hydrogen bonds. Hydrogen bonds are frequently formed between hydroxyl-carbonyl or amide-hydroxyl. Hydroxyl-hydroxyl or amide-hydroxyl bonds are also formed in protein adsorption.

VAN DER WAALS INTERACTION

This interaction force is operative over small distances and only when water has been excluded and the two nonpolar groups come close to each other. Lewin's calculation showed that the van der Waals interaction is negligible compared with the forces involved in the entropy increases, that is, hydrophobic interaction (Lewin, 1974).

CONCLUDING REMARKS

End stage renal disease is a chronic condition in which kidney function is impaired to the extent that the patient's survival requires removal of toxins from the blood by dialysis therapy or kidney transplantation. Due in part to the limited availability of kidneys for transplantation, hemodialysis is the primary clinical treatment for the patients with ESRD. The central element of a hemodialysis machine is the semi-permeable membrane that allows for selective transport of low molecular weight biological metabolites from the blood. One limitation of current dialysis technologies is the inability to efficiently remove middle molecular weight toxins such as β_2-microglobulin (β_2M) and interleukin 6 (IL-6). β_2M is a causative protein of dialysis-related amyloidosis (DRA), a disease arising in patients with chronic kidney failure as a serious complication of long-term hemodialysis treatment. Although attempts have been made to increase the efficiency of middle molecular weight toxin removal by changes in the membrane pore size and the use of innovative materials to adsorb these toxins, removal efficiency is not as high as those achieved by a normal healthy kidney. Traditional membranes have a number of processing and performance limitations, such as a restricted choice of surface chemistries and limited control of porosity. The development of novel engineering membrane technology is needed to remove middle molecule toxins.

Polymeric latex particles have received increasing attention in areas such as solid-phase supports in biological applications. This is because of the well-defined colloidal and surface characteristics of the particles. By using a seed emulsion polymerization method it is possible to synthesize monodisperse latex particles with various particle size ranges and surface chemistries. Functionalized core-shell latex particles can be introduced by multistep emulsion polymerization. The core-shell particles are useful in a broad range of applications because of their improved physical and chemical properties over their single-component counterparts. Through the development of a hemodialysis membrane using monodisperse latex particles, improvements in advanced separation treatment for patients with ESRD may be realized. This requires the maximum removal of middle molecular weight proteins with minimal removal of other beneficial proteins such as albumin. Thus, an understanding of the fundamental interactions between the particles and biopolymers is vital to maximize the performance of this membrane technology. The state of the art related to synthesis and engineering of polymeric latex particles for hemodialysis has been reviewed in this chapter. An experimental study involving fictionalization of latex particles for selective separation of middle molecular weight protein molecules from human blood has been conducted and the details are presented in part II.

REFERENCES

Abel, J.J., Rowntree, L.G., and Turner, B.B. (1990). On the removal of diffusible substances from the circulating blood by means of dialysis (reprinted from the Transactions of the Association of American Physicians, 1913). *Transfusion Science* 11, 164–165.

Andrade, J.D. (1985). *Surface and Interfacial Aspects of Biomedical Polymer*. Plenum, New York.

Argiles, A. (1996). Beta 2-microglobulin amyloidosis. *Nephrology* 2, 373–386.

Baker, R.W. (2004). *Membrane Technology and Applications,* 2nd ed. John Wiley & Sons, New York.

Bangs, L.B. (1988). Latex agglutination tests. *American Clinical Laboratory News* 7, 20–26.

Bardin, T., Zingraff, J., Kuntz, D., and Drueke, T. (1986). Dialysis related amyloidosis. *Nephrology Dialysis Transplantation* 1, 151–154.

Becker, J.W., and Reeke, G.N. (1985). 3 dimensional structure of beta 2-microglobulin. *Proceedings of the National Academy of Sciences of the United States of America* **82**, 4225–4229.

Bellotti, V., Stoppini, M., Mangione, P., Sunde, M., Robinson, C., Asti, L., Brancaccio, D., and Ferri, G. (1998). Beta 2-microglobulin can be refolded into a native state from *ex vivo* amyloid fibrils. *European Journal of Biochemistry* **258**, 61–67.

Bernhard, W.F., Lafarge, C.G., Liss, R.H., Szycher, M., Berger, R.L., and Poirier, V. (1978). Appraisal of blood trauma and blood prosthetic interface during left ventricular bypass in calf and humans. *Annals of Thoracic Surgery* **26**, 427–437.

Bhat, A.A., and Pangarkar, V.G. (2000). Methanol selective membranes for the pervaporative separation of methanol-toluene mixtures. *Journal of Membrane Science* **167**, 187–201.

Bhutto, A.A., Vesely, D., and Gabrys, B.J. (2003). Miscibility and interactions in polystyrene and sodium sulfonated polystyrene with poly(vinyl methyl ether) PVME blends. Part II. FTIR. *Polymer* **44**, 6627–6631.

Blanco, J.F., Nguyen, Q.T., and Schaetzel, P. (2001). Novel hydrophilic membrane materials: Sulfonated polyethersulfone Cardo. *Journal of Membrane Science* **186**, 267–279.

Bosch, T., Lennertz, A., Kordes, B., and Samtelben, W. (1999). Low density lipoprotein hemoperfusion by direct adsorption of lipoproteins from whole blood (DALI apheresis): Clinical esperience from a single center. *Therapeutic Apheresis* **3**, 209–213.

Bowry, S.K. (2002). Dialysis membranes today. *International Journal of Artificial Organs* **25**, 447–460.

Brescia, M.J., Cimino, J.E., Appel, K., and Hurwich, B.J. (1966). Chronic hemodialysis using venipuncture and a surgically created arteriovenous fistula. *New England Journal of Medicine* **275**, 1089–1092.

Brink, L.E.S., Elbers, S.J.G., Robbertsen, T., and Both, P. (1993). The anti-fouling action of polymers preadsorbed on ultrafiltration and microfiltration membranes. *Journal of Membrane Science* **76**, 281–291.

Broadhead, K.W., and Tresco, P.A. (1998). Effects of fabrication conditions on the structure and function of membranes formed from poly(acrylonitrile-vinylchloride). *Journal of Membrane Science* **147**, 235–245.

Brown, J.R. (1976). Structural origins of mammalian albumin. *Federation Proceedings* **35**, 2141–2144.

Bryjak, M., Hodge, H., and Dach, B. (1998). Modification of porous polyacrylonitrile membrane. *Angewandte Makromolekulare Chemie* **260**, 25–29.

Carllister, W.D. (1999). *Materials Science and Engineering: An Introduction*, 5th ed. John Wiley & Sons, New York.

Carter, D.C., and Ho, J.X. (1994). Structure of serum albumin. In *Advances in Protein Chemistry*, Vol. 45, 153–203.

Casadevall, N., and Rossert, J. (2005). Importance of biologic follow-ons: Experience with EPO. *Best Practice & Research Clinical Haematology* **18**, 381–387.

Castino, F., Scheucher, K., Malchesky, P.S., Koshino, I., and Nose, Y. (1976). Microemboli free blood detoxification utilizing plasma filtration. *Transactions American Society for Artificial Internal Organs* **22**, 637–645.

Chen, Y.C., Dimonie, V., and El-Aasser, M.S. (1991a). Interfacial phenomena controlling particle morphology of composite latexes. *Journal of Applied Polymer Science* **42**, 1049–1063.

Chen, Y.C., Dimonie, V., and El-Aasser, M.S. (1991b). Effect of interfacial phenomena on the development of particle morphology in a polymer latex system. *Macromolecules* **24**, 3779–3787.

Chen, Y.C., Dimonie, V., and El-Aasser, M.S. (1992a). Role of surfactant in composite latex particle morphology. *Journal of Applied Polymer Science* **45**, 487–499.

Chen, Y.C., Dimonie, V., and El-Aasser, M.S. (1992b). Theoretical aspects of developing latex particle morphology. *Pure and Applied Chemistry* **64**, 1691–1696.

Chen, Y.C., Dimonie, V.L., Shaffer, O.L., and El-Aasser, M.S. (1993). Development of morphology in latex particles- the interplay between thermodynamic and kinetic parameters. *Polymer International* **30**, 185–194.

Chern, C.S., and Poehlein, G.W. (1990). Polymerization in nonuniform latex particles. 2. Kinetics of 2 phase emulsion polymerization. *Journal of Polymer Science Part a: Polymer Chemistry* **28**, 3055–3071.

Chern, C.S., Lee, C.K., and Chang, C.J. (2003). Synthesis and characterization of amphoteric latex particles. *Colloid and Polymer Science* **281**, 1092–1098.

Cheung, A.K., Agodoa, L.Y., Daugirdas, J.T., Depner, T.A., Gotch, F.A., Greene, T., Levin, N.W., Leypoldt, J.K., and Beck, G.J. (1999). Effects of hemodialyzer reuse on clearances of urea and beta(2)-microglobulin. *Journal of the American Society of Nephrology* **10**, 117–127.

Cousin, P., and Smith, P. (1994). Synthesis and characterization of styrene based microbeads possessing amine functionality. *Journal of Applied Polymer Science* **54**, 1631–1641.

Cowgill, L.D., and Francey, T. (2001). Biocompatibility of adsorptive hem perfusion with or without dialysis in normal and uremic dogs. *Abstract European Renal Association: European Dialysis Transplant Association*, 292.

Cristol, J.P., Canaud, B., Rabesandratana, H., Gaillard, I., Serre, A., and Mion, C. (1994). Enhancement of reactive oxygen species production and cell surface markers expression due to hemodialysis. *Nephrology Dialysis Transplantation* **9**, 389–394.

Dahman, Y., Puskas, J.E., Margaritis, A., Merali, Z., and Cunningham, M. (2003). Novel thymine-functionalized polystyrenes for applications in biotechnology. Polymer synthesis and characterization. *Macromolecules* **36**, 2198–2205.

de Francisco, A.L.M., Ghezzi, P.M., Brendolan, A., Fiorini, F., La Greca, G., Ronco, C., Arias, M., Gervasio, R., and Tetta, C. (2000). Hemodiafiltration with online regeneration of the ultrafiltrate. *Kidney International* **58**, S66–S71.

Delacal, J.C., Urzay, R., Zamora, A., Forcada, J., and Asua, J.M. (1990). Simulation of the latex particle morphology. *Journal of Polymer Science Part A: Polymer Chemistry* **28**, 1011–1031.

Dhondt, A., Vanholder, R., Van Biesen, W., and Lameire, N. (2000). The removal of uremic toxins. *Kidney International* **58**, S47–S59.

Drueke, T.B. (2000). Beta(2)-microglobulin and amyloidosis. *Nephrology Dialysis Transplantation* **15**, 17–24.

Dugaiczyk, A., Law, S.W., and Dennison, O.E. (1982). Nucleotide sequence and the encoded amino acids of human serum albumin messenger RNA. *Proceedings of the National Academy of Sciences of the United States of America: Biological Sciences* **79**, 71–75.

Elaïssari, A., Chauvet, J.P., Halle, M.A., Decavallas, O., Pichot, C., and Cros, P. (1998). Effect of charge nature on the adsorption of single-stranded DNA fragments onto latex particles. *Journal of Colloid and Interface Science* **202**, 251–260.

Fehske, K.J., Muller, W.E., and Wollert, U. (1981). The location of drug binding sites in human serum albumin. *Biochemical Pharmacology* **30**, 687–692.

Floege, J., and Ketteler, M. (2001). Beta(2)-microglobulin derived amyloidosis: An update. *Kidney International* **59**, S164–S171.

Furuyoshi, S., Kobayashi, A., Tamai, N., Yasuda, A., Takada, S., Tani, N., Nakazawa, R., Mimura, H., Gejyo, F., and Arakava, M. (1991). *Blood Purification* **9**, 9.

Ganachaud, F., Mouterde, G., Delair, T., Elaissari, A., and Pichot, C. (1995). Preparation and characterization of cationic polystyrene latex particles of different aminated surface charges. *Polymers for Advanced Technologies* **6**, 480–488.

Gejyo, F., Homma, N., Hasegawa, S., and Arakawa, M. (1993). A new therapeutic approach to dialysis amyloidosis-intensive removal of beta2-microglobulin with adsorbent column. *Artificial Organs* **17**, 240–243.

Gejyo, F., Teramura, T., Ei, I., Arakawa, M., Nakazawa, R., Azuma, N., Suzuki, M., Furuyoshi, S., Nankou, T., Takata, S., and Yasuda, A. (1995). Long-term clinical evaluation of an adsorbent column (BM-O1) of direct hemoperfusion type for beta(2)-microglobulin on the treatment of dialysis-related amyloidosis. *Artificial Organs* **19**, 1222–1226.

Gejyo, F., Yamada, T., Odani, S., Nakagawa, Y., Arakawa, M., Kunitomo, T., Kataoka, H., Suzuki, M., Hirasawa, Y., Shirahama, T., Cohen, A.S., and Schmid, K. (1985). A new form of amyloid protein associated with chronic hemodialysis was identified as beta2 microglobulin. *Biochemical and Biophysical Research Communications* **129**, 701–706.

Gilbert, R.G. (1995). *Emulsion Polymerization: A Mechanistic Approach.* Academic Press, London.

Godjevargova, T., Konsulov, V., and Dimov, A. (1999). Preparation of an ultrafiltration membrane from the copolymer of acrylonitrile-glycidylmethacrylate utilized for immobilization of glucose oxidase. *Journal of Membrane Science* **152**, 235–240.

Godjevargova, T., Konsulov, V., Dimov, A., and Vasileva, N. (2000). Behavior of glucose oxidase immobilized on ultrafiltration membranes obtained by copolymerizing acrylonitrile and N-vinylimidazol. *Journal of Membrane Science* **172**, 279–285.

Godjevargova, Z., Dimov, A., and Petrov, S. (1992). Chemical modification of dialysis membrane based on poly(acrylonitrile methylmethacrylate sodium vinylsulphonate). *Journal of Applied Polymer Science* **44**, 2139–2143.

Goosen, M.F.A., Sefton, M.V., and Hatton, M.W.C. (1980). Inactivation of thrombin by antithrombin III on a heparinized biomaterial. *Thrombosis Research* **20**, 543–554.

Guiver, M.D., Black, P., Tam, C.M., and Deslandes, Y. (1993). Functionalized polysulfone membranes by heterogeneous lithiation. *Journal of Applied Polymer Science* **48**, 1597–1606.

Haag-Weber, M., Cohen, G., and Horl, W.H. (2000). Clinical significance of granulocyte-inhibiting proteins. *Nephrology Dialysis Transplantation* **15**, 15–16.

Hancock, L.F., Fagan, S.M., and Ziolo, M.S. (2000). Hydrophilic, semipermeable membranes fabricated with poly(ethylene oxide)-polysulfone block copolymer. *Biomaterials* **21**, 725–733.

Hansen, F.K., and Ugelstad, J. (1979). Particle nucleation in emulsion polymerization. 2. Nucleation in emulsifier free systems investigated by seed polymerization. *Journal of Polymer Science Part A Polymer Chemistry* **17**, 3033–3045.

Hariss, J.M. (1992). *Poly(ethylene glycol) Chemistry.* Plemum, New York.

Harkins, W.D. (1947). A general theory of the mechanism of emulsion polymerization. *Journal of the American Chemical Society* **69**, 1428–1444.

He, X.M., and Carter, D.C. (1992). Atomic structure and chemistry of human serum albumin. *Nature* **358**, 209–215.

Hergeth, W.D., Bittrich, H.J., Eichhorn, F., Schlenker, S., Schmutzler, K., and Steinau, U.J. (1989). Polymerizations in the presence of seeds. 5. Core shell structure of 2 stage emulsion polymers. *Polymer* **30**, 1913–1917.

Hicke, H.G., Lehmann, I., Malsch, G., Ulbricht, M., and Becker, M. (2002). Preparation and characterization of a novel solvent-resistant and autoclavable polymer membrane. *Journal of Membrane Science* **198**, 187–196.

Higuchi, A., Mishima, S., and Nakagawa, T. (1991). Separation of proteins by surface modified polysulfone membranes. *Journal of Membrane Science* **57**, 175–185.

Higuchi, A., and Nakagawa, T. (1990). Surface modified polysulfone hollow fibers. 3. Fibers having a hydroxide group. *Journal of Applied Polymer Science* **41**, 1973–1979.

Higuchi, A., Sugiyama, K., Yoon, B.O., Sakurai, M., Hara, M., Sumita, M., Sugawara, S., and Shirai, T. (2003). Serum protein adsorption and platelet adhesion on pluronic (TM)-adsorbed polysulfone membranes. *Biomaterials* **24**, 3235–3245.

Hoenich, N.A., and Katopodis, K.P. (2002). Clinical characterization of a new polymeric membrane for use in renal replacement therapy. *Biomaterials* **23**, 3853–3858.

Homma, N., Gejyo, F., Isemura, M., and Arakawa, M. (1989). Collagen binding affinity of beta2 microglobulin, a preprotein of hemodialysis associated amyloidosis. *Nephron* **53**, 37–40.

Huysmans, K., Lins, R.L., Daelemans, R., Zachee, P., and De Broe, M.E. (1998). Hypertension and accelerated atherosclerosis in endstage renal disease. *Journal of Nephrology* **11**, 185–195.

Ishihara, K., Aragaki, R., Ueda, T., Watenabe, A., and Nakabayashi, N. (1990). Reduced thrombogenicity of polymers having phospholipid polar groups. *Journal of Biomedical Materials Research* **24**, 1069–1077.

Izquierdo, M.P.S., Martin-Molina, A., Ramos, J., Rus, A., Borque, L., Forcada, J., and Galisteo-Gonzalez, F. (2004). Amino, chloromethyl and acetal-functionalized latex particles for immunoassays: A comparative study. *Journal of Immunological Methods* **287**, 159–167.

James, R.O. (1985). *Polymer colloids*. Elsevier, Amsterdam.

Jaroniec, M., and Madey, R. (1988). *Physical Adsorption on Heterogenous Solids*. Elsevier, New York.

Jung, B. S., Yoon, J. K., Kim, B., and Rhee, H. W. (2005). Effect of crystallization and annealing on polyacrylonitrile membranes for ultrafiltration. *Journal of Membrane Science,* **246**, 67–76.

Kauzmann, W. (1959). Some factors in the interpretation of protein denaturation. *Advances in Protein Chemistry* **14**, 1–63.

Keusch, P., and Williams, D.J. (1973). Equilibrium encapsulation of polystyrene latex particles. *Journal of Polymer Science Part A: Polymer Chemistry* **11**, 143–162.

Khan, A.R., Baker, B.M., Ghosh, P., Biddison, W.E., and Wiley, D.C. (2000). The structure and stability of an HLA-A*0201/octameric tax peptide complex with an empty conserved peptide-N-terminal binding site. *Journal of Immunology* **164**, 6398–6405.

Kim, J.H., Chainey, M., Elaasser, M.S., and Vanderhoff, J.W. (1989). Preparation of highly sulfonated polystyrene model colloids. *Journal of Polymer Science Part A: Polymer Chemistry* **27**, 3187–3199.

Kim, K.J., Fane, A.G., and Fell, C.J.D. (1988). The performance of ultrafiltration membranes pretreated by polymers. *Desalination* **70**, 229–249.

Kim, Y.W., Ahn, W.S., Kim, J.J., and Kim, Y.H. (2005). *In situ* fabrication of self-transformable and hydrophilic poly(ethylene glycol) derivative-modified polysulfone membranes. *Biomaterials* **26**, 2867–2875.

Kim, Y.W., Kim, J.J., and Kim, Y.H. (2003). Surface characterization of biocompatible polysulfone membranes modified with poly(ethylene glycol) derivatives. *Korean Journal of Chemical Engineering* **20**, 1158–1165.

Ko, M.K., Pellegrino, J.J., Nassimbene, R., and Marko, P. (1993). Characterization of the adsorption fouling layer using globular proteins on ultrafiltration membranes. *Journal of Membrane Science* **76**, 101–120.

Kolarz, B.N., and Jermakowiczbartkowiak, D. (1995). Hyper-crosslinked sorbents for hemoperfusion. *Angewandte Makromolekulare Chemie* **227**, 57–68.

Kolarz, B.N., Trochimczuk, A., Wojaczynska, M., Bartkowiak, D. (1989). Polymer sorbents for hemoperfusion. I. Preparation, structure, and sorption properties of chemically modified polymer sorbents for hemoperfusion. *Polimery Medycynie* **19**, 29–35.

Kondo, A., and Higashitani, K. (1992). Adsorption of model proteins with wide variation in molecular properties on colloidal particles. *Journal of Colloid and Interface Science* **150**, 344–351.

Kondo, A., Oku, S., and Higashitani, K. (1991). Structural changes in protein molecules adsorbed on ultrafine silica particles. *Journal of Colloid and Interface Science* **143**, 214–221.

Korshak, V.V., Leikin, Y.A., Neronov, A.Y., Tikhonova, L.A., Ryabov, A.V., Kabanov, O.V., Gorchakov, V.D., and Evseev, N.G. (1978). *Sorbents Compatible with Blood for Extraction of Exogenous and Endogenous Toxins.* Demande.

Kragh-hansen, U. (1981). Molecular aspects of ligand binding to serum albumin. *Pharmacological Reviews* **33**, 17–53.

Kresheck, G.C., and Klotz, I.M. (1969). Thermodynamics of transfer of amides from an apolar to an aqueous solution. *Biochemistry* **8**, 8–&.

Kurbel, S., Radic, R., Kotromanovic, Z., Puseljic, Z., and Kratofil, B. (2003). A calcium homeostasis model: orchestration of fast acting PTH and calcitonin with slow calcitriol. *Medical Hypotheses* **61**, 346–350.

Kurisawa, M., Terano, M., and Yui, N. (1995). Doublestimuli responsive degradable hydrogels for drug delivery—Interpenetrating polymer networks composed of oligopeptide-terminated poly(ethylene glycol) and dextran. *Macromolecular Rapid Communications* **16**, 663–666.

Lee, C.F. (2000). The properties of core-shell composite polymer latex. Effect of heating on the morphology and physical properties of PMMA/PS core-shell composite latex and the polymer blends. *Polymer* **41**, 1337–1344.

Lee, C.F. (2002). The morphology of composite polymer particles produced by multistage soapless seeded emulsion polymerization. *Colloid and Polymer Science* **280**, 116–123.

Lee, C.F., Chiu, W.Y., and Chern, Y.C. (1995). Kinetic study on the poly(methyl methacrylate) seeded soapless emulsion polymerization of styrene. 2. Kinetic-model. *Journal of Applied Polymer Science* **57**, 591–603.

Lee, C.F., Young, T.H., Huang, Y.H., and Chiu, W.Y. (2000). Synthesis and properties of polymer latex with carboxylic acid functional groups for immunological studies. *Polymer* **41**, 8565–8571.

Lee, J.H., Yoon, J.Y., and Kim, W.S. (1998). Continuous separation of serum proteins using a stirred cell charged with carboxylated and sulfonated microspheres. *Biomedical Chromatography* **12**, 330–334.

Lee, S. H., and Ruckenstein, E. (1988). Adsorption of proteins onto polymeric surfaces of different hydrophilicities- a case study with bovine serum albumin. *Journal of Colloid and Interface Science* **125**, 365–379.

Legido-Quigley, C., Marlin, N., and Smith, N.W. (2004). Comparison of styrene-divinylbenzene based monoliths and vydac nano-liquid chromatography columns for protein analysis. *Journal of Chromatography A* **1030**, 195–200.

Lewin, S. (1974). *Displacement of Water and Its Control of Biochemical Reactions.* Academic Press, New York.

Leypoldt, J.K., Cheung, A.K., and Deeter, R.B. (1998). Effect of hemodialysis resue: Dissociation between clearances of small large solutes. *American Journal of Kidney Diseases* **32**, 295–301.

Li, J., Guo, S.Y., and Li, X.N. (2005). Degradation kinetics of polystyrene and EPDM melts under ultrasonic irradiation. *Polymer Degradation and Stability* **89**, 6–14.

Lin, W.C., Liu, T.Y., and Yang, M.C. (2004). Hemocompatibility of polyacrylonitrile dialysis membrane immobilized with chitosan and heparin conjugate. *Biomaterials* **25**, 1947–1957.

Lonnemann, G., and Koch, K.M. (2002). Beta(2)-microglobulin amyloidosis: effects of ultra-pure dialysate and type of dialyzer membrane. *Journal of the American Society of Nephrology* **13**, S72–S77.

Lu, M., Keskkula, H., and Paul, D.R. (1996). Thermodynamics of solubilization of functional copolymers in the grafted shell of core-shell impact modifiers. 2. Experimental. *Polymer* **37**, 125–135.

Luck, M., Paulke, B.R., Schroder, W., Blunk, T., and Muller, R.H. (1998). Analysis of plasma protein adsorption on polymeric nanoparticles with different surface characteristics. *Journal of Biomedical Materials Research* **39**, 478–485.

Luo, Y., Rong, M.Z., Zhang, M.Q., and Friedrich, K. (2004). Surface grafting onto SiC nanoparticles with glycidyl methacrylate in emulsion. *Journal of Polymer Science Part a-Polymer Chemistry* **42**, 3842–3852.

Magalhaes, I., Pihan, J.C., and Falla, J. (2004). A latex agglutination test for the field determination of abnormal vitellogenin production in male fishes contaminated by estrogen mimics. *Journal of Physics—Condensed Matter* **16**, S2167–S2175.

Malchesky, P.S. (2004). *Extracorpeal Artificial Organs*. Elsevier Academic Press, San Diego.

Malchesky, P.S., Piatkiewicz, W., Varnes, W.G., Ondercin, L., and Nose, Y. (1978). Sorbent membranes-device designs, evaluations and potential applications. *Artificial Organs* **2**, 367–371.

Malmsten, M. (1998). Biopolymers at interface, Marcel Dekker, New York.

McDonald, C. (2003). Personal communications.

McMurry, J. (2004). *Organic Chemistry*. Thompson-Brooks/Cole, Belmont.

Meier, P., von Fliedner, V., Markert, M., van Melle, G., Deppisch, R., and Wauters, J.P. (2000). One-year immunological evaluation of chronic hemodialysis in end-stage renal disease patients. *Blood Purification* **18**, 128–137.

Mikhalovsky, S.V. (1987). The organism detoxification with bioselective carbon sorbents. *Artificial Organs* **11**, 504–504.

Mikhalovsky, S.V. (1989). The detoxication with bioselective carbon sorbents. *Biomaterials Artificial Cells and Artificial Organs* **17**, 157–160.

Min, T.I., Klein, A., El-Aasser, M.S., and Vanderhoff, J.W. (1983). Morphology and grafting in polybutylacrylate-polystyrene core-shell emulsion polymerization. *Journal of Polymer Science Part A: Polymer Chemistry* **21**, 2845–2861.

Miraballes-Martinez, I., and Forcada, J. (2000). Synthesis of latex particles with surface amino groups. *Journal of Polymer Science Part A: Polymer Chemistry* **38**, 4230–4237.

Miraballes-Martinez, I., Martin-Molina, A., Galisteo-Gonzalez, F., and Forcada, J. (2001). Synthesis of amino-functionalized latex particles by a multistep method. *Journal of Polymer Science Part A: Polymer Chemistry* **39**, 2929–2936.

Mok, S., Worsfold, D.J., Fouda, A., and Matsuura, T. (1994). Surface modification of polyethersulfone hollow fiber membranes by gamma ray irradiation. *Journal of Applied Polymer Science* **51**, 193–199.

Musale, D.A., and Kulkarni, S.S. (1997). Relative rates of protein transmission through poly(acrylonitrile) based ultrafiltration membranes. *Journal of Membrane Science* **136**, 13–23.

Musyanovych, A., and Adler, H.J.P. (2005). Grafting of amino functional monomer onto initiator-modified polystyrene particles. *Langmuir* **21**, 2209–2217.

Myers, D. (1999). *Surfaces, Interfaces, and Colloids: Princeples and Applications*, 2nd ed. Wiley-VCH, New York.

Nabe, A., Staude, E., and Belfort, G. (1997). Surface modification of polysulfone ultrafiltration membranes and fouling by BSA solutions. *Journal of Membrane Science* **133**, 57–72.

Nagai, K., Ohashi, T., Kaneko, R., and Taniguchi, T. (1999). Preparation and applications of polymeric microspheres having active ester groups. *Colloids and Surfaces A—Physicochemical and Engineering Aspects* **153**, 133–136.

Nakazawa, R., Azuma, N., Suzuki, M., Nakatani, M., Nankou, T., Furuyoshi, S., Yasuda, A., Takata, S., Tani, N., and Kobayashi, F. (1993). A new treatment for dialysis related amyloidosis with beta2-microglobulin adsorbent column. *International Journal of Artificial Organs* **16**, 823–829.

Nelliappan, V., El-Aasser, M.S., Klein, A., Daniels, E.S., Roberts, J.E., and Pearson, R.A. (1997). Effect of the core/shell latex particle interphase on the mechanical behavior of rubber-toughened poly(methyl methacrylate). *Journal of Applied Polymer Science* **65**, 581–593.

Nestor, J., Esquena, J., Solans, C., Levecke, B., Booten, K., and Tadros, T.F. (2005). Emulsion polymerization of styrene and methyl methacrylate using a hydrophobically modified inulin and comparison with other surfactants. *Langmuir* **21**, 4837–4841.

Nie, F.Q., Xu, Z.K., Ming, Y.Q., Kou, R.Q., Liu, Z.M., and Wang, S.Y. (2004). Preparation and characterization of polyacrylonitrile-based membranes: Effects of internal coagulant on poly(acrylonitrile-co-maleic acid) ultrafiltration hollow fiber membranes. *Desalination* **160**, 43–50.

Niemeyer, C.M. (2001). Nanoparticles, proteins, and nucleic acids: biotechnology meets materials science. *Angewandte Chemie–International Edition* **40**, 4128–4158.

Nilsson, J.L. (1990). Protein fouling of UF membranes - causes and consequences. *Journal of Membrane Science* **52**, 121–142.

Nilsson, K.G.I. (1989). Preparation of nanoparticles conjugated with enzyme and antibody and their use in heterogeneous enzyme immunoassays. *Journal of Immunological Methods* **122**, 273–277.

Nir, S. (1977). van der Waals interactions between surfaces of biological interest. *Progress in Surface Science* **8**, 1–58.

Nissenson, A.R., Fine, R.N., and Gentile, D.E. (1995). *Clinical Dialysis*, 3rd ed. Appleton & Lange, Connecticut.

National Kidney Foundation. (2002). Kidney disease: Are you at increased risk for chronic kidney disease? http://www.kidney.org/atoz/atozitem.cfm?id=134.

Norde, W., and Favier, J.P. (1992). Structure of adsorbed and desorbed proteins. *Colloids and Surfaces* **64**, 87–93.

Norde, W. (1998). Biopolymers at interfaces: chap2. driving forces for protein adsorption at solid surfaces, Marcel Dekker, New York.

Obrien, T. F., Baxter, C. R., and Teschan, P. E. (1959). Prophylactic daily hemodialys*Transactions American Society for Artificial Internal Organs,* **5**, 77–77.

Odian, G. (1991). *Principles of Polymerization*, 3rd ed. John Wiley & Sons, New York.

Oh, J.T., and Kim, J.H. (2000). Preparation and properties of immobilized amyloglucosidase on nonporous PS/PNaSS microspheres. *Enzyme and Microbial Technology* **27**, 356–361.

Okubo, M., and Nakagawa, T. (1992). Preparation of micron-size monodisperse polymer particles having highly crosslinked structures and vinyl groups by seeded polymerization of divinylbenzene using the dynamic swelling method. *Colloid and Polymer Science* **270**, 853–858.

Onen, F., Turkay, C., Meydan, A., Doknetas, H.S., Sumer, H., Hocaoglu, L., Icagasioglu, S., and Bakici, M. Z. (1998). Prevalence of rheumatoid factor (RF) and anti-native-DNA antibodies (anti-n DNA) in different age subpopulations. *Journal of Medical Sciences* **28**, 85–88.

Oscik, J. (1982). *Adsorption*. John Wiley & Sons. New York.

Panichi, V., Bianchi, A.M., Andreini, B., Casarosa, L., Migliori, M., De Pietro, S., Taccola, D., Giovannini, L., and Palla, R. (1998). Biocompatibility evaluation of polyamide hemofiltration. *International Journal of Artificial Organs* **21**, 408–413.

Park, J.G., Kim, J.W., and Suh, K.D. (2001). Chloromethyl functionalized polymer particles through seeded polymerization. *Colloids and Surfaces A—Physicochemical and Engineering Aspects* **191**, 193–199.

Percival, D.A. (1996). The measurement of hormones and bacterial antigens using rapid particle-based immunoassays. *Pure and Applied Chemistry* **68**, 1893–1895.

Peters, T. (1985). Serum albumin. *Advances in Protein Chemistry* **37**, 161–245.

Pieracci, J., Crivello, J.V., and Belfort, G. (2002). Increasing membrane permeability of UV-modified poly(ether sulfone) ultrafiltration membranes. *Journal of Membrane Science* **202**, 1–16.

Piskin, E., Tuncel, A., Denizli, A., and Ayhan, H. (1994). Monosize microbeads based on polystyrene and their modified forms for some selected medical and biological applications. *Journal of Biomaterials Science—Polymer Edition* **5**, 451–471.

Premstaller, A., Oberacher, H., and Huber, C.G. (2000). High-performance liquid chromatography-electrospray ionization mass spectrometry of single- and double-stranded nucleic acids using monolithic capillary columns. *Analytical Chemistry* **72**, 4386–4393.

Premstaller, A., Oberacher, H., Walcher, W., Timperio, A.M., Zolla, L., Chervet, J.P., Cavusoglu, N., van Dorsselaer, A., and Huber, C.G. (2001). High-performance liquid chromatography-electrospray ionization mass spectrometry using monolithic capillary columns for proteomic studies. *Analytical Chemistry* **73**, 2390–2396.

Putnam, F.W. (1984). *The Plasma Proteins*, 2nd ed. Academic Press, London.

Quinton, W., Dillard, D., and Scribner, B.H. (1960). Cannulation of blood vessels for prolonged hemodialysis. *Transactions American Society for Artificial Internal Organs* **6**, 104–113.

Radomska-Galant, I., and Basinska, T. (2003). Poly(styrene/alpha-tert-butoxy-omega-vinylbenzylpolyglycidol) microspheres for immunodiagnostics—Principle of a novel latex test based on combined electrophoretic mobility and particle aggregation measurements. *Biomacromolecules* **4**, 1848–1855.

Ramakrishnan, S., McDonald, C. J., Prud'homme, R. K., and Carbeck, J. D. (2004). Latex composite membranes: structure and properties of the discriminating layer. *Journal of membrane Science* **154**, 1–13.

Ray, S.K., Sawant, S.B., Joshi, J.B., and Pangarkar, V.G. (1999). Methanol selective membranes for separation of methanol-ethylene glycol mixtures by pervaporation. *Journal of Membrane Science* **154**, 1–13.

Reb, P., Margarit-Puri, K., Klapper, M., and Mullen, K. (2000). Polymerizable and nonpolymerizable isophthalic acid derivatives as surfactants in emulsion polymerization. *Macromolecules* **33**, 7718–7723.

Riceevans, C.A., and Diplock, A.T. (1993). Current status of antioxidant therapy. *Free Radical Biology and Medicine* **15**, 77–96.

Roe, C.P. (1968). Surface chemistry aspects of emulsion polymerization. *Industrial and Engineering Chemistry* **60**, 20–&.

Ronco, C., Brendolan, A., Winchester, J.F., Golds, E., Clemmer, J., Polaschegg, H.D., Muller, T.E., La Greca, G., and Levin, N.W. (2001a). First clinical experience with an adjunctive hemoperfusion device designed specifically to remove beta(2)-microglobulin in hemodialysis. *Blood Purification* **19**, 260–263.

Ronco, C., Ghezzi, P.M., Hoenich, N.A., and Delfino, P.G. (2001b). *Membranes and Filters for Hemodialysis: Database 2001*. Karger, Basel.

Roselaar, S.E., Nazhat, N.B., Winyard, P.G., Jones, P., Cunningham, J., and Blake, D.R. (1995). Detection of oxidants in uremic plasma by electron spin resonance spectroscopy. *Kidney International* **48**, 199–206.

Samtleben, W., Blumenstein, M., Bosch, T., Lysaght, M.J., and Schmidt, B. (1996a). Plasma therapy at klinikum grosshadern: A 15-year retrospective. *Artificial Organs* **20**, 408–413.

Samtleben, W., Gurland, H.J., Lysaght, M.J., and Winchester, J.F. (1996b). *Plasma Exchange and Hemoperfusion*. Kluwer Academic, Dordrecht.

Santos, R.M., and Forcada, J. (1997). Acetal-functionalized polymer particles useful for immunoassays. *Journal of Polymer Science Part A. Polymer Chemistry* **35**, 1605–1610.

Saper, M.A., Bjorkman, P.J., and Wiley, D.C. (1991). Refined structure of the human histocompatibility antigen HLA-A2 at 2.6Å resolution. *Journal of Molecular Biology* **219**, 277–319.

Sarobe, J., and Forcada, J. (1998). Synthesis of monodisperse polymer particles with chloromethyl functionality. *Colloids and Surfaces A—Physicochemical and Engineering Aspects* **135**, 293–297.

Sarobe, J., Molina-Bolivar, J.A., Forcada, J., Galisteo, F., and Hidalgo-Alvarez, R. (1998). Functionalized monodisperse particles with chloromethyl groups for the covalent coupling of proteins. *Macromolecules* **31**, 4282–4287.

Scharnagl, N., and Buschatz, H. (2001). Polyacrylonitrile (PAN) membranes for ultra- and microfiltration. *Desalination* **139**, 191–198.

Schmidt, E. (1972). Synthetic latex foam rubber and method of making same. U.S. Patent 3 673 133.

Shimizu, N., Sugimoto, K., Tang, J. W., Nishi, T., Sato, I., Hiramoto, M., Aizawa, S., Hatakeyama, M., Ohba, R., Hatori, H., Yoshikawa, T., Suzuki, F., Oomori, A., Tanaka, H., Kawaguchi, H., Watanabe, H., and Handa, H. (2000). High-performance affinity beads for identifying drug receptors. *Nature Biotechnology* **18**, 877–881.

Shirahama, H., Suzuki, K., and Suzawa, T. (1989). Bovine hemoglobin adsorption onto polymer lattices. *Journal of Colloid and Interface Science* **129**, 483–490.

Singer, J.M., and Plotz, C.M. (1956). Latex Fixation Test. 1. Application to the serologic diagnosis of rheumatoid arthritis. *American Journal of Medicine* **21**, 888–892.

Singh, N.P., Bansal, R., Thakur, A., Kohli, R., Bansal, R. C., and Agarwal, S.K. (2003). Effect of membrane composition on cytokine production and clinical symptoms during hemodialysis: a crossover study. *Renal Failure* **25**, 419–430.

Smith, W.V., and Ewart, R.H. (1948). Kinetics of emulsion polymerization. *Journal of Chemical Physics* **16**, 592–599.

Snyder, H.W., Cochran, S.K., Balint, J.P., Bertram, J.H., Mittelman, A., Guthrie, T.H., and Jones, F.R. (1992). Experience with protein α-immunoadsorption in treatment resistant adult immune thrombocytopenic purpura. *Blood* **79**, 2237–2245.

Soderquist, M.E., and Walton, A.G. (1980). Structural changes in proteins adsorbed on polymer surfaces. *Journal of Colloid and Interface Science* **75**, 386–397.

Solomon, T.W., and Fryhle, C.B. (2000). *Organic Chemistry*, 7th ed. John Wiley & Sons, New York.

Song, Y.Q., Sheng, J., Wei, M., and Yuan, X.B. (2000). Surface modification of polysulfone membranes by low-temperature plasma-graft poly(ethylene glycol) onto polysulfone membranes. *Journal of Applied Polymer Science* **78**, 979–985.

Suzawa, T., Shirahama, H., and Fujimoto, T. (1982). Adsorption of bovine serum albumin onto homo-polymer and co-polymer lattices. *Journal of Colloid and Interface Science* **86**, 144–150.

Takenaka, T., Itaya, Y., Tsuchiya, Y., Kobayashi, K., and Suzuki, H. (2001). Fitness of biocompatible high-flux hemodiafiltration for dialysis-related amyloidosis. *Blood Purification* **19**, 10–14.

Tamai, H., Hasegawa, M., and Suzawa, T. (1989). Surface characterization of hydrophilic functional polymer latex particles. *Journal of Applied Polymer Science* **38**, 403–412.

Tanford, C. (1967). *Physical Chemistry of Macromolecules*. John Wiley & Sons, New York.

Tangpasuthadol, V., Pongchaisirikul, N., and Hoven, V.P. (2003). Surface modification of chitosan films—effects of hydrophobicity on protein adsorption. *Carbohydrate Research* **338**, 937–942.

Trinh, C.H., Smith, D.P., Kalverda, A.P., Phillips, S.E.V., and Radford, S.E. (2002). Crystal structure of monomeric human beta 2 microglobulin reveals clues to its amyloidogenic properties. *Proceedings of the National Academy of Sciences of the United States of America* **99**, 9771–9776.

Trotta, F., Drioli, E., Baggiani, C., and Lacopo, D. (2002). Molecular imprinted polymeric membrane for naringin recognition. *Journal of Membrane Science* **201**, 77–84.

Tuncel, A., Tuncel, M., Ergun, B., Alagoz, C., and Bahar, T. (2002). Carboxyl carrying large uniform latex particles. *Colloids and Surfaces A—Physicochemical and Engineering Aspects* **197**, 79–94.

Ugelstad, J., Kaggerud, K.H., and Fitch, R.M. (1980). *Polymer Colloid II*. Plenum, New York.

USRDS (2003). United State Renal Date System (USRDS) 2003 annual report.

Van Noordwijk, J. (2001). *Dialyzing for Life: The Development of the Artificial Kidney*. Kluwer Academic, Dordrecht.

Vanholder, R., Desmet, R., Vogeleere, P., and Ringoir, S. (1995). Middle molecules toxicity and removal by hemodialysis and related strategies. *Artificial Organs* **19**, 1120–1125.

Vincent, C., Chanard, J., Caudwell, V., Lavaud, S., Wong, T., and Revillard, J.P. (1992). Kinetics of I-125 beta2 microglobulin turnover in dialyzed patients. *Kidney International* **42**, 1434–1443.

Ward, R.A., Schmidt, S., Hullin, J., Hillebrand, G.F., and Samtleben, W. (2000). A comparison of on-line hemodiafiltration and high-flux hemodialysis: A prospective clinical study. *Journal of the American Society of Nephrology* **11**, 2344–2350.

Wavhal, D.S., and Fisher, E.R. (2002). Hydrophilic modification of polyethersulfone membranes by low temperature plasma induced graft polymerization. *Journal of Membrane Science* **209**, 255–269.

Westhuyzen, J., Foreman, K., Battistutta, D., Saltissi, D., and Fleming, S.J. (1992). Effect of dialyzer reprocessing with renalin on serum beta2 microglobulin and complement activation in hemodialysis patients. *American Journal of Nephrology* **12**, 29–36.

Williams, K.M., Arthur, S.J., Burrell, G., Kelly, F., Phillips, D.W., and Marshall, T. (2003). An evaluation of protein assays for quantitative determination of drugs. *Journal of Biochemical and Biophysical Methods* **57**, 45–55.

Winchester, J.F., Ronco, C., Brady, J.A., Cowgill, L.D., Salsberg, J., Yousha, E., Choquette, M., Albright, R., Clemmer, J., Davankov, V., Tsyurupa, M., Pavlova, L., Pavlov, M., Cohen, G., Horl, W., Gotch, F., and Levin, N.W. (2002). The next step from high-flux dialysis: application of sorbent technology. *Blood Purification* **20**, 81–86.

Winchester, J.P., Ronco, J.A., Brendolan, A., Davankov, V., Tsyurupa, M., Pavlova, L., Clemme, J., Polaschegg, H.D., Muller, T.E., La Greca, G., and Levin, N.W. (2001). Rationale for combined hemoperfusion/hemodialysis in uremia. *Contribution of Nephrology* **133**, 174–179.

Winslow, R.M., Vandergriff, K.D., and Motterlini, R. (1993). Mechanism of hemoglobin toxicity. *Thrombosis and Haemostasis* **70**, 36–41.

Yang, S.C., Ge, H.X., Hu, Y., Jiang, X.Q., and Yang, C.Z. (2000). Formation of positively charged poly(butyl cyanoacrylate) nanoparticles stabilized with chitosan. *Colloid and Polymer Science* **278**, 285–292.

Zollars, R.L. (1979). Kinetics of the emulsion polymerization of vinyl acetate. *Journal of Applied Polymer Science* **24**, 1353–1370.

Zou, D., Ma, S., Guan, R., Park, M., Sun, L., Aklonis, J.J., and Salovey, R. (1992). Model filled polymers. 5. Synthesis of cross-linked monodisperse polymethacrylate beads. *Journal of Polymer Science Part A. Polymer Chemistry* **30**, 137–144.

Zunszain, P.A., Ghuman, J., Komatsu, T., Tsuchida, E., and Curry, S. (2003). Crystal structural analysis of human serum albumin complexed with hemin and fatty acid. *BMC Structural Biology, BiomMed Central* doi:10.1186/1472-6807-3-6. The electronic version of this article is the complete one and can be found online at: http://www.biomed-central.com/1472-6807/3/6

5 Synthesis and Engineering of Polymeric Latex Particles for Hemodialysis

Part II—An Experimental Study

S. Kim, H. El-Shall, R. Partch,
T. Morey, and B. Koopman

CONTENTS

INTRODUCTION

The flexibility in latex surface properties is a significant advantage to this material's ability to separate proteins since the selectivity is strongly dependent on the charge interactions of both the protein and the latex particles under the conditions of separation (Menon and Zydney, 1999; Chun and Stroeve, 2002). First, all sorbents were designed to have negative surface properties because most plasma proteins in blood are negative and should not be removed with charge interaction between proteins and sorbents at physiological condition (pH 7.4). It was reported that a negatively charged surface was more blood compatible than a positive one (Srinivasan and Sawyer, 1971). The toxin protein, β_2-microglobulin (β_2M) can be selectively adsorbed on the surface of latex particles by the design of a suitable partially hydrophobic surface where only β_2M protein can be anchored. Albumin adsorption on the latex particle is not allowed because charge repulsion is more dominant than hydrophobic interaction with side-on mode. Figure 5.1 shows the schematic representation of the selective adsorption of β_2M protein on the engineered latex particles.

Hydrophilic/hydrophobic microdomain structures were proven to be more blood compatible (Higuchi, Ishida, and Nakagawa, 1993; Deppisch et al., 1998). The optimization of a suitable hydrophobic-to-hydrophilic ratio is also important for the biocompatibility. The monomers, such as styrene (St), methyl methacrylate (MMA) and acrylic acid (AA) are widely used hydrophobic and hydrophilic monomer models in emulsion polymerization process.

In summary, the background and fundamental literature survey about the history, material properties and limitation of existing hemodialysis membranes; latex particle preparation, surface chemistry, and manufacturing process; target proteins and protein adsorption; and finally the hypothesis for a toxin removal with high separation efficiency have been suggested. The materials and characterization methodology for achievement of suggested hypotheses are described in.

EXPERIMENTAL METHODOLOGY

MATERIAL

Styrene (St) monomer used for preparing seed and core particles, and acrylic acid (AA), and methyl methacrylate (MMA) monomers for shell formation, were purchased from Fisher Scientific and used without purification. Sodium persulfate (SPS) and sodium bicarbonate (SBC) were obtained from Fisher Scientific and used as received. Divinylbenzene (DVB), crossliking agent, was purchased from Aldrich. The anionic surfactant, Aerosol MA80-I (sodium di(1,3-dimethylbutyl) sulfosuccinate), was kindly donated by Cytec. Ion exchange resin, AG 501-x8 Resin (20–50

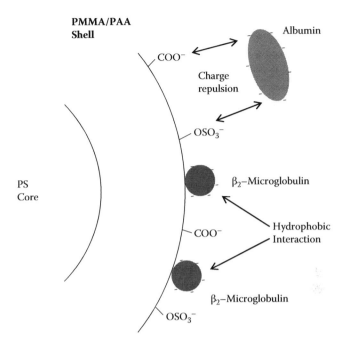

FIGURE 5.1 Schematic representation of the protein adsorption on the core-shell latex particles at pH 7.4.

mesh), was purchased from Bio-Rad Laboratories, Inc. This is mixed-bed resin having H^+ and OH^- as the ionic forms and is 8% crosslinked. Bovine serum albumin (BSA) (heat shock treated) was purchased from Fisher Scientific. The isoelectric point for BSA is 4.7–4.9 and molecular weight is 66,300 Da. β_2-microglobulin (β_2M) was purchased from ICN Biomedicals, Inc. β_2M was separated from patients with chronic renal disease. This was lyophilized from ammonium bicarbonate. The molecular weight by SDS-PAGE was approximately 12,000 Da. BSA and β_2M proteins were used as received without further purification. In order to investigate the protein adsorption evaluation, the bicinchoninic acid protein assay kits were purchased from Sigma (Cat # BCA-1) and Pierce Biotechnology (Cat # 23225). The chemical structures for the main chemicals are shown in Figure 5.2.

LATEX PARTICLE PREPARATION

The experimental setup and the particle preparation scheme in various types and size ranges of latex particles are shown in Figure 5.3 and Figure 5.4.

SEED LATEX PARTICLE PREPARATION

Polystyrene seed particles were synthesized using a typical semi-continuous emulsion polymerization. Polystyrene latex particles were prepared in a four-necked Paar glass vessel equipped with mechanical glass stir, thermometer, glass funnel for nitrogen

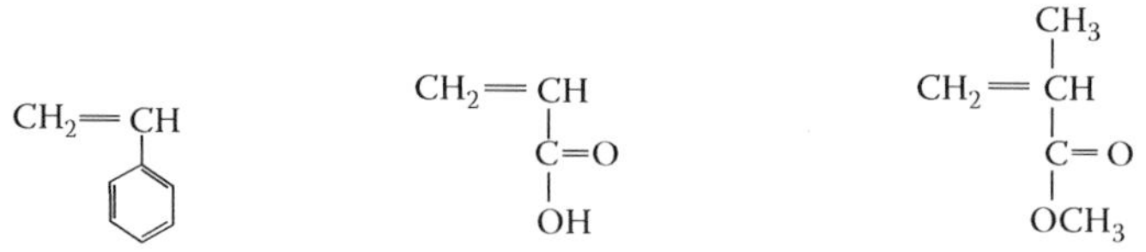

$$CH_2\!=\!CH$$

Styrene (Mw = 104.15)

$$CH_2\!=\!CH$$
$$C\!=\!O$$
$$OH$$

Acrylic acid (Mw = 72.06)

$$CH_2\!=\!CH$$
$$\overset{CH_3}{\underset{}{\mid}}$$
$$C\!=\!O$$
$$OCH_3$$

Methyl methacrylate (Mw = 100.12)

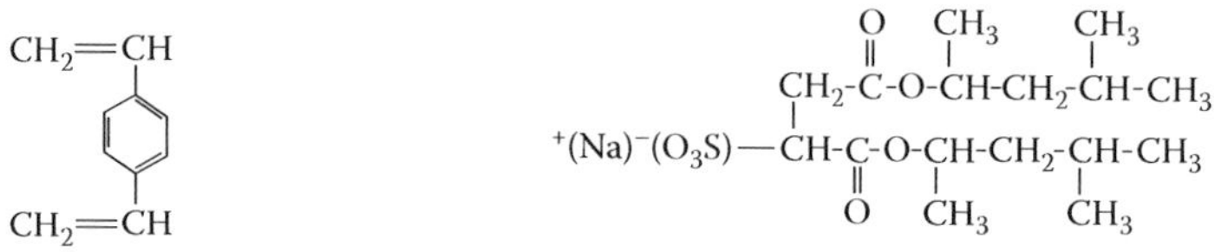

$$CH_2\!=\!CH$$

$$CH_2\!=\!CH$$

Divinylbenzene (Mw = 130.2)

$$CH_2\text{-}\overset{O}{\overset{\parallel}{C}}\text{-O-}\overset{CH_3}{\underset{}{\mid}}CH\text{-}CH_2\text{-}\overset{CH_3}{\underset{}{\mid}}CH\text{-}CH_3$$
$$^+(Na)^-(O_3S)\!-\!CH\text{-}\overset{O}{\overset{\parallel}{C}}\text{-O-}\overset{CH_3}{\underset{}{\mid}}CH\text{-}CH_2\text{-}\overset{CH_3}{\underset{}{\mid}}CH\text{-}CH_3$$

Anionic surfactant (Aerosol MA80–I) (Mw = 388.5)

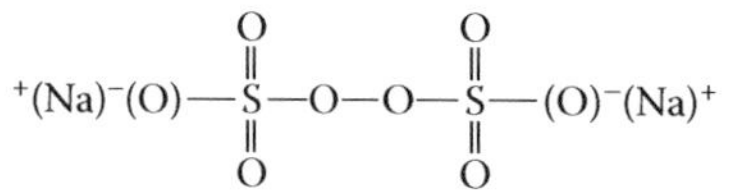

$$^+(Na)^-(O)\!-\!\overset{O}{\overset{\parallel}{\underset{\underset{O}{\parallel}}{S}}}\!-\!O\!-\!O\!-\!\overset{O}{\overset{\parallel}{\underset{\underset{O}{\parallel}}{S}}}\!-\!(O)^-(Na)^+$$

Initiator, sodium persulfate (Mw = 238.11)

FIGURE 5.2 Chemical structure of main chemicals.

purge, and tube for monomer feeding. The reactor was placed in a silicon oil bath on a hot plate, which accommodated homogenous and stable temperature control. The vessel was firstly charged with de-ionized (DI) water, emulsifier, sodium persulfate (SPS) initiator, sodium bicarbonate (SBC), and styrene (St) and divinylbenzene (DVB) monomers under nitrogen atmosphere, and then heated to 72 ± 2°C for 1 h. Then, additional St and DVB monomers, SPS initiator, and SBC were fed separately using fine fluid metering pumps (Model RHSY, Fluid Metering Inc., NY) under

1. Motor
2. Mechanical stirrer bar
3. Thermometer
4. Inlet of N_2
5. Inlet for initiator and monomer
6. Glass reactor (1L)
7. Oil bath
8. Hot plate
9. Pump for initiator
10. Pump for monomer
11. Tubes for reactant feed

FIGURE 5.3 Experimental setup for semicontinuous emulsion polymerization.

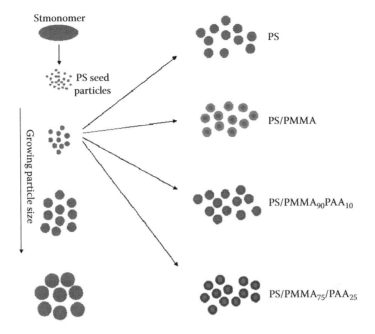

FIGURE 5.4 The particle preparation scheme in various types and size ranges of latex particles.

nitrogen atmosphere for 2 h. After complete feeding of all reactants, the reaction was continued for another hour at $80 \pm 1°C$, and then cooled to room temperature.

SEEDED LATEX PARTICLE PREPARATION

The particle growth was done by seeded continuous emulsion polymerization. The vessel was charged with the polystyrene seeds and DI water to the desired solid content, and heated to $72 \pm 2°C$ for 1 h under nitrogen atmosphere. Then, St and DVB monomers and a solution of SPS and SBC in DI water were continuously added using a metering pump with a precisely controlled feed rate. After feeding of all reactants, the reaction was continued for 3 h, keeping a temperature of $80 \pm 2°C$, then to room temperature.

CORE-SHELL LATEX PARTICLE PREPARATION

Monodisperse and spherical polystyrene seeds were used for the synthesis of core-shell structured latex particles. Selected polystyrene seeds with 280 nm of mean diameter were charged into a 500-mL glass flask under nitrogen atmosphere. DI water was added to make a desired solid content. The emulsion was heated to $80 \pm 2°C$ for 1 h. Monomers, initiator, and SBC for shell were added by metering pumps with fine controlled feed rate. Then the reaction was sustained for an additional 3 h at $90 \pm 2°C$, and then cooled to room temperature.

CHARACTERIZATION METHODOLOGY

Synthesized latex particles were purified with ion exchange resin. 200 grams (10% w/w polystyrene solid content) of emulsion were mixed with 4 g of ion exchange resin and stirred for 40 min to remove unreacted monomers, initiators and other impurities. Then the emulsion was diluted with DI water to a specific degree of solid content as needed.

DEGREE OF CONVERSION

The degree of monomer-to-polymer conversion was determined gravimetrically (Lee and Chiu, 1995) before the latex cleaning process. The synthesized emulsion raw materials were weighed and dried in a conventional oven at 120°C for 30 min to evaporate unreacted monomers and water. Remaining solid was weighed and the degree of conversion calculated. This process was repeated three times to obtain a mean value of the degree of conversion.

FOURIER TRANSFORM INFRARED (FTIR) SPECTROSCOPY

A molecule absorbs only selected frequencies (energy) of infrared radiation, which matches the natural vibration frequencies of the molecule, and can serve for identification of the chemical structure of the sample (Pavin, Lampman, and Kriz, 1996).

The synthesis of latex particles such as polystyrene (PS) homopolymer and PS/PMMA, $PS/PMMA_{90}PAA_{10}$, $PS/PMMA_{75}PAA_{25}$ core-shell copolymers, were verified with FTIR. Purified latex particles were dried in a vacuum oven at 40°C for 24 h before analyzing by FTIR (Nicolet Magma, USA). KBr was also dried in a vacuum oven at 120°C for 3 h. 10 milligrams of dried latex particles were weighed and mixed with 250 mg KBr. Transmission spectra using drift mode were plotted with 128 scans and 4 cm^{-1} resolution.

QUASIELASTIC LIGHT SCATTERING (QELS)

Latex particle size was measured by Brookhaven ZetaPlus particle size analyzer. The solid concentration of latex particles was 0.001% (w/v).

SCANNING ELECTRON MICROSCOPE (SEM)

Particle size and surface morphology were characterized by FE-SEM (JEOL JSM-6335F, Japan). Very dilute latex particles with water were dried on a silicon wafer at room temperature and coated with carbon as thin as possible. Secondary electron image mode with 15 KV of accelerating voltage was used. Magnification range was used between 10,000× and 70,000×.

ZETA POTENTIAL MEASUREMENT

Zeta potential measurement was carried out using Brookhaven ZetaPlus zeta-potential analyzer. Synthesized and purified latex particles were diluted with DI water to 0.01 wt % and the dispersions adjusted to six different pH values (2.05, 3.45, 4.81,

5.65, 6.43 and 7.47). Each dispersion was transferred to a standard cuvette for zeta potential measurement. At least ten measurements for each sample were taken and averaged.

PROTEIN ADSORPTION

Protein adsorption on latex particles was performed with bovine serum albumin (BSA) as a standard model protein and β_2-microglobulin (β_2M) as a target protein in two types of buffer solution: phosphate buffer (PB) and phosphate buffered saline (PBS). 0.345 g of sodium phosphate monobasic ($NaH_2PO_4 \cdot H_2O$) was dissolved in 500 mL DI water to make 5 mM PB solution. 0.345 grams of sodium phosphate monobasic ($NaH_2PO_4 \cdot H_2O$) and 4.178 g (143 mM) sodium chloride (NaCl) were dissolved in 500 mL DI water to make the PBS solution. The synthesized latex particles were diluted with each buffer to be 0.5% (w/w) solid content and adjusted to the desired pH values such as 3.2, 4.8 and 7.4, and mixed with a selected protein, where BSA concentration were 0.05, 0.1, 0.3, 0.5 and 0.7 mg/mL and β2M were 0.015, 0.030, 0.045 and 0.060 mg/mL. Each mixture was gently rotated in the incubator at 37°C for 12 h. Thereafter, the latex-protein mixture was centrifuged at 13,000 rpm for 15 min. The amount of protein adsorbed was determined by quantifying the free protein in the supernatant after centrifugation using the bicinchoninic acid (BCA) assay method (Lowry et al., 1951; Smith et al., 1985; Wiechelman, Braun, and Fitzpatrick, 1988; Brown, Jarvis, and Hyland, 1989; Baptista et al., 2003). The BCA assay kit consists of Reagent A containing bicinchoninic acid, sodium carbonate, sodium tartrate, and sodium bicarbonate in 0.1 N NaOH with pH = 11.25 and Reagent B containing 4% (w/v) copper (II) sulfate pentahydrate. The BCA working reagent was prepared by mixing 50 parts Reagent A with 1 part Reagent B. 100 μL of protein supernatant was then mixed with 2 ml BCA working reagent in UV cuvette. Incubation was allowed for color development at room temperature for 2 h. The absorbance on a UV-VIS spectrophotometer (Perkin-Elmer Lambda 800, USA) at 562 nm was read. The unknown concentration of sample protein was determined using a standard curve, which was established with already-known concentrations of a protein. The adsorbed amount per unit surface area was determined by the mass balance of the protein.

An adsorption kinetic experiment was also performed with BSA and β_2M in 5 mM phosphate buffer at 37°C and pH = 7.4. The concentrations of proteins were 0.7 mg/mL and 0.06 mg/mL for BSA and β_2M, respectively.

BLOOD BIOCOMPATIBILITY BY HEMOLYSIS TEST

Hemolysis is the destruction of red blood cells (RBC), which leads to the release of hemoglobin into the blood plasma. Healthy people donated samples of whole blood. The red blood cells were separated by centrifuging the normal whole blood at 1500 rpm for 15 min and washing with isotonic phosphate buffer solution (PBS) at pH 7.4 to remove debris and serum protein. This process was repeated three times. Prepared latex particles were re-dispersed in PBS by sonification to obtain homogeneously well-dispersed latex particles. 100 mL of the mixture of RBCs (3 parts) and

PBS (11 parts) was added to the 1.0 mL of 0.5% (w/w) particle suspension. PBS was used as a negative control (0% hemolysis) and DI water was used as a positive control to produce 100% hemoglobin released from completely destroyed RBCs. The mixture was incubated in a water bath with gentle shaking for 30 min at 37°C and then centrifuged at 1500 rpm for 15 min. 100 mL of the supernatant was mixed with 2 mL of the mixture of ethanol (99%) and hydrochloric acid (37%) (EtOH/HCl = 200/5, w/w) to prevent precipitation of hemoglobin. In order to confirm the free particles, the mixture was centrifuged again at 13,000 rpm at room temperature. The supernatant was then transferred to the UV cuvette. The amount of hemoglobin release was determined by monitoring the UV absorbance at the wavelength of 397 nm.

RESULTS AND DISCUSSION

POLYMERIZATION OF POLYSTYRENE SEED LATEX PARTICLES

Polymer particles used as lubricants, packing material for chromatography columns and medical diagnostic standards are required to have uniform particle shape and size. Uniform particle size is the first requirement for latex composite membranes formed by close-packed particle arrays with the interstitial spaces serving as pores for eluent size discrimination (Jons, Ries, and McDonald, 1999; Ramakrishnan et al., 2004). Although suspension polymerization has been the conventional way to produce latex particles, their uniformity is insufficient in the present work. The so-called seed emulsion polymerization method comprising a vinyl monomer to be absorbed into fine monodiserse seed particles and polymerizing the monomer to increase the sizes does achieve the required uniformity. In order to obtain monodisperse larger particles by this method, the procedure of absorption of monomer into fine polymer particles and polymerization of the monomer is repeated.

Cross-linked polystyrene seed latex particles were synthesized using batch semicontinuous emulsion polymerization. The polymerization conditions and seed particle sizes are summarized in Table 5.1. The emulsion polymerization of latex particles has to be carried out in a narrow range of surfactant concentration, where particles are stable (Nestor et al., 2005). Wide-range changes in surfactant concentration cause flocculation or phase separation, resulting in a broad range of particle

TABLE 5.1
The Polymerization Recipe of PS Seed Latex Particles

PS Seeds	MA-80 (g)	St (g)	DVB (g)	H_2O (g)	SPS (g)	SB (g)	Mean Particle Diameter (nm)	Conversion (%)
PS_{S1}	2.59	99.48	0.52	156.4	1.9	1.12	126	96.5
PS_{S2}	2.33	99.48	0.52	156.4	1.9	1.12	171	95.3
PS_{S3}	2.07	99.48	0.52	156.4	1.9	1.12	182	95.7
PS_{S4}	1.81	99.48	0.52	156.4	1.9	1.12	216	97.1

size distribution and an unstable emulsion system. The surfactant amount was carefully added from 1.81 g to 2.59 g in 0.26-g increments to know the relationship between the surfactant concentration and particle size. Other reactants were fixed in weight. As expected, the seed particle size increased as the amount of surfactant decreased. The amount of emulsifier affects the number of micelles formed and their size. Large amounts of emulsifier produce larger numbers of smaller-sized particles (Odian, 1991). Monomer-to-polymer conversion was obtained by calculation using gravimetric data and was more than 95% for all seed polymers ($PS_{S1} \sim PS_{S4}$).

The particle morphology and size of representative polystyrene seed particles can be seen in Figure 5.5. The surface morphology and shape of seed particles were very smooth and spherical, respectively. The particle size distribution of each PS latex particle obtained also showed high uniformity from the SEM.

Seeded emulsion polymerization has been conducted for several decades, and various mechanisms have been proposed. Grancio and Williams (1970) suggested that the growing polymer particles consist of an expanding polymer-rich core surrounded

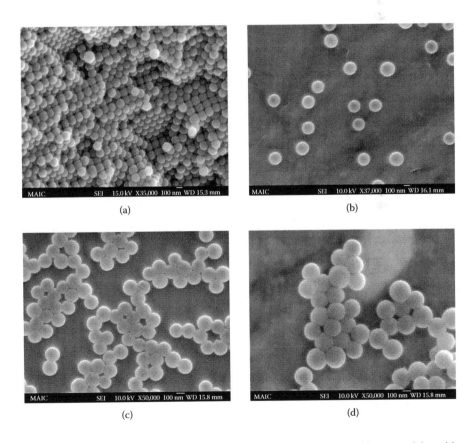

(a)

(b)

(c)

(d)

FIGURE 5.5 Scanning electron micrograph (SEM) of polystyrene seed latex particles with various amounts of surfactant : (a) PS_{S1} (2.59 g), (b) PS_{S2} (2.33 g), (c) PS_{S3} (2.07 g), (d) PS_{S4} (1.81 g).

by a monomer-rich shell, with the outer shell providing a major locus of polymerization. Seeded emulsion polymerization is normally used for preparing latex particles
with less than 1 μm in size (Gandhi et al., 1990; Park et al, 1990; Zou et al., 1990;
1992; Cha and Choe, 1995). Such latexes can be obtained from pre-prepared seed
particles using a further growth step by first swelling the seed latex particles with
the additional monomer required to grow the particles to the desired size and then
effecting polymerization. Several size ranges of monodisperse PS latex particles
were prepared by the multistep seeded emulsion polymerization method. In order
to obtain larger seeded latex particles, PS_{S3} particles were used as seeds because of
their perfect spherical shape and smooth surface as well as their high uniformity in
particle size distribution. The seeds introduced in this process serve as nucleation
sites for particle growth.

One of the recipes used in seeded emulsion polymerization is shown at Table 5.2.
The seed size of PS_{S3} was 182 nm. The expected latex particle size by calculation
was 260 nm in diameter after seeded emulsion polymerization was done. A simple
geometric calculation allows prediction of the ultimate size of latex particle reached
after the polymerization. The solid content used for seed was about 40% and the
solid content after polymerization was set to about 30%.

Cross-linked PS latex particles with size ranges from 258 nm ~ 790 nm in diameter were obtained using sequential multistep growth emulsion polymerization
reactions (Table 5.2). Especially, the latex particles polymerized by the recipe in
Table 5.2 are shown in Figure 5.6. From the SEM characterization, no newly nucle-
ated particles formed; thus product monodispersity was retained. Particles with less
than 500 nm diameter in size were very spherical in and of desired narrow size distribution. However, as the particles were grown larger than 500 nm diameter, they
became nonspherical and had uneven surfaces.

The amount of surfactant as well as monomer used in an emulsion system is the
primary determinant of the particle diameter (Odian, 1991). The dependence on the
particle size and the amount of surfactant is shown in Figure 5.7. As the surfactant-
to-monomer ratio decreases, mean particle diameter increases. Generally, lower

TABLE 5.2

The Recipe of Seeded Emulsion Polymerization of PS Particles

Stream	Component	Weight (g)	Normalized Wt (g)	Feed Rate (g/min)
Initial charge	Seed latex: 0.18 μm	100 @ 39.24%	124.07 @ 39.24%	
	H_2O	147.0	205.3	
A	St	71.0	99.2	0.44
	DVB	0.6	0.8	
B	H_2O	75.0	104.7	0.46
	SBC	0.7	1.0	
	SPS	0.7	1.0	

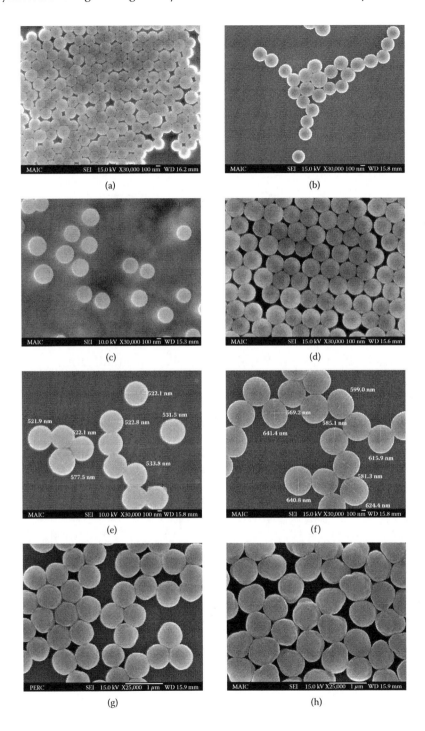

FIGURE 5.6 SEM of PS latex particles with various size ranges: (a) 258 nm, (b) 320 nm, (c) 370 nm, (d) 410 nm, (e) 525 nm, (f) 585 nm, (g) 640 nm, (h) 790 nm.

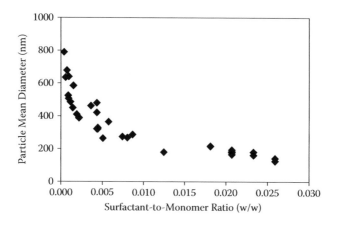

FIGURE 5.7 Dependence of the particle size on surfactant-to-monomer ratio.

surfactant concentrations form fewer micelles resulting in larger particles (Evans and Wennerstron, 1999).

Polystyrene (PS) Core and Polymethyl Methacrylate (PMMA)/ Polyacrylic Acid (PAA) Shell Latex Particles

Seed particles with 258 nm size (Fig. 5.6(a)) were used to form core-shell structures. Methyl methacrylate (MMA) and acrylic acid (AA) monomers were introduced to make the latex particle surface more hydrophilic than bare polystyrene particles. MMA is more hydrophilic than styrene and the surface carboxyl group of AA may have many promising applications in biomedical and biochemical fields. The recipes for core-shell structure are shown in Table 5.3.

Three types of core shell structures were prepared. The $PMMA_{75}PAA_{25}$ shell is a copolymer consisting of a PMMA-to-PAA ratio of 75% to 25% by weight. The $PMMA_{90}PAA_{10}$ shell is a copolymer with a PMMA-to-PAA ratio of 90% to 10% by weight. $PMMA_{100}$ is the PMMA homopolymer shell on the PS core. The particle sizes of these core-shell particles and bare PS particles were prepared to be about 370 nm in

TABLE 5.3
The Preparation Recipe of PS Core with Various Shell Latex Particles

Shell	MMA (g)	AA (g)	PS Seed (g)	H₂O (g)	SPS (g)	SB (g)	Monomer Feed Rate (g/min)
$PMMA_{75}PAA_{25}$	24.0	8.0	50	177	0.2	0.2	0.53
$PMMA_{90}PAA_{10}$	28.8	3.2	50	177	0.2	0.2	0.64
$PMMA_{100}$	32.0	0.0	50	177	0.2	0.2	0.27

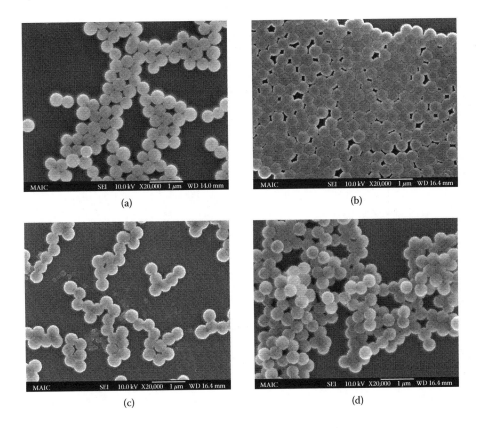

FIGURE 5.8 Scanning electron micrograph of latex particles (a) PS, (b) PS/PMMA$_{75}$PAA$_{25}$, (c) PS/PMMA$_{90}$PAA$_{10}$, (d) PS/PMMA$_{100}$.

diameter to give latex particles in the same experimental conditions for characterization. SEM micrographs of PS and core-shell particles are shown at Figure 5.8.

FOURIER TRANSFORM INFRARED SPECTROSCOPY (FTIR)

Exemplary FTIR spectra of polymerized latex particles are shown in Figure 5.9. There are a number of characteristic peaks for PS (Bhutto et al., 2003; Li et al., 2005): C-H aromatic peaks between 3002 cm^{-1} and 3103 cm^{-1}, aliphatic C-H peaks at 2900 cm^{-1} and 2850 cm^{-1}, respectively. There is a carbonyl (C=O) characteristic peak at 1730 cm^{-1} for PS/PMMA$_{100}$ core-shell latex particles.

The broad OH group peak at 3400 cm^{-1} appears in the spectra of particles with PAA in the shell, the intensity of which depends on the ratio of PMMA to PAA. This peak is not seen for PS and PS core with only PMMA in the shell.

SURFACE PROPERTIES OF LATEX PARTICLES

The surface electrical property of the latex prepared in this work was determined by zeta potential measurement. The results as a function of pH and ion strength for

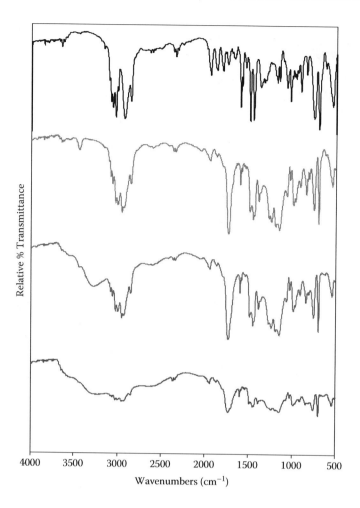

FIGURE 5.9 FTIR spectroscopy of polymerized latex particles.

all latex particles are shown in Figure 5.10 and Figure 5.11. PB stands for phosphate buffer in water and PBS for phosphate buffered saline.

All latex particles exhibited the expected negative surface charge due to the sulfate groups (Van den Hul et al., 1970) originating from initiators. Carboxylate groups from polyacrylic acid also caused negative surface charge for the latex particles PS/PMMA$_{75}$PAA$_{25}$ and PS/PMMA$_{90}$PAA$_{10}$. Sulfate groups have pK_a values in the range 1 to 2, while the carboxylate pK_a values are between 4 and 6 (Ottewill and Shaw, 1967).

Since the zeta potential values of all latex particles synthesized were negative between pH = 2.0 and pH = 7.8, the isoelecric point (IEP) of these latex particles would be less than pH = 2.0. The zeta potential of the latex particles in phosphate buffer (PB) was more negative than that in phosphate buffered saline (PBS). This is

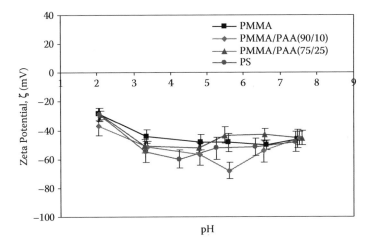

FIGURE 5.10 Zeta potential of latex particles in phosphate buffer (PB).

because high concentrations of sodium and chloride electrolytes in PBS compresses the electrical double layer (Hunter, 1981; Burns and Zydney, 2000).

PROTEIN ADSORPTION STUDY

An adsorption study of blood proteins on monodisperse latex particles is of great importance because many biomedical applications, such as artificial tissues and organs, drug delivery systems, biosensors, solid-phase immunoassays, immunomagnetic cell separations and immobilized enzymes or catalysts (Guiot and Couvreur, 1986; Bangs, 1987; Rembaum and Tokes, 1988; Berge et al., 1990), need the

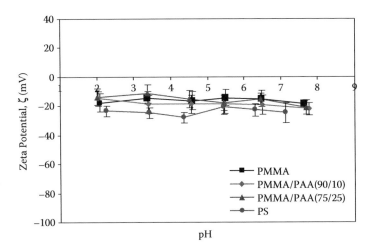

FIGURE 5.11 Zeta potential of latex particles in phosphate buffered saline (PBS).

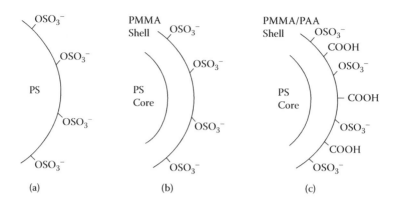

FIGURE 5.12 Schematic representation of latex particles (a) PS, (b) PS/PMMA, (c) PS/ PMMAPAA.

combination of blood proteins and latex particles. Although many researchers have systematically studied adsorption of proteins, the results and explanations were not sufficient to explain the quantitative effects of interaction forces and to determine which will be dominant in given systems and conditions. It also has been difficult to investigate the intrinsic properties of the protein in a thorough and systematic way. This is because of the specificity and complexity of protein structure.

In order to elucidate adsorption of blood proteins on latex particles, properties such as surface electrical charge and hydrophobicity of the adsorbents, and environmental conditions like pH, ionic strength and temperature are considered as experimental factors (Norde, 1998). Figure 5.12 shows the schematic representation of the functionalized surface of latex particles prepared for protein adsorptions. The surface properties on latex particles are closely related to hydrophobic, hydrogen bonding, and van der Waals interaction forces with blood proteins to be adsorbed (Andrade, 1985).

ADSORPTION ISOTHERM

The protein adsorption isotherm is generally defined as the relationship between the amounts of protein adsorbed by a solid (latex particles) at a constant temperature and as a function of the concentration of protein. In the present study, adsorption isotherms were studied to determine the amount of protein adsorbed onto latex particles having various surface hydrophobicity as well as adsorption affinity with target proteins.

Calibration curves and an equation for each adsorption test were firstly needed to evaluate the equilibrium concentration of target proteins. Figure 5.13 shows the UV intensity as a factor of BSA concentration. The UV absorbance of the protein was monitored at 562 nm because the UV intensity of protein is the maximum at this wavelength. Figure 5.14 shows the calibration curve used in adsorption isotherm experiments of BSA onto PS latex particles. The BCA assay was used for evaluation of protein because of the technical advantages, such as high sensitivity, ease of use, stability for color complex, and less susceptibility to any detergents. Five different

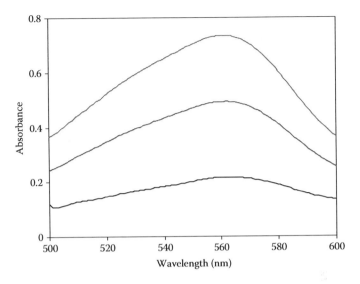

FIGURE 5.13 UV intensity with various BSA concentrations. (A) 1.0 mg/ml, (B) 0.6 mg/ml, (C) 0.2 mg/ml.

BSA concentrations (0.05, 0.1, 0.3, 0.5 and 0.7 mg/ml) were used for this curve as well as for measuring adsorption isotherms.

All isotherms were fitted to the Langmuir-Freundlich isotherms using the nonlinear regression method. Although the shapes of the isotherms were Langmuir-type, the initial slope of the adsorption curve very rapidly increased and the Langmuir isotherm model for all adsorption experiments failed to fit the experimental data:

$$q = q_m \frac{kC_{eq}^{\frac{1}{n}}}{1 + kC_{eq}^{\frac{1}{n}}} \tag{5.1}$$

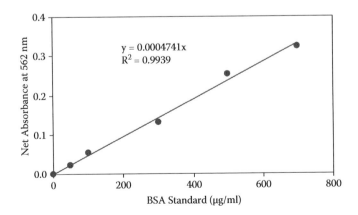

FIGURE 5.14 Standard curve of net absorbance versus BSA sample concentration.

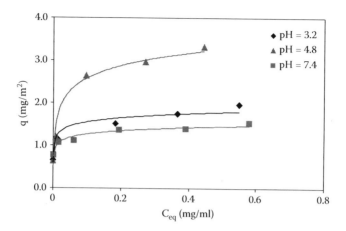

FIGURE 5.15 Representative adsorption isotherms of BSA in PBS on PS latex particles.

where q is the amount of adsorbed BSA per unit surface area, q_m is the adsorbed amount in equilibrium, C_{eq} is the equilibrium concentration of BSA, k is the adsorption constant, and n is the exponential factor for a heterogeneous system. n values are empirical and usually from zero to no more than unity (Jaekel and Vereecken, 2002). The values of n in our system were between 0.45 and 0.55, and hence n was fixed to its average 0.5. Actual q_m values were evaluated after fixing the n value.

Representative adsorption isotherms of BSA on PS latex particles at pH values of 3.2, 4.8 and 7.4 and in either PB or PBS media are shown in Figure 5.15. These pH values correspond to the acidic pH lower than the isoelectric point (IEP) of BSA, the IEP, and the alkaline pH higher than the IEP, respectively. There are plateaus in all isotherms, which suggests that an equilibrium adsorption of BSA onto PS latex particles occurs in both PB and PBS. At pH 4.8, BSA has almost no net electrical charge. At pH 3.2 and 7.4, BSA has positive and negative charge, respectively. The adsorbed amount (q_m) in this equilibrium was maximum at pH 4.8, in accordance with data accumulated by other researches (McLaren, 1954; Watkins and Robertson, 1977; Yoon et al., 1996). This is probably because BSA molecules form the most compact structures without repulsion force between proteins at the IEP, and more BSA molecules can adsorb on a given surface area. The difference of adsorbed amount in PB and in PBS is small. This suggests that the conformation of BSA is not affected very much by ionic strength.

The adsorbed amount of BSA decreased at pH values above and below the IEP. The decrease in adsorption efficiency may be caused by an increase in conformational size of protein molecules and the lateral electrostatic repulsions between adjacent adsorbed BSA molecules. In addition, the adsorbed amount of BSA at pH 3.2 was larger than that at pH 7.4 and still less than that at 4.8. This is because there are electrostatic interactions between the positive BSA molecules and negative PS latex particles at pH 3.2, and electrostatic repulsion is dominant between negative BSA and negative PS latex particles at pH 7.4. The amount adsorbed increased with

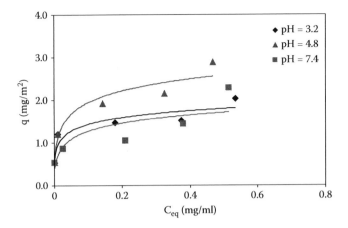

FIGURE 5.16 Representative adsorption isotherms of BSA in PB or PBS on PS/PMMA$_{100}$ latex particles.

increasing ionic strength at pH 3.2 and 7.4. This may be because the high concentration of electrolytes reduces the electrostatic repulsion between BSA and PS latex particles.

These results also indicate that the assumption in the Langmuir model, that there are no interactions between molecules adsorbed to adjacent sites, is difficult to meet in practice.

Representative adsorption isotherms of BSA on PS/PMMA$_{100}$ core-shell particles at 3.2, 4.8, and 7.4 pH are shown in Figure 5.16. Plateaus in the curves at all pH ranges indicate an equilibrium adsorption of BSA onto these adsorbents in both PB and PBS.

At pH 4.8, the adsorbed amount (q_m) of BSA on PS/PMMA$_{100}$ core-shell particles was 2.2 mg/m^2 in PB and 2.4 mg/m^2 in PBS, which are maximum values but less than those on PS latex particles. Suzawa, Shirahama, and Fujimoto (1982) have obtained similar results. This can also be explained by compact conformation of BSA molecules, similar to the PS latex example.

The adsorbed amount of BSA increased more rapidly with increasing ionic strength (in PBS) at pH 7.4 than occurred on PS particles. This may be explained by the low concentration of electrolytes enhancing the electrostatic repulsion between positive BSA and positive PS/PMMA$_{100}$ core-shell particles. As ionic strength is increased in PBS, however, a high concentration of electrolyte decreases electrostatic repulsion, promoting more rapid protein adsorption. It is known that protein adsorption is greater on a hydrophobic surface than on a hydrophilic one (Suzawa and Murakami, 1980; Suzawa et al., 1982; Lee and Ruckenstein, 1988). The amount of adsorbed protein with PS latex particles and PS/PMMA core-shell particles is greater than the amount of protein adsorption on PS latex particles because the PS polymer is more hydrophobic than PMMA.

Unlike PS and PS/PMMA$_{100}$ latex particles, the adsorbed amount of BSA on PS/PMMA$_{90}$PAA$_{10}$ and PS/PMMA$_{75}$PAA$_{25}$ core-shell particles is higher at pH 3.2 than

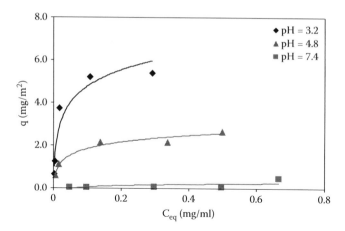

FIGURE 5.17 Representative adsorption isotherms of BSA in PBS on PS/PMMA$_{90}$PAA$_{10}$ and on PS/PMMA$_{75}$ PAA$_{25}$ particles.

at pH 4.8 and 7.4. Representative curves are shown in Figure 5.17. Furthermore, the adsorbed amount of BSA increased at pH 3.2 and pH 4.8 as the amount of PAA in the shell increased from 10 to 25% (w/w). However, the adsorbed amount of BSA dramatically decreased at pH 7.4.

Carboxylic acid groups as well as sulfate groups are distributed on the surface of PS/PMMA$_{90}$PAA$_{10}$ and PS/PMMA$_{75}$PAA$_{25}$ core-shell particles and seem to largely affect the BSA adsorption. The isotherms showed very strong dependence of adsorption on pH. At pH 3.2, over one unit lower than the IEP of 4.5 for PAA, the carboxylic acid groups are in the protonated, -COOH, form, and hydrogen bonding with protein should be enhanced. At pH 7.4, only 0.7 mg/m^2 BSA adsorbed on these core-shell particles, presumably due to anionic carboxylate-anionic protein repulsion (Gebhardt and Fuerstenau, 1983).

The equilibrium concentration values (q_m) of BSA adsorption on latex particles are calculated and listed in Table 5.4.

TABLE 5.4
The Equilibrium Concentration Values of BSA Adsorption Calculated by the Langmuir-Freundlich Isotherm Model

		PS		PS/PMMA$_{100}$		PS/PMMA$_{90}$PAA$_{10}$		PS/PMMA$_{75}$PAA$_{25}$	
	pH	PB	PBS	PB	PBS	PB	PBS	PB	PBS
q_m (mg/m^2)	3.2	1.74	2.60	1.40	1.69	3.61	5.19	10.83	10.77
	4.8	2.99	2.83	2.21	2.37	2.83	2.26	6.62	6.81
	7.4	1.36	2.07	1.30	1.62	0.43	0.43	0.75	0.57

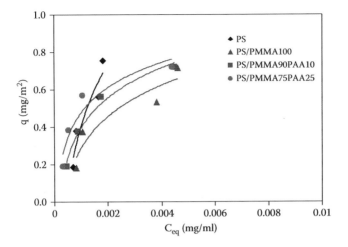

FIGURE 5.18 Adsorption isotherms of β_2M in PB at 37°C and pH = 7.4.

The adsorption isotherms for the β_2-microglobulin (β_2M) on latex particles are shown in Figure 5.18. There are steep initial slopes, indicating high affinity between β_2M and latex particles under all conditions tested. The complete plateau regions were not experimentally evaluated so we did not calculate the layer thickness of β_2M molecules.

The equilibrium concentration values (q_m) of β_2M adsorption on latex particles are calculated and listed in Table 5.5. The difference in maximum adsorption amount of β_2M on all latex particles was not much deviated and their values were between 0.69 and 0.8 mg/m^2.

There are several possibilities to explain the large adsorbed amount of BSA at acidic pH, including the end-on conformation of BSA, the flexibility of BSA molecules, the tilting of protein molecules due to the asymmetry of the charge distribution, or multilayer formation (Peula and Delasnieves, 1993).

The thickness (δ) of the adsorbed BSA monolayer was calculated using equation (5.2) (Chiu, Nyilas, and Lederman, 1976) to determine the adsorption mode of BSA

TABLE 5.5

The Equilibrium Concentration Values of β_2M Adsorption Calculated by the Langmuir-Freundlich Isotherm Model

	PS	PS/PMMA$_{100}$	PS/PMMA$_{90}$PAA$_{10}$	PS/PMMA$_{75}$PAA$_{25}$
q_m (mg/m^2)	0.80	0.69	0.72	0.72

TABLE 5.6

Calculated Absorbed BSA Layer Thickness, Å

	PS		PS/PMMA$_{100}$		PS/PMMA$_{90}$PAA$_{10}$		PS/PMMA$_{75}$PAA$_{25}$	
pH	PB	PBS	PB	PBS	PB	PBS	PB	PBS
3.2	21	32	17	21	44	64	133	132
4.8	37	35	27	29	35	28	81	83
7.4	17	25	16	20	5	5	9	7

onto latex particles (Shirahama and Suzawa, 1985):

$$\delta = \frac{3\sqrt{3} \bullet q_m}{\pi \bullet \rho_{BSA}} \ldots \tag{5.2}$$

where

$$\frac{3\sqrt{3}}{\pi}$$

is the packing factor, q_m is the plateau value of the adsorbed amount, and ρ_{BSA} is the density of the BSA molecule, which corresponds to the reciprocal of its known partial specific volume. The calculated δ values in the range of 2.7 ~ 8.3 nm at pH 4.8 are given in Table 5.6. Only δ values near the BSA IEP, pH 4.8, are important, because intra- and intermolecular electrostatic repulsion of protein molecules is minimized near the IEP (Shirahama and Suzawa, 1985). Compared to the hydrodynamic dimensions of BSA, $140 \times 38 \times 38$ Å (Fair and Jamieson, 1980), these values of adsorbed layer thickness indicate that the BSA molecules exist between the side-on (38 Å) and end-on (140 Å) mode That is, only up to a monolayer is adsorbed.

In the Langmuir isotherm model, k represents the ratio of adsorption to the desorption rate constant. This definition is not applicable in the Langmuir-Freundlich isotherm model but its meaning is almost the same (Yoon et al., 1996). In this sense, k is closely related to the affinity between protein molecules and latex particles. When the equilibrium concentration, C_{eq}, reaches zero, the value of $C_{eq}^{1/n}$ will be very small ($1 >> kC_{eq}^{1/n}$) and then equation (5.1) becomes

$$q = q_m k C_{eq}^{1/n} \tag{5.3}$$

k is closely related to the initial slope of the isotherm showing the affinity between BSA and latex particles (Yoon et al., 1996).

A key point in the characterization of adsorption processes on solid-liquid interfaces is to determine the change in Gibbs's free energy (ΔG°), one of the corresponding thermodynamic parameters, during adsorption. The traditional method to evaluate adsorption thermodynamic constants is based on measurement of adsorption isotherms under static conditions (Malmsten, 1998). The Langmuir-Freundlich

TABLE 5.7

The Values of Gibbs Free Energy Change of BSA Adsorption

		PS		PS/PMMA$_{100}$		PS/PMMA$_{90}$PAA$_{10}$		PS/PMMA$_{75}$PAA$_{25}$	
	pH	PB	PBS	PB	PBS	PB	PBS	PB	PBS
$\Delta G°_{ads}$ (kJ/mol)	3.2	−28.1	−26.8	−28.5	−23.1	−27.3	−26.2	−24.1	−23.1
	4.8	−19.7	−21.4	−22.0	−20.9	−23.8	−23.0	−23.5	−21.5
	7.4	−2.8	−2.6	−3.2	−2.6	−8.9	−19.9	−24.5	−24.3

adsorption equilibrium constant, k, can be relatively easily found from the slope of the isotherm. Then, this enables calculation of the free-energy change, $\Delta G°_{ads}$, during adsorption (equation 5.4).

$$\Delta G°_{ads} = -RT \ln k \qquad (5.4)$$

where R is the general gas constant ($R = 8.314$ J·mol^{-1}K^{-1}) and T is the absolute temperature. All calculated values are in Table 5.7.

Table 5.8 shows the calculated values of Gibbs free energy change of β_2M adsorption. PS/PMMA$_{75}$PAA$_{25}$ core-shell latex particles have the largest negative Gibbs free energy of β_2M adsorption.

The Gibbs free energy for β_2M adsorption on all latexes was more negative than that of BSA (Figure 5.19). This indicates that β_2M molecules are much more favorably adsorbed on latex particles than BSA, and also that the affinity of β_2M is much larger than BSA. This can be explained by the rate of diffusion and accessibility of β_2M molecules being faster and easier than BSA because of smaller molecular weight and size. Also, hydrophobic forces (less electrostatic repulsion) may contribute to β_2M adsorption over BSA because the IEP of β_2M (5.7) is closer to physiological condition (pH 7.4) than that of BSA (4.8).

KINETICS OF ADSORPTION

Protein adsorption is a complex process known to have large biological impact and is currently not well understood quantitatively because of extreme sensitivity to pH,

TABLE 5.8

The Values of Gibbs Free Energy Change of β_2M Adsorption

	PS	PS/PMMA$_{100}$	PS/PMMA$_{90}$PAA$_{10}$	PS/PMMA$_{75}$PAA$_{25}$
$\Delta G°_{ads}$ (kJ/mol)	−36.26	−34.66	−36.59	−39.03

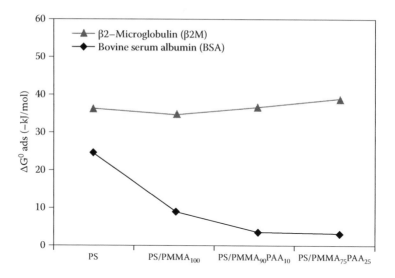

FIGURE 5.19 Gibbs free energy of adsorption of proteins on latex particles in PB at 37°C and pH = 7.4.

concentration of other electrolytes and molecules, temperature, etc. Accurate knowledge of the adsorption kinetics under a given set of conditions is a prerequisite for elucidating the mechanisms of many fundamental biological processes to be predicted at the molecular level. The adsorption kinetics approach is used in protein adsorption studies because of the uncertainty related to the time needed for the equilibrium to be established. It is generally accepted that the process of protein adsorption comprises the following steps: (a) transport toward the interface, (b) attachment at the interface, (c) eventual structural rearrangements in the adsorbed state, (d) detachment from the interface and transport away from the interface. Each of these steps can determine the overall rate of the adsorption process.

The quantitative analysis of the protein adsorption kinetics requires that the protein amount adsorbed is known as a function of time. Kinetic tests were carried out in phosphate buffer (PB) at pH 7.4 and 37°C, which are similar with physiological conditions. The protein concentrations were 0.7 mg/mL for BSA and 0.06 mg/mL for β_2M, which were the maximal concentrations employed in adsorption isotherm tests. A time course of the adsorption was made by measuring the concentration of protein in solution at different incubation times. Representative results are shown in Figure 5.20. Slight differences in the curve plateau values for adsorption on the different latexes were evident, but overall the shapes of the curves were similar. This data indicates that the adsorption of protein on latex particles can be expressed as a second-order model (Özacar, 2003) shown by the following equation:

$$\frac{dq_t}{dt} = k(q_e - q_t)^2 \cdots \cdots \qquad (5.5)$$

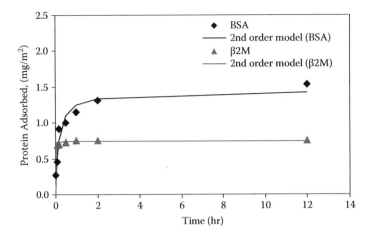

FIGURE 5.20 Representative kinetics of protein adsorption in PB on four types of latex particles.

Integration and applying the boundary conditions, gives

$$q_t = \frac{t}{\frac{1}{kq_e^2} + \frac{t}{q_e}} \cdots \cdots \tag{5.6}$$

where q_t and q_e stand for the amount of protein (mg/m²) adsorbed at time t and at equilibrium, respectively, k is the equilibrium rate constant of second-order adsorption (m²/mg·h), and t is the incubation time (h).

The amount of BSA adsorbed at equilibrium was higher than that of β_2M on PS and PS/PMMA$_{100}$ latex particles but less than that of β_2M on PS/PMMA$_{90}$PAA$_{10}$ and PS/PMMA$_{75}$PAA$_{25}$ latex particles. The variation in β_2M adsorption amount at equilibrium was nearly negligible for all latex particles, and the plateau region for β_2M was reached earlier than BSA on all latex particles. The K values in the β_2M adsorption process on latex particles were much larger than in BSA adsorption, indicating the β_2M molecules have higher adsorption rates than BSA molecules.

TABLE 5.9

Fitting Parameters of Second-Order Kinetic Model

		PS	PS/PMMA$_{100}$	PS/PMMA$_{90}$PAA$_{10}$	PS/PMMA$_{75}$PAA$_{25}$
BSA	q_e (mg/m²)	1.44	1.75	0.46	0.60
	K (h⁻¹)	6.7	35.3	9.7	17.9
β_2M	q_e (mg/m²)	0.74	0.76	0.75	0.75
	K (hr⁻¹)	754.9	443.8	569.5	516.8

There are several explanations for these phenomena. First, the smaller size of β_2M molecules can more easily diffuse to the substrate. Second, reduced hydrophobic sites of the latex particle surface by sulfate groups (SO_4^-) and ionized carboxylic groups (COO^-) may not be as powerful fixation groups for larger BSA molecules as for smaller β_2M. Third, the conformational change by intra- and intermolecular repulsion would be faster in β_2M due to smaller molecular weight. Finally, the ionic repulsion between β_2M and latex particles is less than that between BSA and latex particles because the IEP (5.7) of β_2M is closer to the tested condition (pH 7.4) than the IEP (4.8) of BSA.

Blood Compatibility

Blood compatibility can be defined as the property of a material or device that permits it to function in contact with blood without inducing adverse reactions. Before the *in vivo* application of any materials for humans is allowed, biocompatibility tests are required.

An important function of red blood cells is to defend the body against infections and other foreign materials. Red blood cells (erythrocytes) are the most numerous types in the blood. These cells average about 4.8–5.4 million per cubic millimeter but these values can vary depending on such factors as health and altitude. They live about 120 days and then are ingested in the liver and spleen. They manufacture hemoglobin and are responsible for the transport of oxygen and carbon dioxide.

Studies on *in vitro* hemolysis serve as screening methods for the toxicity of intravenous formulations and this data allows estimation of the concentration of hemoglobin released from red blood cells damage (ruptured) by foreign material.

The latex particle concentrations used in the present tests were 0.5, 2.5 and 5.0% (w/w) and incubation time with moderate shaking was 30 min at 37°C. Test results are shown in Figure 5.21. Blood cell rupture by PS latex particles was largely increased from about 30 to 91% as the solid content of PS increased from 0.5 to 5.0%. However, hemolysis caused by the other core-shell latex particles, such as

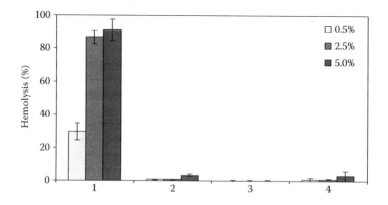

FIGURE 5.21 Hemolysis of latex particles, n = 5.

PS/PMMA$_{100}$, PS/PMMA$_{90}$PAA$_{10}$, and PS/PMMA$_{75}$PAA$_{25}$, was less than 3.5%. The PS/PMMA$_{90}$PAA$_{10}$ particles at 5.0% solid concentration have less than 0.2% hemolysis, which is the lowest value among the latex particles. The lowest hemolysis values indicate that the PS/PMMA$_{90}$PAA$_{10}$ core-shell latex particles are most blood compatible at all solid contents. Other core-shell particles such as PS/PMMA$_{100}$ and PS/PMMA$_{75}$PAA$_{25}$ also have good blood compatibility.

SUMMARY

The goal of this study was to synthesize monodisperse polymeric latex particles with tailored properties, and to investigate the fundamental interactions between the synthesized latex particles and target proteins. An understanding of the fundamental mechanism of selective adsorption is strongly required in order to maximize the separation performance of membranes made of these materials so that it may applicable to hemodialysis therapy for end stage renal disease (ESRD) patients. For the achievement of above research goal, several investigations have been performed. Analysis of the obtained data leads to several conclusions.

Monodisperse polystyrene seed latex particles were synthesized with the size ranges between 126 ± 7.5 and 216 ± 5.3 nm. Divinylbenzene (DVB) was used as a crosslinking agent to make the latex particles more hydrodynamically stable.

800-nm particles were synthesized using a semicontinuous seeded emulsion polymerization process. The particles with a mean diameter less than 500 nm are very spherical in shape and highly uniform in particle size distribution. However, as the particles grow larger than 500 nm in mean diameter, they become non-spherical with an uneven surface. This irregularity of the particle surface can be attributed to the non-homogeneous monomer swelling into the shell of a growing polymer, which can be controlled by factors such as temperature, agitation speed, initiator feeding rate, and surfactant amount required to stabilize the colloidal system. It was determined that as the surfactant-to-monomer ratio decreased, the mean particle diameter increased.

Methyl methacrylate (MMA) and acrylic acid (AA) monomers were introduced to make the latex particle surfaces more hydrophilic than bare polystyrene particles. The prepared core-shell latex particles were made of PMMA$_{100}$, PMMA$_{90}$PAA$_{10}$ and PMMA$_{75}$PAA$_{25}$. PMMA$_{100}$ is the PMMA homopolymer shell on a PS core. The PMMA$_{90}$PAA$_{10}$ shell is a copolymer with a PMMA-to-PAA ratio of 90 to 10% by weight. The PMMA$_{75}$PAA$_{25}$ shell is a copolymer consisting of a PMMA-to-PAA ratio of 75 to 25% by weight. The particle size of these core-shell particles was found to be about 370 nm in mean diameter.

The bicinchoninic acid (BCA) assay technique was used to determine the protein adsorption and all isotherms were fitted to Langmuir-Freundlich isotherms using the nonlinear regression method. The hydrophobic interaction between PS latex particle and BSA protein was dominant for the adsorption process. Conformational change of BSA at each pH also affected the adsorption amount, which was highest at pH 4.8 because of low ionic repulsion making compact conformation of BSA. The smallest amount of BSA adsorption was at pH 7.4, where charge repulsion between negative particles and proteins is present at this pH. Lateral repulsion between proteins was dominant for protein adsorption even though there always exists the hydrophobic interaction between solid latex particle and protein. The charge repulsion was reduced by the high concentration of electrolytes in the PBS medium. The BSA adsorption process on PS/PMMA core-shell particles was very similar to the case of PS latex particles where hydrophobic interaction between latex particle and protein was dominant. However, the adsorbed BSA amount was less than that of PS in PBS.

There were steep initial slopes for adsorption isotherms of β_2M on latex particles in all isotherms, indicating a high affinity type adsorption. The complete plateau regions were not seen for all isotherms, indicating that β_2M proteins were not saturated and there are still available sites for β_2M adsorption on latex particles. The initial slope of β_2M isotherms for PS was steeper than for the other latex particles (PS/PMMA$_{100}$, PS/PMMA$_{90}$PAA$_{10}$ and PS/PMMA$_{75}$PAA$_{25}$).

The Gibbs free energy of β_2M was more negative in phosphate buffer (PB) than that of BSA in physiological conditions (37°C and pH 7.4), indicating that β_2M molecules are more favorably adsorbed on the latex particles than BSA, and that the affinity of β_2M was much larger than BSA.

In the biocompatibility test, the number of red blood cells (RBCs) ruptured by PS latex particles ranged from 30 to 91% as the solid content of PS increased from 0.5 to 5.0%, respectively. However, hemolysis of the other core-shell latex particles, such as PS/PMMA$_{100}$, PS/PMMA$_{90}$PAA$_{10}$, and PS/PMMA$_{75}$PAA$_{25}$ was less than 3.5%. The PS/PMMA$_{90}$PAA$_{10}$ particles at all applied solid concentrations had less than 0.2% hemolysis and were the most blood compatible among the prepared latex particles. Other core-shell particles such as PS/PMMA$_{100}$ and PS/PMMA$_{75}$PAA$_{25}$ also had excellent blood compatibility with low RBC breakage.

REFERENCES

Andrade, J.D. (1985). Surface and interfacial aspects of biomedical polymer, Plenum, New York.

Bangs, L.B. (1987). Uniform latex particles. In *Technical Bulletin*. Seradyn Inc., 1987.

Baptista, R.P., Santos, A.M., Fedorov, A., Martinho, J.M.G., Pichot, C., Elaissari, A., Cabral, J.M.S., and Taipa, M.A. (2003). Activity, conformation and dynamics of cutinase adsorbed on poly(methyl methacrylate) latex particles. *Journal of Biotechnology* 102, 241–249.

Berge, A., Ellingsen, T., Skyeltorp, A.T., and Ugelstad, J. (1990). *Scientific Methods for the Study of Polymer Colloids and Their Applications*. Kluwer Academic, Dordrecht.

Bhutto, A.A., Vesely, D., and Gabrys, B.J. (2003). Miscibility and interactions in polystyrene and sodium sulfonated pol ethyl ether) PVME blends. Part II. FT IR. *Polymer* **44**, 6627–6631.

Brown, R.E., Jarvis, K.L., and Hyland, K.J. (1989). Protein measurement using bicinchoninic acid- elimination of interfering dubstances. *Analytical Biochemistry* 180, 136–139.

Burns, D.B., and Zydney, A.L. (2000). Buffer effects on the zeta potential of ultrafiltration membranes. *Journal of Membrane Science* 172, 39–48.

Cha, Y.J., and Choe, S. (1995). Characterization of crosslinked polystyrene beads and their composite in SBR matrix. *Journal of Applied Polymer Science* 58, 147–157.

Chiu, T.H., Nyilas, E., and Lederman, D.M. (1976). Thermodynamics of native protein foreign surface interactions. 4. Calorimetric and microelectrophoretic study of human fibrinogen sorption onto glass and LTI-carbon. *Transactions American Society for Artificial Internal Organs* 22, 498–513.

Chun, K.Y., and Stroeve, P. (2002). Protein transport in nanoporous membranes modified with self-assembled monolayers of functionalized thiols. *Langmuir* 18, 4653–4658.

Deppisch, R., Storr, M., Buck, R., and Gohl, H. (1998). Blood material interactions at the surfaces of membranes in medical applications. *Separation and Purification Technology* 14, 241–254.

Evans, D.F., and Wennerstron, H. (1999). *The Colloidal Domain: Where Physics, Chemistry, Biology, and Technology Meet*, 2nd ed. Wiley-VCH, New York.

Fair, B.D., and Jamieson, A.M. (1980). Studies of protein adsorption on polystyrene latex surfaces. *Journal of Colloid and Interface Science* 77, 525–534.

Gandhi, K., Park, M., Sun, L., Zou, D., Li, C.X., Lee, Y.D., Aklonis, J.J., and Salovey, R. (1990). Model filled polymers. 2. Stability of polystyrene beads in a polystyrene matrix. *Journal of Polymer Science Part B—Polymer Physics* 28, 2707–2714.

Gebhardt, J.E., and Fuerstenau, D.W. (1983). Adsorption of polyacrylic acid at oxide water interfaces. *Colloids and Surfaces* 7, 221–231.

Grancio, M.R., and Williams, D.J. (1970). Morphology of monomer polymer particle in styrene emulsion polymerization. *Journal of Polymer Science Part A1—Polymer Chemistry* 8, 2617–2629.

Guiot, P., and Couvreur, P. (1986). *Polymeric Nanoparticles and Microspheres*. CRC, Boca Raton (FL).

Higuchi, A., Ishida, Y., and Nakagawa, T. (1993). Surface modified polysulfone membranes—separation of mixed proteins and optical resolution of tryptophan. *Desalination* 90, 127–136.

Hsieh, H.P., Liu, P.K.T., and Dillman, T.R. (1991). Microporous ceramic membranes. *Polymer Journal* 23, 407–415.

Hunter, R.J. (1981). Zeta potential in colloid science: principles and applications, Academic Press, London.

Jaekel, U., and Vereecken, H. (2002). Transport of solutes undergoing a Freundlich type nonlinear and nonequilibrium adsorption process. *Physical Review E* 65, 041402.

James, R.O. (1985). Polymer colloids, Elsevier, Amsterdam.

Jons, S., Ries, P., and McDonald, C.J. (1999). Porous latex composite membranes: fabrication and properties. *Journal of Membrane Science* 155, 79–99.

Jung, B.S., Yoon, J.K., Kim, B., and Rhee, H. W. (2005). Effect of crystallization and annealing on polyacrylonitrile membranes for ultrafiltration. *Journal of Membrane Science* 246, 67–76.

Lee, C.F., and Chiu, W.Y. (1995). Kinetic study on the poly(methyl methacrylate) seeded soapless emulsion polymerization of dtyrene. 1. Experimental investigation. *Journal of Applied Polymer Science* 56, 1263–1274.

Lee, S.H., and Ruckenstein, E. (1988). Adsorption of proteins onto polymeric surfaces of different hydrophilicities—A case study with bovine serum albumin. *Journal of Colloid and Interface Science* 125, 365–379.

Li, J., Guo, S.Y., and Li, X.N. (2005). Degradation kinetics of polystyrene and EPDM melts under ultrasonic irradiation. *Polymer Degradation and Stability* 89, 6–14.

Lowry, O.H., Rosebrough, N.J., Farr, A.L., and Randall, R.J. (1951). Protein measurement with the folin phenol reagent. *Journal of Biological Chemistry* 193, 265–275.

Malmsten, M. (1998). *Biopolymers at Interface*. Marcel Dekker, New York.

McLaren, A.D. (1954). The adsorption and reactions of enzymes and proteins on daolinite. *Journal of Physical Chemistry* 58, 129–137.

Menon, M.K., and Zydney, A.L. (1999). Effect of ion binding on protein transport through ultrafiltration membranes. *Biotechnology and Bioengineering* 63, 298–307.

Nestor, J., Esquena, J., Solans, C., Levecke, B., and Tadros, T.F. (2005). Emulsion polymerization of styrene and methyl methacrylate using a hydrophobically modified inulin and comparison with other surfactants. *Langmuir* **21**, 4837–4841.

Norde, W. (1998). *Biopolymers at Interfaces: Chapter 2. Driving Forces for Protein Adsorption at Solid Surface*. Marcel Dekker, New York.

Odian, G. (1991). *Principles of Polymerization,* 3rd ed. John Wiley & Sons, New York.

Ottewill, R.H., and Shaw, J.N. (1967). Studies on preparation and characterisation of monodisperse polystyrene latices. 2. Electrophoretic characterisation of surface groupings. *Kolloid-Zeitschrift and Zeitschrift Fur Polymere* 218, 34.

Ozacar, M. (2003). Equilibrium and kinetic modelling of adsorption of phosphorus on calcined alunite. *Adsorption—Journal of the International Adsorption Society* 9, 125–132.

Park, M., Gandhi, K., Sun, L., Salovey, R., and Aklonis, J.J. (1990). Model filled polymers. 3. Rheological behavior of polystyrene containing cross-linked polystyrene beads. *Polymer Engineering and Science* 30, 1158–1164.

Pavin, D.L., Lampman, G.M., and Kriz, G.S. (1996). *Introduction to Spectroscopy: A Guide for Students of Organic Chemistry,* 2nd ed. Harcourt Brace College, Fort Worth.

Peula, J.M., and Delasnieves, F.J. (1993). Adsorption of monomeric bovine serum albumin on sulfonated polystyrene model colloids. 1. Adsorption isotherms and effect of the surface charge density. *Colloids and Surfaces A—Physicochemical and Engineering Aspects* 77, 199–208.

Ramakrishnan, S., McDonald, C.J., Prud'homme, R.K., and Carbeck, J.D. (2004). Latex composite membranes: structure and properties of the discriminating layer. *Journal of Membrane Science* 231, 57–70.

Rembaum, A., and Tokes, Z.A. (1988). *Microspheres: Medical and Biological Applications.* CRC, Baca Raton (FL).

Shirahama, H., and Suzawa, T. (1985). Adsorption of bovine serum albumin onto styrene acrylic acid copolymer latex. *Colloid and Polymer Science* 263, 141–146.

Smith, P.K., Krohn, R.I., Hermanson, G.T., Mallia, A.K., Gartner, F.H., Provenzano, M.D., Fujimoto, E.K., Goeke, N.M., Olson, B.J., and Klenk, D.C. (1985). Measurement of protein using bicinchoninic acid. *Analytical Biochemistry* 150, 76–85.

Srinivasan, S., and Sawyer, P.N. (1971). *Correlation of the Surface Charge Characterizations of Polymer with Their Antithrombogenic Characteristics.* Marcel Dekker, New York.

Suzawa, T., and Murakami, T. (1980). Adsorption of bovine serum albumin on synthetic polymer lattices. *Journal of Colloid and Interface Science* 78, 266–268.

Suzawa, T., Shirahama, H., and Fujimoto, T. (1982). Adsorption of bovine serum albumin onto homo-polymer and co-polymer lattices. *Journal of Colloid and Interface Science* 86, 144–150.

United States Renal Date System (USRDS). (2003). Annual report.

Vanderhu, H., and Vanderho, J. (1970) Inferences on mechanism of emulsion and polymerization of styrene from characterization of polymer end groups. *Chemical Process Engineering* **51**, 89.

Watkins, R.W., and Robertson, C.R. (1977). Total internal reflection technique for examination of protein sdsorption. *Journal of Biomedical Materials Research* 11, 915–938.

Wiechelman, K.J., Braun, R.D., and Fitzpatrick, J.D. (1988). Investigation of the bicinchoninic acid protein assay. Identification of the groups responsible for color formation. *Analytical Biochemistry* 175, 231–237.

Yoon, J.Y., Park, H.Y., Kim, J.H., and Kim, W.S. (1996). Adsorption of BSA on highly carboxylated microspheres—Quantitative effects of surface functional groups and interaction forces. *Journal of Colloid and Interface Science* 177, 613–620.

Zou, D., Derlich, V., Gandhi, K., Park, M., Sun, L., Kriz, D., Lee, Y.D., Kim, G., Aklonis, J.J., and Salovey, R. (1990). Model filled polymers. 1. Synthesis of cross-linked monodisperse polystyrene beads. *Journal of Polymer Science Part A. Polymer Chemistry* 28, 1909–1921.

Zou, D., Ma, S., Guan, R., Park, M., Sun, L., Aklonis, J.J., and Salovey, R. (1992). Model filled polymers. 5. synthesis of cross-linked monodisperse polymethacrylate beads. *Journal of Polymer Science Part A. Polymer Chemistry* **30**, 137–144.

6 Product Engineering of Nanoscaled Materials

W. Peukert and A. Voronov

CONTENTS

INTRODUCTION

A major trend in Chemical Engineering and Particle Technology is the shift from commodities toward high-end products with specific properties and functionality. Whereas in the past research was mainly directed toward better understanding of unit operations, modern trends are characterized by approaches for product formulation and means to tailor specific functions and product properties. This general trend is complemented by efforts for miniaturization that led to the important development of nanotechnology. The aim to build materials from smaller and smaller building blocks, that is, nanoparticles, raises the question of how to control self-assembly. Since nanoparticles are controlled by surface forces rather than by volume forces, control of particulate interfaces is the critical issue in nanoparticle technology and in product engineering of nanoscaled systems. Whereas great effects on the nanoscale have been discovered by basic sciences, the transfer from lab to industrial practice is often less developed, if not missing at all. A major challenge for nano-technology in general is therefore the processing aspect. Handling of nanoparticles, including dispersion, stabilization, or separation, is thus a key issue in the author's research.

CONCEPTS OF PRODUCT ENGINEERING

For particulate materials, the product properties depend on the chemical composition and on the dispersity of the material. The dispersity is characterized by the particle size distribution, the shape and morphology of particles and their interfacial properties. This relation was called by Rumpf [1] "property function," and control of the property function is known as product engineering or product design.

PROPERTY FUNCTION

> Product property = f (dispersity, chemical composition)
> Dispersity: particle size and shape and their respective distribution, particle
> morphology, particle surface properties

The property function relates the particulate structure (size, shape, morphology, surface) to the product properties (structure-property-correlation). Examples of property functions are the taste of chocolate, the color of pigments, the strength of cements, or the band gap of nanoparticles. Particle ensembles in the form of agglomerates, thin films, or filter cakes are also included in this consideration. Modern products are often characterized by several properties, which have to be achieved simultaneously: transparent *and* scratch-resistant coatings, sun blockers, which are transparent for visible light *and* UV-absorbing may serve as examples. Thus, a multifunctional product space evolves, which must be created through design of the related structures using the respective process technology.

The process function (process-structure correlation) as defined by Krekel and Polke [2] relates the process parameters to the product property.

PROCESS FUNCTION

> Dispersity = f (process parameters, educt concentrations)

Process parameters are the type of unit operations, their interconnection in the process, the process conditions under which the unit operations are operated (e.g., temperature, pressure, mass flow rates, etc.), and the materials processed. Structure—property as well as process—structure correlations must be known in order to run the process and to achieve the desired goal, that is, to produce well-defined, often multifunctional, product properties. Usually, process chains (with or without recirculation loops) have to be developed during which both handling and end-use properties must be optimized. A simple example to illustrate these ideas is crystallization: a chemical synthesis leads to a molecule, say an organic pharmaceutical active component. First, a supersaturation needs to be generated in order to trigger the phase transition. For substances with relatively high solubility, evaporation or cooling might be a good solution, depending on the solubility as a function of temperature and pressure. For sparingly soluble molecules, fast mixing with an antisolvent can be the method of choice, especially when the final particle size will be in the nanometer range. In any case, the one-phase molecular mixture is transferred to a two-phase suspension in which rheology, particle stability, and settling of the particles are of concern. The next step can be filtration as one specific solution for solid-liquid separation. Now we get a filter cake, which needs to be dried. The dried powder is again finely dispersed in a hammer mill and finally fed into a tabletting machine. The dispersity and the handling properties are changing along the process chain. A prohibitive high filter resistance or poor flowability of the dried powder may not only increase the production costs, but also change the product characteristics in a prohibitive way, for example, by forming hardly dispersible solid bridges between primary particles with negative impact on solubility and dispersion. The final product quality is multifunctional: mechanical stability of the tablet on one side while simultaneously maintaining a sufficient dissolution in body fluids, as well as solubility of the primary particles are just a few important aspects.

While these developments of the process chain have been done mostly empirical in the past, modern developments use more and more computer-based methods, including flow-sheeting software for design of the individual process steps as well as for the whole process. While these computational tools are well developed for fluids processing, it is still in its infancy for solids processing. A larger coordinated project in Germany led to the development of SolidSim, a new flow-sheeting software tool that incorporates major unit operations, including crystallization, agglomeration, comminution, gas-solid as well as solid-liquid separation, and transport [3].

After these general remarks, we now move on to specific boundary conditions for nanoparticle systems. All nanoparticle applications have in common that the interfacial and surface properties of the particles play a central role. The ratio of van der Waals adhesion forces to particle weight scales with particle diameter x^{-2} and is, for instance at 1 μm, on the order of 10^6 (in case of smooth particles). To produce well-defined property functions, the particle interactions have to be carefully controlled. Macroscopic properties can only be tailored by microscopic design of the interfaces. Surface chemistry and physics determine on the one hand the particulate interactions with fluid or solid phases. The types of interactions are van der Waals forces, polar interactions, hydrogen bonds, and even chemical bonds. On the other hand, particle interactions control particle and structure formation as will be discussed in this chapter. For product engineering of nanoparticulate systems, we start conceptually at the particle surface, which "transports" the respective particle interactions, thus leading to the desired structure. Vice versa, structure formation can only be understood by considering the relevant interactions, which are determined by the particle surface. This concept is illustrated in Figure 6.1 for oxide particles in aqueous solution where particle interactions can be understood in the view of well-known DLVO (Deryagin Landau Verwey Overbeek) theory as a superposition of van der Waals and electrostatic double-layer forces. This general concept can also be applied to more complicated and

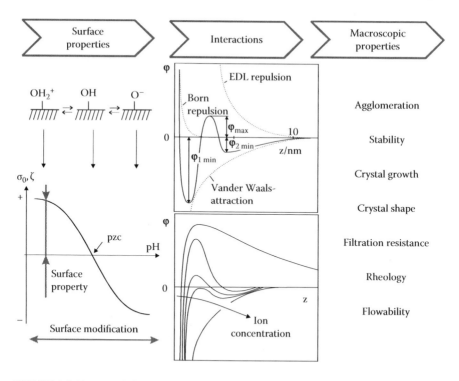

FIGURE 6.1 From particle surfaces to macroscopic properties.

less understood types of interactions, for example, interactions between polymer-modified particles.

This approach is based, however, on some wide-ranging preconditions. In order to bridge the gap between the microscopic molecular nature of a particle surface and the macroscopic properties, we need a multiscale approach covering several orders of magnitude in space and time. On the most basic level, quantum mechanics prevails. However, it is often possible by using the Hellman-Feynman theorem [4] to transfer the intrinsic quantum mechanical nature of surfaces to the physics of molecular interactions described by classical force laws. This theorem states that once the electron density distributions have been determined, the intermolecular interactions can be calculated on the basis of classical electrostatics. The contact value theorem is quite analogous: the force between two surfaces is determined by the density distribution of the molecules and particles in the space between them [5]. By using these classical interaction forces, molecular dynamics and Monte Carlo simulations are nowadays able to describe and even predict mesoscopic phenomena. Thus, we assume for molecular systems (without chemical reactions) the additivity of forces. Macroscopic properties therefore evolve from the summation of the interparticle forces plus hydrodynamic forces and from external fields such as gravity, centrifugal forces in sedimentation, and solid-liquid separation, as well as electromagnetic forces.

POSSIBILITIES TO MODIFY PARTICLE SURFACES

In general, there are three major classes of possibilities to modify the surfaces of particles:

1. Chemical modification by means of chemisorption or physisorption of chemical components, including ions, molecules, as well macromolecules, that is, polymers and biopolymers
2. Change of surface structure, that is, crystallinity and roughness, by either coating processes or by tailoring growth processes leading to a specific surface morphology
3. Process technology, that is, by generating specific forces acting on the particle surfaces, for example, hydrodynamics and/or by time-temperature history

CHEMICAL MODIFICATION

Particulate surfaces may be changed chemically through sorption of ions, molecules, polymers, or even biopolymers such as proteins. A simple example is illustrated in Figure 6.2 for alumina particles in aqueous solution. The alumina surface was modified through pH adjustment, that is, through adsorption of potential determining ions (H^+/OH^- in the case of alumina). The change in the electrochemical behavior of the suspensions can be seen from the change in ζ-potential measured by electroacoustics. It is not important whether the ζ-potential is positive or negative, only the magnitude counts. Higher ζ-potentials correspond with higher surface charge and lead to higher repulsive forces between the particles. This leads in turn to a higher stability against aggregation.

Due to the changed surface chemistry, the rheology is altered rather dramatically. The shear stress and suspension viscosity is changed, depending on the shear rate, by

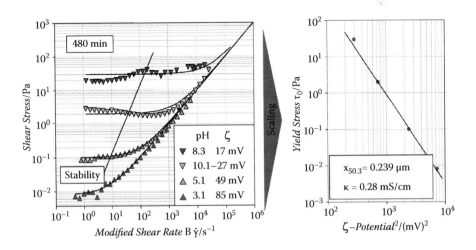

FIGURE 6.2 Influence of surface charge and ζ-potential of alumina (changed by adjusting the pH) on the rheology of alumina suspensions (6).

three orders of magnitude. In this example, the shear rate is modified by a shift factor B, which shifts all curves in the hydrodynamic range into one single master curve. The changes at low shear stresses, including the yield stress where particle-particle interactions prevail, is purely due to particle interactions since the adsorbed ions neither change particle size nor particle shape. The yield stress scales with the square of the ζ-potential. In accordance with particle interactions, the microstructure in the suspension is changed by forming more or less aggregated particulate structures.

These effects on rheology are on the one hand relevant for general considerations concerning multiscale processing and modeling. The described concept bridges many time and length scales from molecular level toward macroscopic effects. On the other hand, these considerations led specifically to a breakthrough in milling. Over many years it was postulated that milling in stirred media mills is limited to particle sizes larger than approximately 0.5 μm. Here and there were indications that this limit might be not an absolute one; for example, results from Kodak indicated that dispersion of aggregated particles may be possible down to a few 10 nm [7]. Recent systematic studies in stirred media mills did show that nanomilling down to mean particle sizes of 10 nm is possible if the suspension stability is safeguarded and thus suspension rheology is carefully controlled. A key element was electrostatic stabilization, which could be controlled on-line in the milling circuit by means of electroacoustic spectroscopy [6,8,9]. Most trials in the past did use polymeric stabilization without having direct access to the state of particle surfaces during the process, so that the state of the particle surfaces could not be controlled on-line during the process.

Similar effects can be obtained when the surface chemistry of the particles is changed by the binding of organic molecules, including polymers and biopolymers; this aspect will be explained later in this contribution in greater detail. Of course, coating procedures working in the gas phase might also be employed. These may comprise gas-solid reactions, including CVD (chemical vapor deposition), PVD (physical vapor deposition), and ALD (atomic layer deposition) to change the surface chemistry; these methods will not be treated in this chapter.

CHANGE OF STRUCTURE

Another possibility to change the particle interactions is the structure of the surface. A prominent example is the attachment of nanoparticles. This concept is known in nature as the Lotus effect; technical realizations of widespread application are toner particles and stabilization of emulsions. The Lotus flower grows in swamps but its surface remains clean. That is why the Lotus flower is known in the Indian culture as a symbol of purity. The flower can grow organic nanostructured organic crystals on the surface of the leaves [10]. These crystals influence both the adhesion force and the wetting behavior of water. The water has high a contact angle so that the almost round water droplets collect the adhering particles by rolling over the surface.

Powder flow and particle adhesion can be strongly influenced by the surface roughness through the distance dependent van der Waals interactions. A striking example of this effect is shown in Figure 6.3. This figure shows two heaps of powders. One consists of nanoparticulate titania with particle diameters around 10 nm, the other of

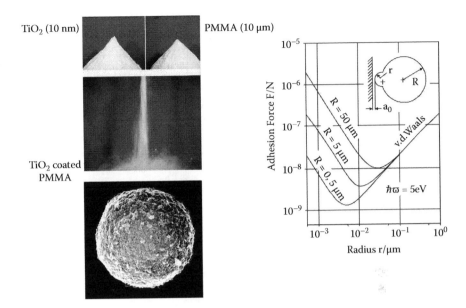

FIGURE 6.3 Influence of particle roughness on the adhesion forces and powder flow [11–14].

PMMA beads with diameters around 10 μm. Obviously, both powders exhibit fairly poor flowability. By applying a mechanical method known as MechanoFusion, the Hosokawa Company succeeded in coating the larger particles with the smaller ones [11]. A SEM picture of a coated particle is shown in the lower part of Figure 6.3. The flowability of this functionalized powder is completely different from that of the primary particles. Now, the powder flows almost like water. This effect is explained in the diagram on the right side of the figure. It shows that the van der Waals adhesion force of a rough particle on a flat wall can be reduced by orders of magnitude by an asperity, which is modeled in the example by a half sphere. In this case, obviously a roughness r of around 10 nm leads to a minimal adhesion force.

A coating procedure based on triboelectrical charging of the particles in the liquid phase (liquid nitrogen) was investigated for pharmaceutical powders in order to improve powder flow. For instance, lactose coated with nanoparticles in the range of 10–50 nm did considerably improve the fluidization behavior, which is of importance in aerosol inhalers for drug delivery to the lung [15]. Although this coating procedure is completely different from the former one, the basic physical effects are similar.

Surface structures can also be influenced by the coating processes, both from the gas and liquid phases. The acting supersaturation, which is the driving force for the surface growth, determines the mechanisms of growth and the growth kinetics. In crystallization, the following growth mechanisms in integration controlled growth are distinguished: mononuclear and polynuclear as the two limiting cases for low and high supersaturations, respectively, and the birth and spread model as an intermediate case. These mechanisms determine the resulting surface roughness. For instance, under conditions where the lateral diffusion of preformed species on the surface is very fast in comparison with the integration step, fairly smooth layers are obtained (mononuclear case). Integration occurs in this case at a few energetically favored sites such as edges and kinks. In contrast, when the adsorbed species on the surface are immobilized fast and thus integrated into the surface at arbitrary locations so that they are not diffusing over the surface, then a much rougher surface will be the result (polynuclear case) [16]. Of course, these effects depend also on each specifically indexed surface in the case of crystals.

Process Technology

A third possibility to change particle interactions is by process technology itself. A quite large and interesting class of applications is related to the gas-phase synthesis of nanoparticles. During gas-phase synthesis, nanoparticles are formed through nucleation with subsequent parallel or sequential steps of growth, agglomeration, and sintering. The temperature-time history in the reactor determines the sintering rate between particles. The relative rates of collision and coalescence determine the state of sintering with the two boundary conditions of aggregation without any sintering or complete coalescence. In both cases, spherical particles are to be expected. In between, the structure of the particles is determined by the temperature-time profile the particles have experienced in the reactor. Figure 6.4 shows examples of sintered nanoparticles and of model structures obtained by computer simulation [17,18]. The simulation comprises a Monte Carlo approach to obtain fractal aggregates with

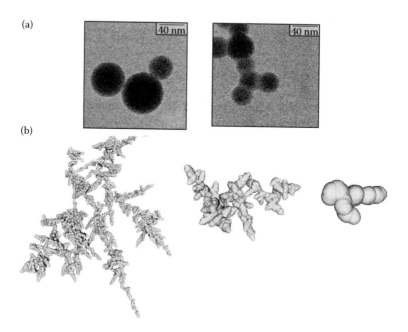

FIGURE 6.4 Sintered nanoparticles in experiments for measuring the sintering kinetics (a), and model structures obtained through computer modeling (b).

subsequent growth and sintering of the aggregate structures. The sintering kinetics can be obtained from the measurements as well as from independent simulations of the viscous sintering between to spherical nanoparticles.

The latter approach uses a volume-of-fluid method in combination with Hamaker summation. On-going research is directed toward model-based property character-ization, for example, of the hydrodynamic [83] and optical properties of the particles. Further possibilities to change the structure of surfaces can be induced by external forces such as hydrodynamic shear or mechanical stressing in ball or stirred media mills. In the latter case, amorphous structures can be created on crystalline surfaces. In order to increase for instance, flowability, nonspherical particles can be rounded by mechanically treating particles.

Due to limited space, we explain only one case in some greater detail, namely the surface modification by polymers.

STRUCTURE FORMATION

Structure formation is a key concept in product engineering (see Figure 6.5). Properties follow functions. In general, the formed structure depends on transport mechanisms and interactions. The transport mechanisms may be diffusive or con-vective. In addition, field forces such as gravity or electric fields contribute also to the mass transfer. The study of transport mechanisms has a long tradition in chemical engineering; the book *Transport Phenomenon*, by Bird, Stuart, and Lightfoot, first published in 1960 and may serve as an example [19]. This field is well developed although many open questions remain to be solved. Intermolecular and interpar-ticle interactions are much less well understood than transport phenomena but are of key importance in evolving fields such as nanoparticle technology. In the case of purely thermodynamically determined systems, the transport mechanisms can be

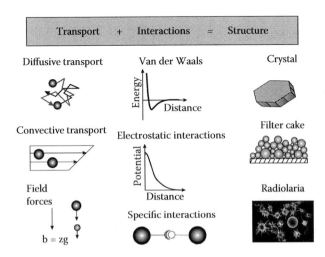

FIGURE 6.5 Principles of structure formation.

neglected. The structure is then only dependent on the interactions and can in principle be determined by minimizing the total energy of the system. The equilibrium shapes of crystals, the crystalline structure of highly charged colloidal suspensions or ordered arrays of optical microlenses or photonic crystals may serve as examples. The types of interactions are dispersive, electrostatic, magnetic forces, structural (entropic) forces in fluids, as well as forces due to material bridges. Material bridges form between particles in close contact in the presence of supersaturation, and due to ripening or sintering, depending on solubility in the liquid phase or the temperature in gas-phase systems, respectively. Specific interactions in biological matter are, of course, also included.

Of special interest are particulate thin films, which can be formed in dip- or spin-coating experiments or by means of spraying techniques. Depending on the application, there is an increasing demand for open or dense structures with defined pore size distributions. Dense coatings are necessary for passivation purposes, whereas porous coatings are required, for instance, when used as a carrier matrix in catalytic applications. Particulate structures on flexible substrates may open up new options in optics and electronics for transparent conductive films or for printable electronics. In any case, the question is to what extent the microstructure can be influenced by means of tailored particle interactions. The microstructure of dip-coated samples is influenced by the rate of evaporation and the rate of aggregation. The former depends on the selection of process parameters such as withdrawal velocity and drying conditions. The latter is associated with the stability and dispersity of the sol, thus with the physical-chemical properties of the coating bath. Investigations are focused on particulate surface properties and their influence on the structure-formation process. The structure can be characterized by means of atomic force microscopy (AFM) for the surface and by means of porosimetry and SAXS [20] for the volume, see Figure 6.6.

The use of stamps allows the formation of structured surfaces [21,22]. This technique is known as microcontact printing (μCP); it uses elastomeric stamps (PDMS) with typical dimensions in the μm-range. This method can be used to generate surface patterns with tailored wetting behavior by transfer of specific surface-active components, to generate metal films, to grow crystals at defined locations, and to selectively adsorb particles. This technique can be also used for stamping particles (Figure 6.6). The particles can be transferred by sucking suspensions through acting capillary forces into the grooves of the stamp and thus to generate particulate patterns on surfaces [23].

CASE STUDY: TAILORING OF NANOPARTICLES SURFACES BY THE USE OF POLYMERS

Polymer molecules tethered at one of their ends to a surface or interface find many applications in a variety of fields, including colloidal stabilization [24–25], biocompatible materials [26–30], and drug carriers [31–35]. The main role of the tethered polymers is to change the interactions of the modified surface or interface with the environment. The most common application is when one desires to cover the surface or interface with a protective "steric" layer [36,37]. This is achieved by choosing polymer molecules for which the solvent is "good." The polymer chains prefer,

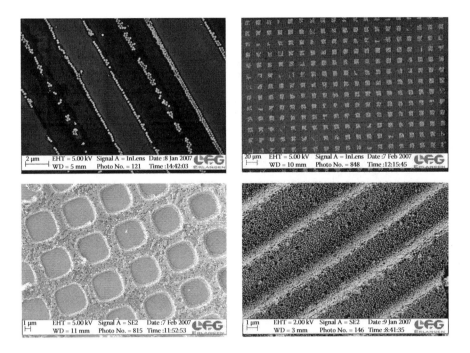

FIGURE 6.6 Particulate thin-film structures obtained by means of μCP.

effectively, to be surrounded by solvent molecules rather than by polymer segments of the same kind. In order to be better dissolved in the solvent, they tend to stretch out of the surface/interface as the surface coverage increases. At high enough surface coverage, the polymers form what is called a polymer brush, where the chain molecules are highly stretched out of the surface. These highly stretched polymers are the ones that have the potential of forming a very effective steric barrier that protects the surface/interface. The behaviour of tethered polymer layers has been studied at length in the last 20 years. There are a large variety of different theoretical studies that include full-scale computer simulation [38–41], molecular theories [42], and analytical approaches [43–45]. There are also many experimental studies; most of the early work concentrated on measuring forces between tethered layers [36,46–48] and the structure of the layers using scattering techniques [49–51].

Recently, strategies have been developed to tailor nanoparticle surfaces by the use of polymers. Tethering of polymer chains to nanoparticles can be reversible or irreversible and usually results in a core-shell type of material architecture. This approach leads to the formation of "polymer brushes" ("hairy nanoparticles"), which can effectively protect the particulate surface and increase its compatibility with the environment (for example, polymer matrices).

The polymer chains can be chemically bonded to the surface or may be just adsorbed onto the surface. *Physisorption* on a solid surface is usually achieved by block copolymers with one block interacting strongly with the substrate and another

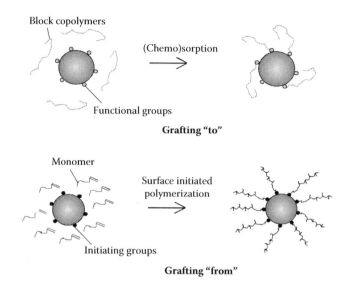

FIGURE 6.7 Schematic representation of polymer layers synthesized by grafting to and grafting from methods.

block interacting weakly. Covalent attachment can be accomplished by either "grafting to" or "grafting from" approaches (Figure 6.7). In a grafting to approach, preformed end-functionalized polymer macromolecules react with an appropriate substrate to form polymer layers. The "grafting from" approach is a more promising method that permits one to vary grafting density. The initiators are immobilized onto the surface, followed by *in situ* surface-initiated polymerization (SIP) to generate tethered polymer layers. It is more attractive due to the controlled density of initiator on the surface and a well-defined mechanism of initiation.

Physisorption

Physisorption of block copolymers or graft copolymers occurs in the presence of selective solvents or selective surfaces, giving rise to selective solvation and selective adsorption, respectively. The polymer layer structure depends therefore on the selectivities of these media and the nature of the copolymers, the architecture of copolymers, the length of each block, and the interactions between blocks and surface. In the case of a selective solvent [43], an ideal solvent is a precipitant for one block that forms an "anchor" layer on the surface, and a good solvent for other block, which forms polymer brushes in the solution. In the case of selective surface [52,53], one block is preferentially adsorbed on the surface and another one forms a polymer brush. Preparation of polymer layers by adsorption of block copolymers from selective solvent is not difficult. However, the polymer layers exhibit thermal and solvolytic instabilities due to the weak interaction between the substrate and block copolymers [54]. The interactions in most cases are van der Waals forces or hydrogen bonding.

Desorption could occur upon exposure to other good solvents or the adsorbed polymers are displaced by other polymers or other low molecular weight compounds. Also, it is not always easy to synthesize block copolymers, which are suitable for physisorption. Some of these difficulties could be overcome by covalently tethering polymer chains to substrates.

GRAFTING TO

The grafting to approach uses preformed end-functionalized polymers. They react with a suitable substrate surface under appropriate conditions and form tethered polymer layers. The covalent bond between surface and polymer chain makes the layer resistant to the chemical environment. This method has been often used in the preparation of polymer brushes. End-functionalized polymers with a narrow molecular weight distribution can be synthesized by living anionic, cationic, and radical polymerizations. The substrate surface also can be modified to introduce suitable functional groups by coupling agents of self-assembled monolayers. In general, only a small amount of polymer can be immobilized on the surface by the grafting to approach. Macromolecular chains must diffuse through the already existing polymer layer to get to the reactive sites on the surface. This barrier becomes more pronounced as the thickness of the tethered layer increases. The developed polymer layer has, therefore, a low grafting density and a low thickness. The grafting to of preformed polymer onto the surface of nanoparticles would not result in a compact core-shell system for steric reasons. Therefore, to achieve a design of a dense polymer shell, the grafting from approach must be used.

GRAFTING FROM AND SIP FROM NANOPARTICLES

The grafting from approach has attracted attention in the preparation of tethered polymers on a solid surface. This is a useful synthetic way to design precisely and functionalize the surface with well-defined polymer and copolymer layers. The grafting from approach has been already widely applied to surface modification of flat substrate surfaces, but high-density polymer brushes tethered to nanoparticles also show very interesting properties as colloids. In grafting from, the initiators are usually immobilized on the substrate with a following *in situ* SIP to generate tethered polymers. The mechanisms of SIP include free radical, cationic [55], ring opening metathesis polymerization [56], atom transfer free-radical polymerization (ATRP) [57], polymerizations using 2,2,6,6-tetramethyl-1-piperidyloxy free radical (TEMPO) [58], and anionic polymerization [59]. All these methods are suitable for polymerization of different types of monomers on a solid substrate, resulting in different polymer growth kinetics and grafting densities. For each polymerization mechanism, particular conditions and parameters have been optimized for controlled polymer brush growth (length, molecular weight, density, etc.).

Nanoparticle surfaces have the advantages of high surface/volume ratios, which allow a variety of *in situ* analytical methods to investigate surface chemistry at various stages of polymerization. The most common approach for the synthesis of modified particulate materials consisting of polymer-coated nanoparticles is by preparing core-shell structures. The polymer shell determines the external chemical properties of material and interaction with the environment. Material physical properties are influenced by both the size and shape of the nanoparticles and the surrounding organic layer [60]. As an example of the work done in this area, polymers were grafted from nanometer- and micrometer-sized silica particles [61–64]. In every case, a key step is the attachment of initiating groups onto the particle surface. The attachment of functional groups can be performed by ionic or covalent bonds.

Free radical polymerization allows the formation of high molecular weight with a high grafting density. Recently, silane-coupling agents such as alkoxysilanes [64–66] or chlorosilanes [62,63] have been used to modify silica surface with subsequent attachment of surface initiator. For example, poly(methyl methacrylate) with a molecular weight of 8.7×10^5 g/mol was grafted from the modified with a 4,4′-azobis (4-cyanopentanoic acid) through attached aminophenyltrimethoxysilane silica particles [67]. Surface-anchored peroxide initiators were also used for the grafting of poly (methyl methacrylate) [66]. The mechanism of the polymerization reactions showed many similarities but some differences to polymerization in solution.

Among the living free radical polymerization methods, ATRP shows the most promising possibilities to control polymerization on nanoparticles surface. Surface-bound 4-chlormethylphenyl type initiator have been used for the polymerization of water-soluble acrylates and methacrylates [68]. The ATRP initiator 3-(2-bromoprpi-onyloxy)propyl dimethylethoxysilane has been used to polymerize from CdS/SiO_2 nanoparticle surface [69]. The SiO_2 shell was necessary to improve the stability of the semiconductor nanoparticles with temperature. Gold nanoparticles have been coated with polymer using emulsion polymerization. The initiator groups were introduced onto the gold surface by the reduction of Au-salt in the presence of initiator carrying disulfide [70]. Surface-initiated ATRP resulted in an Au nanoparticle core of several nanometers and a polymer shell of variable thickness of high-density brush.

Among the polymerization techniques in solution, a living anionic polymerization may show the best possible control of polymer architecture. Monodispersed homopolymers, complex block, graft, and star architectures have been accessible by anionic methods [71]. It has been used to grow polymer brushes from various small particles such as silica, graphite, and carbon black. The results on living anionic polymerization on clay [72] and silica nanoparticles [73] has recently been reported.

Different mechanisms of SIP are thus possible on a variety of nanoparticles. Comparison can be made to their solution or bulk polymerization counterparts in which the main differences are in (a) polymerization rate, (b) termination methods and mechanisms, (c) consideration of interfacial properties and reaction conditions, and (d) methods of analysis and correlation with theoretical trends.

ANALYSIS OF TETHERED POLYMER LAYERS

Modern analytical tools permit detailed study of polymerization mechanisms and the physical properties of tethered polymers *in situ*. Surface-sensitive spectroscopic and microscopic techniques like AFM, SEM, ellipsometry, and x-ray reflectometry allow investigation of polymer layers and kinetics of growth under glass, melt, or swelling conditions. The most effective analysis approach with molecular weight parameters and macromolecular characteristics are by degrafting the polymers and post-polymerization analysis. The grafted polymer layers on nanoparticle surfaces can be compared to polymer films on flat surfaces where the surface-sensitive spectroscopic and microscopic measurements are complementary to analysis methods for colloidal particles.

Adaptive Nanoparticle Surfaces Through Controlled Reorganization of Invertible Polymer Layers

In many studies, special attention is devoted to tuning the interface for special applications. A predictable surface response or variable surface response under different conditions can be developed by careful design of the top surface layer [74–80]. The structure and characteristics of the interface are of great importance in understanding the materials properties in processing and use. Further advances in materials science require dual surface properties. A given material, depending on the conditions under which it is utilized, has to be hydrophobic and hydrophilic, acidic and basic, adhesive and repellent, etc.

Random or statistical copolymers possessing monomers of different nature and being introduced on the surface of nanoparticles can be certainly considered for building the adaptive/responsive smart particulate materials, since the different units of the copolymer macromolecule may segregate to the surface in response to some stimuli.

Recently, a new range of amphiphilic polyesters with both hydrophilic and hydrophobic functionalities alternately distributed in the polymer backbone was synthesized by the polycondensation of poly(ethylene glycol) with aliphatic dicarboxylic acids [81]. These polyesters are soluble in aqueous and organic media where they reveal inverse behavior that could be correlated to the chemical structure (Figure 6.8).

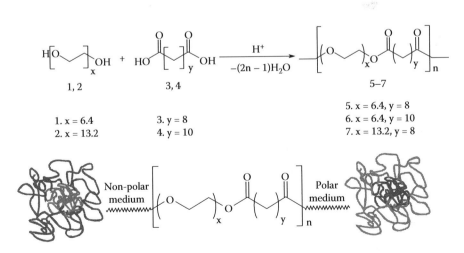

FIGURE 6.8 Synthesis of amphiphilic polyesters and scheme of their invertible behavior in media of various polarity.

TABLE 6.1

Properties and Characteristics of Synthesized Polyesters

Polyester Sample	Molecular Weight Mw, kDa	Critical Micelle Concentration cmc, mmol/l	Surface Tension at the cmc, mN/m	Radius R_h of Macromolecule, nm	
				In Toluene	In Acetone
5	22.6	1.3×10^{-4}	45.0	9.0 ± 0.6	2.8 ± 0.2
6	21.1	1.4×10^{-4}	48.9	6.9 ± 0.7	3.3 ± 0.1
7	9.2	2.6×10^{-4}	47.5	7.4 ± 0.6	2.0 ± 0.1

Their invertibility has been shown by the change in hydrodynamic radius R_h in solvents of different polarity (Table 6.1) as well as by immediate extraction of nonsoluble in polar and nonpolar medium dyes into the solvent phase of polyester solution (Figure 6.9). Figure 6.9 shows the observed color changes for both the dye solutions. This behavior can be explained by the solubilization of the dye molecules by the hydrocarbon groups of polyester and hence extraction of Sudan Red B to aqueous polymer solution. Increasing the polymer concentration in water leads to an increase in the total amount of architectural units being able to extract the hydrophobic dye. A similar experiment was performed with Malachite Green, extracted to toluene solutions of polyesters 5–7. The UV/VIS spectra results demonstrated the highest loading capacity for polyester 7. The longer hydrophilic poly(ethylene glycol) part of polyester 7 being solved in toluene can obviously encapsulate more hydrophilic dye molecules than polyesters 5 and 6. Therefore, the incorporation of poly(ethylene glycol) with various lengths to the polyester macromolecule allows for the adjustment of solubilization activity in nonpolar media.

The authors suppose that invertible coatings from amphiphilic polyesters depending on the effect of environment may be obtained on nanoparticulate surfaces (fillers, pigments, semiconductors like SiO_2, TiO_2, ZnO). Recently, they suggested a simple

| Sudan red B in water | Addition of the amphiphilic polyester 7 → | Encapsulation of the dye | Malachite green in toluene | Addition of the amphiphilic polyester 5–7 → | Encapsulation of the dye |

FIGURE 6.9 Photographs of Sudan Red B and Malachite Green sample, respectively, in water and toluene before and after the addition of amphiphilic polyesters. (There is a rapid change of a color to red and green, respectively, in the presence of polyester.)

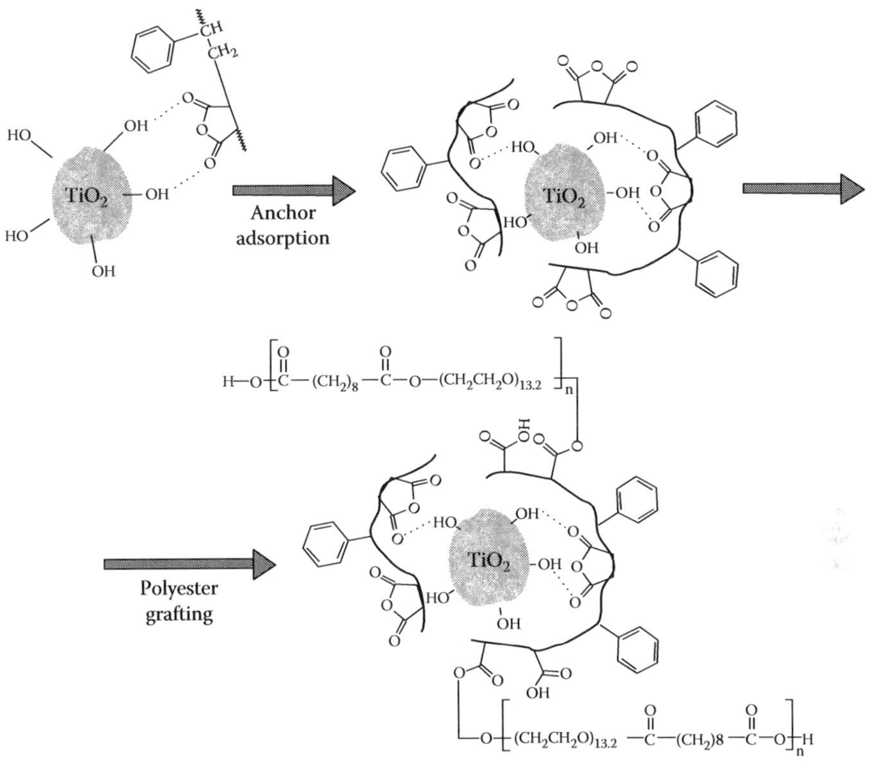

FIGURE 6.10 Reaction scheme of the grafting of invertible coating onto TiO$_2$ surface via PSM coupling.

method for fabricating, on the solid surface, self-adjustable invertible polymer coatings made of graft copolymer with the anchor poly(styrene-*alt*-maleic anhydride) (PSM) and tethered amphiphilic polyester (Figure 6.10) [82]. The invertible properties of the polymer coatings were successfully confirmed on disperse (titanium dioxide colloidal particles) as well as flat (silicon wafers) solid substrates.

In the absence of the polymer coating, the titania particles precipitate in toluene rapidly, since they do not form any stable suspension in a nonpolar solvent. On the contrary, the titania particles surface-modified with the invertible polyester form in toluene a suspension with enhanced stability. The range of stability experimental tests showed that addition of even small amounts of polar solvent, for example, acetone in a nonpolar dispersing medium (toluene), increases the suspension stability (Figure 6.11). A plot demonstrates the tendency to lower the volume mean particle size with an increase in acetone content in the mixture, which reaches its minimum value at the volumetric ratio between polar and nonpolar constituents of the dispersion medium of 1:1. This means that the wettability of the polyester coatings can be switched, depending upon the environment. For the modified titania particles, an

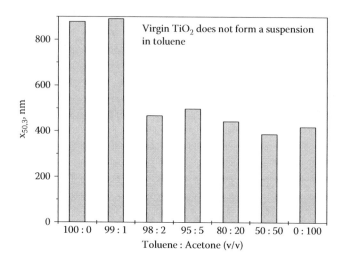

FIGURE 6.11 The volume mean particle size of modified with a grafted invertible coating on titanium dioxide dispersed in toluene, acetone, and its mixture.

essential increase in the colloidal stability after the modification with the invertible polymer chains was observed.

Drying/redispersion cycles demonstrate the ability of the modified titania particles, after their removal from one type of dispersion and subsequent drying, to be redispersed in another dispersing medium strongly differing in polarity from the previous one. Photographs presented in Figure 6.12 support the experimental results described above. Figure 6.12A shows that nonmodified titanium dioxide, after immersing into toluene and vigorous stirring, settles down as soon as the stirring was stopped. On the contrary, the toluene suspensions of titania with the grafted invertible tails is stable enough (Figure 6.12B). The powder of modified titania can be removed from toluene, dried in vacuum, and redispersed in water to form a stable suspension (Figure 6.12C). The same procedure has been repeated

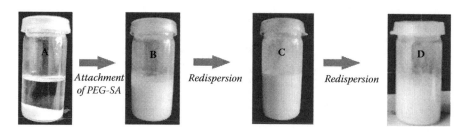

FIGURE 6.12 Photographs of nonmodified titania in toluene (A), polyester (PEG-SA) modified titania in toluene (B), modified titania redispersed in water after drying (C), modified titania again redispersed in toluene after drying (D).

in another sequence: the modified titanium dioxide is removed from an aqueous dispersion and subsequently redispersed in toluene without any loss of its colloid stability (Figure 6.12D).

An environmentally induced switching of the surface properties has also been proven on a flat surface via measuring the wetting contact angles and by scanning force microscopy of silicon wafers covered by PSM with the tethered polyester chains.

Experiments give strong support for the capability of such polymer coatings to switch their environmentally faced appearance, that is, to behave as self-adjustable invertible interfaces due to the ability of the tethered amphiphilic oligoester chains to change their conformations in response to changes in the environment. The elaborated process may be utilized for the creation of smart fillers, for example, as well as for paint coatings or highly filled polymer composites. This adaptivity of the particles can be understood as first steps toward intelligent particles.

REFERENCES

[1]   H. Rumpf, *Staub - Reinhaltung der Luft,* 1967, 27, 1, 3–13.

[2]   J. Krekel, R. Polke, *Chem. Ing. Technol.* 1992, 64, 6, 528–535.

[3]   J. Werther, E.U. Hartge, Gruhn G., *Chem. Ing. Technol.* 2004, 76, 709–713.

[4]   B.M. Deb, *Rev. Mod. Phys.*, 1973, 45, 1, 22–43.

[5]   J. Israelachvili, *Intermolecular and Surface Forces*, Academic Press, London, 1997.

[6]   F. Stenger, W. Peukert, *Chem. Eng. Technol.* 2003, 26, 2, 177–183.

[7]   J.D. Mendel, D. Bugner, A.D. Bermel, *J. Nanoparticle Res.*, 1999, 1, 421–424.

[8]   S. Mende, J. Schwedes, F. Stenger, W. Peukert, *Powder Technol.*, 2003, 132, 1, 64–73.

[9]   F. Stenger, S. Mende, J. Schwedes, W. Peukert, *Chem. Eng. Sci.*, 2005, 60, 4557–4565.

[10] W. Barthlott, W. Wollenweber, *Planta*, 1997, 202, 1–8.

[11] Hosokawa Micron Ltd., Osaka, Japan.

[12] H. Krupp, *Adv. Coll. Interf. Sci.*, 1967, 1, 111.

[13] H. Rumpf, *Chem. Ing. Technol.*, 1974, 46, 1, 1.

[14] H. Zhou, W. Peukert, *Langmuir*, 2008, 24, 4, 1459–1469.

[15] G. Huber, K.E. Wirth, *Powder Technol.*, 2003, 134, 3, 181–192.

[16] A. Mersmann, *Crystallization Technology Handbook*, Marcel Dekker, New York, 1995.

[17] H. Schmid, B. Al-Zaitone, C. Artelt, W. Peukert, *Chem. Eng. Sci.*, 2006, 61, 1, 293–305.

[18] H.-J. Schmid, S. Teijwani, C. Artelt, W. Peukert, *J. Nanoparticle Res.*, 2004, 6, 613–626.

[19] R.B. Bird, W.E. Steward, E.N. Lightfoot, *Transport Phenomena*, John Wiley & Sons, New York, 1960.

[20] L. Günther, W. Peukert, G. Goerik, N. Dingenhouts, *J. Colloid & Interf. Sci.*, 206, 294, 309–320.

[21] A.Kumar, A. Biebuyck, G.M. Whitesides, *Langmuir*, 1994, 10, 1498.

[22] Y. Xia, J. Tien, D. Qin, G.M. Whitesides, *Langmuir*, 1996, 12, 4033.

[23] B. Winzer, W. Peukert, *Chem. Ing. Technol.*, 2006, 78, 93–97.

[24] D.H. Napper. *Polymeric Stabilization of Colloidal Dispersions*, Academic Press, 1983.

[25] D.F. Evans, H. Wennerström, *The Colloidal Domain: Where Physics, Chemistry, Biology and Technology Meet*. VCH Publisher, New York, 1994.

[26] J.M. Harris, S. Zalipsky. *Poly(ethylene glycol): Chemistry and Biological Applications.* American Chemical Society, Washington, D.C., 1997.

[27] K.D. Park, W.K. Suzuki, W.K. Lee, Y.H. Kim, Y. Sakurai, T. Okano, *ASAIO J.*, 1996, M876.

[28] K.D. Nelson, R. Eisenbaumer, M. Pomerantz, R.C. Eberhart, *ASAIO J.*, 1996, M884.

[29] C.R. Jenney, J.M. Anderson, *J. Biomed. Mater. Res.*, 1999, 44, 206.

[30] T.A. Horbett, J.L. Brash, *Proteins at Interfaces II: Fundamentals and Applications.* American Chemical Society, Washington, D.C., 1995.

[31] T.M. Allen, C. Hansen, F. Martin, C. Redemann, A. Yauyoung. *Biochim. Biophys. Acta*, 1991, 1066, 29.

[32] V.P. Torchilin, V.G. Omelyanenko, M.I. Papisov, A.A. Bogdanov, V.S. Trubetskoy, J.N. Herron, C.A. Gentry. *Biochim. Biophys. Acta*, 1991, 1195, 11.

[33] G.S. Kwon. *Crit. Rev. er. Drug Carrier Syst.*, 1998, 15, 481.

[34] D. Lasic, F. Martin, *Stealth Liposomes.* 1995, CRC Press, Boca Raton, FL.

[35] L. Fisher, D. Lasic, *Curr. Opin. Colloid Interf. Sci.*, 1998, 3, 509.

[36] A. Halperin, M. Tirrell, T.P. Lodge, *Adv. Polym. Sci.*, 1992, 100, 31.

[37] S.T. Milner, *Science*, 1991, 251, 905.

[38] G.S. Grest, M. Murat, *Macromolecules*, 1993, 26, 3108.

[39] P.-Y. Lai, K. Binder, *J. Chem. Phys.*, 1991, 95, 9288.

[40] A. Chakrabarti, R. Toral, *Macromolecules*, 1990, 23, 2016.

[41] G. S. Grest, M. Murat, in *Monte Carlo and Molecular Dynamics Simulations in Polymer Science*, K. Binder, Clarendon Press, Oxford 1994.

[42] I. Szleifer, M. A. Carignano, *Adv. Chem. Phys.*, 1996, XCIV, 165.

[43] S. Alexander, *J. Phys.*, 1977, 38, 983.

[44] S. T. Milner, T. A. Witten, M. E. Cates, *Macromolecules*, 1988, 22, 2610.

[45] A. Halperin, in *Soft Order in Physical Systems*, Y. Rabin, R. Bruinsma, Plenum Press, New York, 1994.

[46] H. J. Taunton, C. Toprakcioglu, L. J. Fetters, J. Klein, *Nature*, 1988, 332, 712.

[47] H. J. Taunton, C. Toprakcioglu, L. J. Fetters, J. Klein, *Macromolecules*, 1990, 23, 571.

[48] M. Tirrell, S. Patel, G. Hadziioannou, *Proc. Natl. Acad. Sci.*, 1987, 84, 4725.

[49] P. Auroy, Y. Mir, L. Auvray, *Phys. Rev. Lett.*, 1992, 69, 93.

[50] B. J. Factor, L.-T. Lee, M. S. Kent, F. Rondelez, *Phys. Rev. E*, 1993, 48, 2354.

[51] A. Karim, S.K. Satija, J. F. Douglas, J. F. Ankner, L. J. Fetters, *Phys. Rev. Lett.*, 1994, 74, 3407.

[52] J. Marra, M.L. Hair, *Colloids Surf.*, 1989, 34, 215.

[53] D. Guzonas, D. Boils, M.L. Hair, *Macromolecules,* 1991, 24, 3383.

[54] G.J. Fleer, M.A. Cohen-Stuart, J.M.H. Scheutjens, T. Cosgrove, B. Vincent, *Polymers at Interfaces.* Chapman and Hall, London, 1993.

[55] R. Jordan, A. Ulman, *J. Am. Chem. Soc.*, 1998, 120, 2, 243.

[56] M. Weck, J.J. Jackiw, R.R. Rossi, P.S. Weiss, R.H. Grubbs, *J. Am. Chem. Soc.*, 121, 16, 4088.

[57] D.M. Jones, A.A. Brown, W.T.S. Huck, *Langmuir*, 2002, 18, 4, 1265.

[58] M. Husseman, E.E. Malmstrom, M. McNamara et al. *Macromolecules*, 1999, 32, 5, 1424.

[59] R. Advincula, Q. Zhou, M. Park et al., *Langmuir*, 2002, 18, 22, 8672.

[60] G. Kickelbick, *Prog. Polym. Sci.*, 2003, 28, 83.

[61] R. Laible, K. Hamann, *Angew. Makromol. Chem.*, 1975, 48, 97.

[62] X. Huang, M.J. Wirth, *Anal. Chem.*, 1997, 69, 4477.

[63] O. Prucker, J. Rühe, *Macromolecules*, 1998, 31, 602.

[64] T. von Werne, T.E. Patten, *J. Am. Chem. Soc.*, 1999, 121, 7409.

[65] G. Boven, M.L.C.M. Oosterling, G. Chella, A.J. Schouten, *Polymer*, 1990, 31, 2377.

[66] N. Tsubokawa, H. Ishida, *J. Polym. Sci., Part A: Polym. Chem.*, 1992, 30, 2241.

[67] G.V. Schulz, G., Harborth, *Makromol. Chem.*, 1948, 1, 106.

[68] C. Perruchot, M.A. Khan, A. Kamitsi et al., *Langmuir*, 2001, 17, 15, 4479.

[69] J. Chen, H. Iwata, N. Tsubokawa et al., *Polymer*, 2002, 43, 2201.

[70] T.K. Mandal, M.S. Fleming, D.R. Walt, *Nano Lett.*, 2002, 2, 1, 3.

[71] N. Hadjichristiadis, H. Iatrou, S. Pispas et al., *J. Polym. Sci. A: Polym. Chem.*, 2000, 38, 3211.

[72] X. Fan, Q. Zhou, C. Xia et al., *Langmuir*, 2002, 18, 11, 4511.

[73] Q. Zhou, S. Wang, X. Fan et al., *Langmuir*, 2002, 18, 8, 3324.

[74] T.P. Russel, *Science*, 2002, 297, 964.

[75] C. Creton, *MRS Bull.*, 2003, 28, 434.

[76] V.V. Tsukruk, K. Wahl, *Microstructure and Microtribology of Polymer Surfaces,* Washington, ACS Symposium Series, 2000.

[77] I. Luzinov, S. Minko, V.V. Tsukruk, *Prog. Polym. Sci.* 2004, 29, 635.

[78] I. Sanchez I., *Physics of Polymer Surfaces and Interfaces.* Manning, New York, 1993.

[79] T.M. Birshtein, V.M. Amoskov, *Comput. Theor. Polym. Sci.,* 2000, 10, 1/2, 159.

[80] Ionov L., Minko S., Stamm M., Gohy J.F., Jerome R., Scholl A. J., *Am. Chem. Soc.,* 2003, 25, 8302.

[81] A. Voronov, A. Kohut, W. Peukert et al., *Langmuir*, 2006, 22, 5, 1946–1948.

[82] A. Kohut, W. Peukert, A. Voronov et al., *Langmuir*, 2006, 22, 5, 6498–6506.

[83] C. Binder, C. Feichtinger, H. Schmid, N. Thürey, W. Peukert, U. Rüde, *J. Colloid & Interf. Sci.*, 2006, 301, 1, 155–167.

7 Surface Engineering Quantum Dots at the Air-Water Interface

J. Orbulescu and R.M. Leblanc

CONTENTS

INTRODUCTION

It has been shown that quantum dots are submicron-sized structures that are able to confine excitons in one or more directions. Electron confinement in all three dimensions is called "quantum confinement," and the particles quantum dots (QDs) with a main characteristic of discrete levels of energy not continuous such as in metals.

Langmuir films are formed by spreading an amphiphilic molecule at the air-water interface and its interface properties are studied upon compression of the film. The main requirement of the molecule is that the solvent in which it is dissolved should be immiscible with water and fairly volatile. This requirement must be fulfilled such that the solvent can evaporate within 10–15 min and the monolayer would be formed only by the amphiphilic molecules. Compression can be applied to the monolayer and the change in the surface pressure as a function of the molecular area can be followed. Compression is a very convenient method to control the packing of a monolayer, or in other words, the intermolecular distance.

This aspect is important because arrays of QDs can be formed by various techniques (for example, lithographic methods) and upon deposition on metallic substrates, new electronic properties given by the array could arise.

In recent years, there has been great progress in synthetic methods of quantum dot production, which made large-scale preparation of QDs with high quality and narrow size distribution possible. Luminescent QDs have been successfully attached to dendrimer, protein, sugar, and other biologically active agents.[1–4] Among the various types of modifications, cadmium selenide (CdSe) QDs have been studied most thoroughly due to their possible applications in biological disciplines. However, consideration of QD behavior at biological interfaces, for example, cell membranes, is one of the most important issues requiring further study prior to implementation in real applications. Currently, few studies have been conducted regarding surface chemistry properties of quantum dots at biological interfaces.

Quantum dots of II–VI semiconductors (cadmium sulfide [CdS], CdSe, and cadmium telluride [CdTe]) in the 1–12-nm-size range have attracted a great deal of research interest in the past few years in the fields of physics, chemistry, biology, and engineering.[5–13] Dramatically different from the bulk state, the properties of QDs are dependent upon quantum confinement effects in all three spatial dimensions. These QDs demonstrate great potential in applications for optical devices, optical switches, and fluorescence labeling.[14]

The first article focusing on the Langmuir films of QDs was published by Dabbousi et al.[15] with the focus of controlling the QD array deposited as a Langmuir-Blodgett film. The synthesis of QDs was based on the routine of Murray, Norris, and

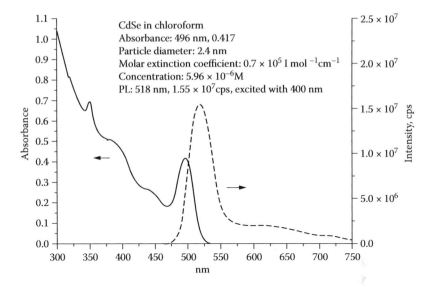

FIGURE 7.1 UV-Vis absorption and fluorescence spectra of CdSe QDs.

Bawendi[16]. Following synthesis, the QDs were transferred into methanol to wash the excess TOPO capping and then dried and transferred to HPLC grade chloroform that was spread on the water surface.

Following this study, there was not much research published for a few years. Sui and co-workers[17] focused on the influence of ligand length on the packing of the QD monolayer. In this case, the synthetic routine was carried out using the method reported by Peng and co-workers.[18,19] By controlling the temperature and reaction time, these experiments are reproducible, and the QDs have the same size distribution (see Figure 7.1).

SYNTHESIS OF CdSe NANOPARTICLES

0.0514 mg of CdO, 0.2232 mg of TDPA, and 3.7768 mg of TOPO were placed into a 100-mL flask. The mixture was heated to 300°C (under argon gas flow) during dissolution of CdO into TDPA and TOPO. This solution was kept at 300°C for 30 min and subsequently cooled down to 150°C. A solution of 0.041 mg of selenium in 2 mg of TOPO was then injected into this solution. Approximately 2 min after injection, the solution underwent a color change from clear to yellow. Immediately after observation of the color change, the heated solution was quickly injected into cool chloroform. The chloroform solution was evaporated under vacuum conditions to produce the raw CdSe QDs (yellow powder). These powders were subsequently washed with methanol several times. After TOPO was removed from solution, the powder was dissolved in chloroform and filtered to provide a clean QD solution.

MODIFICATION OF QDs (QD-C$_x$)

Alkane thiols were directly added into pure CdSe QD solutions (molar ratio, QDs:thiol = 1:300). After agitation of the solution for 24 h, copper powder was added to the solution to remove excess thiol (concentrations of QDs and modified QDs were calculated from the solution UV-Vis absorption) (Figure 7.2).

QDs (CdSe) have been modified with surfactants of varying chain lengths (C6SH, C8SH to C18SH). These QD-surfactants can form a stable monolayer at the air-water interface.

SURFACE PRESSURE-AREA ISOTHERMS

These modified quantum dots can form a stable monolayer (Langmuir film) at the air-water interface. The surface pressure-area isotherm gives the most important indicator of the monolayer properties. A standard surface pressure-area isotherm includes three parts: the gaseous phase at large area per molecule, the liquid expanded and liquid condensed phases during which surface pressure increases with compression, and the solid phase. By extrapolating the solid or liquid condensed phase part of the isotherm at zero surface pressure, a limiting molecular area is measured, which is an important parameter for the Langmuir film studies. The limiting molecular area corresponds to the smallest area per nanoparticle at the air-water interface in a 2-D arrangement. Surface pressure-area isotherms of modified QDs are shown in Figure 7.3. Collapse surface pressures of the modified QDs Langmuir films were all higher than 25 mN/m. There are two different behaviors of modified QDs Langmuir films that could be identified from these isotherms. QD-C6 (QDs modified with hexane thiol), QD-C8, QD-C10, and QD-C12 showed similar limiting molecular areas when compared with pure QDs at approximately 500 Å^2/molecule. From QD-C14 to QD-C18, the limiting molecular area grew larger with increasing alkane thiol chain length. QD-C14 showed a limiting molecular area of 1200 Å^2/molecule, whereas the limiting molecular area of QD-C16 was observed to be approximately 9300 Å^2/molecule. A limiting molecular area around 9600 Å^2/molecule was obtained from the surface pressure-area isotherm of QD-C18.

A possible explanation for the different behaviors of modified QDs may be found in the roles of TOPO on the QD surface. The TOPO molecule contains three hydrocarbon chains that produce its hydrophobic property, and it also has a large dipole

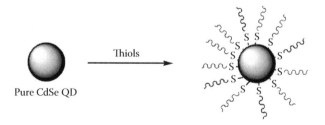

FIGURE 7.2 Illustration of modification of QDs with thiols.

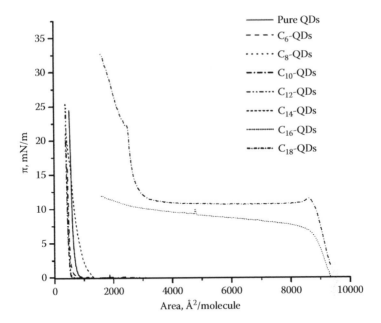

FIGURE 7.3 Surface pressure-area isotherms of pure and modified CdSe QDs.

moment derived from the phosphorus-oxygen bond.[16,20,21] TOPO acts as both the solvent and the capping molecules in the preparation of the CdSe QDs. After preparation, most of the TOPO molecules were removed by methanol because of good solubility in this solvent. Hydrophobic TOPO is the capping agent responsible for CdSe's solubility in nonpolar organic solvents. Therefore, TOPO molecules could not be completely washed away. In fact, the capping layer of modified QDs still contains a small amount of TOPO mixed among the alkane thiols. The limiting molecular area, obtained from these isotherms, is determined by the diameter of modified QDs. Its diameter includes the diameter of the CdSe core and the length of the longest chains. Since the length of TOPO is close to C12 and greater than C6 and C8 (from the CPK model), the modified QDs (QD-C6 to QD-C10) showed limiting molecular areas similar to that of pure QDs that was also covered with some TOPO molecules. The structure of the modified QDs is illustrated in Figure 7.4. From C12 to C18, the length of the alkane thiols are longer than TOPO, and the modified QDs show increased area with increasing thiol molecule chain lengths.

In Situ UV-Vis Spectroscopy

The UV-Vis absorption of modified CdSe QDs at the air-water interface was also obtained. Figure 7.5 show UV-Vis absorption of QD-C12 at different surface pressures. Increasing the surface pressure revealed a linear increase in the UV-Vis absorption intensity. All of the modified QDs from QD-C6 to QD-C18 demonstrated the same linear relationship. Therefore, there appears to be a strong linear relationship between visible absorbance (503 nm) and surface pressure.

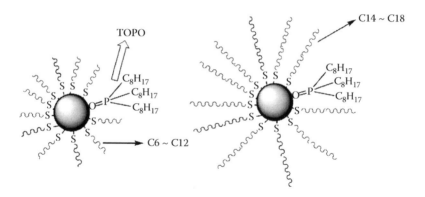

FIGURE 7.4 Illustration of the structure of CdSe QD-thiols.

A different approach was to use QD monolayers and the influence of the nature of the surfactant, particle size, surface pressure, and monolayer mixing could have on the QDs engineered at the air-water interface.

For this, the amounts of 0.0514 g CdO, 0.2232 g TDPA, and 3.7768 g TOPO were loaded into a 25-mL flask with a condenser assembly and drying tube attached to it. The reaction mixture was heated to 310°C under Ar flow and agitation. When a

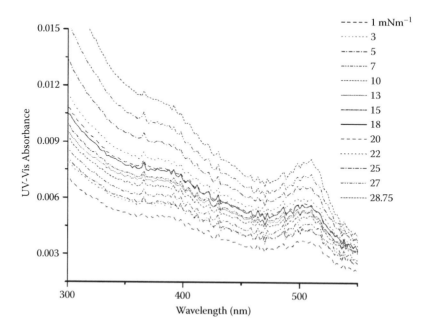

FIGURE 7.5 UV-Vis absorption of CdSe QD-C12 at air-water interface for different surface pressures.

clear and colorless solution was obtained, the temperature was lowered to 270°C, followed by injection of 0.0411 g Se powder dissolved in 2.4 mL trioctylphosphine. Nucleation and growth of the QDs occurred at 250°C until the desired particle size was reached. Immediately, the resulting solution was cooled to 60°C and the QDs were collected and purified by precipitating/dispersing three times with 28 mL anhydrous methanol/0.5 mL chloroform. Further purification was achieved by washing the nanoparticles three times with a mixture consisting of 8 mL of anhydrous methanol and 2 mL chloroform, followed by dispersion of the nanocrystals in 5 mL chloroform and filtration through a 0.45-μm pore size polypropylene syringe filter, from Sigma-Aldrich.

MODIFICATION OF CDSE QDS: TOPO/ODT SURFACE CAP EXCHANGE

To replace the TOPO surface cap on the CdSe QDs by an octadecyl thiol (ODT) cap, 1 mL of the TOPO-capped QDs in chloroform was reacted with 0.287 g (1 M) ODT for ~19 h. The reaction mixture was kept under gentle agitation, in the dark, and at room temperature. Purification of the resulting ODT-capped CdSe QDs was achieved by first evaporating all of the chloroform present in the mixture utilizing argon flow. The QDs were then washed three times with 10 mL of the anhydrous methanol/chloroform (8:2, v/v) mixture. Finally, the QDs were dispersed in 1 mL chloroform and filtered as described above.

MOLAR ABSORPTIVITY OF TOPO-CAPPED QDS

The molar absorptivity (ε) of CdSe QDs at λ_{max} has been found to be dependent on nanoparticle size.[22–25] Linear,[24] cubic,[22,23] and intermediate between quadratic and cubic[25,26] functions are implied or reported in the literature as shown in Figure 7.6b. Molar absorptivity of QDs has been calculated from data collected by atomic absorption spectroscopy,[22,25,26] osmotic methods,[23] absorption cross section,[24] and/or controlled etching.[25,26] These literature ε values are included in the calculations to determine the concentration of QD samples prepared for this study.

In addition, it was found that UV-Vis spectroscopy, surface pressure-area (π-A) isotherms, monolayer properties, QD size, and Langmuir-Blodgett (LB) films could be used collectively to determine the molar absorptivity of QDs. Proposed calculations and procedures are indicated in Gattas-Asfura and colleagues.[27] Based on these combined results, the molar absorptivity values for the 23, 41, 75, and 103 Å CdSe QDs used in this study are 6.6×10^4, 2.0×10^5, 6.3×10^5, and 1.2×10^6 M^{-1} cm^{-1}, respectively. Although in close agreement with previously published calibration data, the derived ε values tend to follow a quadratic dependence (Figure 7.6b). This suggests that molar absorptivity is particle size-dependent and proportional to the QD surface area (Figure 7.6a).

TOPO-CAPPED AND ODT-CAPPED CDSE QDS

The replacement of TOPO by ODT molecules on the surface of the QDs had a minimal effect (±1–3 nm) on λ_{max} in the absorption and emission spectra of the nanocrystals. However, this surface cap exchange resulted in 86% quenching of PL

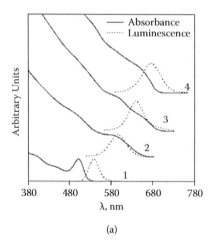
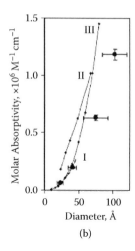

(a)   (b)

FIGURE 7.6 (a) Absorbance (solid lines) and PL (dashed lines, λ_{ex} = 400 nm) spectra of TOPO-capped CdSe QDs with average diameters of [1] 23 Å, [2] 41 Å, [3] 75 Å, and [4] 103 Å. (b) Molar absorptivity values of TOPO-capped CdSe QDs as a function of particle size. The error bars consist of the standard deviations for the respective axes. The dashed lines in (b) indicate the corresponding literature values.

intensity from the QDs. Introduction of S-Cd bonds on the surface of the CdSe QDs has been correlated to changes in PL properties arising from modifications of electronic states.[28–31] PL intensity was 98% quenched when an equimolar mixture of *N,N*-diisopropylethylamine (DIEA) and ODT was utilized to exchange the surface TOPO cap on the QDs. The non-nucleophilic base is utilized to facilitate binding of ODT to the QDs. However, utilization of DIEA during modification of the 103 Å QDs could result in a smaller average particle size, 79 Å, as imaged by the TEM. This suggests that chemical "etching" of the nanoparticles could take place with prolonged reaction time.

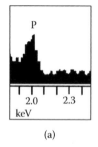
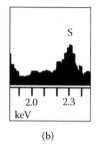

(a)   (b)

FIGURE 7.7 Energy-dispersive x-ray spectra of (a) TOPO-capped and (b) ODT-capped CdSe QDs (79 Å). Intensity of *y*-axes = 3.00 k counts.

It has been proposed that every Cd atom on the surface of the CdSe QDs may be coordinated to a TOPO molecule.[32] Therefore, the degree of quenching of PL intensity could be correlated to the extent of surface ligand exchange. These changes in the QD's optical properties provide evidence of an efficient surface cap exchange. To confirm these results, energy-dispersive x-ray analysis was performed on purified dried films of QDs. Appearance of a peak at 2.3 keV revealed the presence of sulfur atoms in the purified ODT-capped nanocrystals (Figure 7.7). The phosphorus peak at 2.0 keV decreased to the noise level of the spectrum. This indicates that ODT replaced most of the TOPO molecules on the surface of the QDs. The results compare to reports in the literature in which thiol-containing organic molecules could replace 85–90% of surface TOPO/TOPSe molecules on CdSe QDs.[31]

SURFACE CHEMISTRY

The limiting nanoparticle or limiting molecular area is defined as the smallest area occupied by the nanoparticle or molecule, respectively, in 2D. It is obtained by the extrapolation of the linear portion of the π-A isotherm to zero surface pressure. First, the π-A isotherms for the TOPO and ODT molecules were obtained. TOPO and ODT have limiting molecular areas of 135.0 and 22.5 Å^2/molecule, respectively, as shown in Figure 7.8A and Figure 7.8B. ODT and stearic acid are very similar in structure, both having a C18 hydrocarbon chain but a different polar group. Both compounds have comparable limiting molecular areas of 22.5 and 21.8 Å^2/molecule, respectively (Figure 7.8A and Figure 7.8C). No signs of ODT's thiol group oxidation were detected at the air-water interface. The ODT π-A isotherm has an additional feature. After the collapse of the monolayer film at about 15 mN/m, ODT presents a plateau region with almost constant surface pressure, followed by another rapid increase in surface pressure (Figure 7.8A). The limiting molecular area extrapolated from this second lifting point is 7.7 Å^2/molecule. This indicates the capacity of ODT to form a stable "multilayer" film at about 1/3 of its limiting molecular area. Self-assembly of ODT into a trilayer film has been reported on liquid mercury.[33]

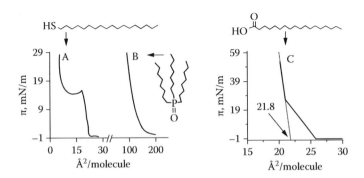

FIGURE 7.8 Characteristic π-A isotherms for (A) ODT, (B) TOPO, and (C) stearic acid molecules.

TOPO-Capped

The π-A isotherms of QDs were recorded utilizing only fresh and purified QD samples. This practice was utilized to minimize the presence of 3D QD aggregates at the air-water interface. Considering a spherical nanocrystal, the area of a single TOPO-capped CdSe QD in 2D at the air-water interface could be predicted theoretically. Based on simple molecular modeling calculations, it has been reported that TOPO can form a monolayer on the surface of the nanocrystals.[32] Also, TOPO monolayers on metal surfaces have been measured at about 7 Å thick.[20] The area per QD in 2D then should be as follows: $\pi(r + 7)^2$, where r is the QD radius in Å. The QDs used in this study had average r values of 11.5, 20.5, 37.5, and 51.5 Å. The respective theoretical area/QD values in 2D are 1075, 2376, 6221, and 10,751 Å^2/QD. The corresponding experimental values of the limiting nanoparticle areas extrapolated from the π-A isotherms (Figure 7.9) of the QDs are 1086, 2375, 6370, and 10,757 Å^2/QD, respectively. The predicted and experimentally derived limiting nanoparticle areas match very closely, with an error <3%. These results indicate minimum interdigitation of TOPO surface molecules, confirming that TOPO forms close-packed monolayers on the surface of the QDs. These results also prove the accuracy of the molar absorptivity values that were derived previously for the respective TOPO-capped CdSe QDs.

To verify that there is minimum interdigitation among the TOPO cap on the QDs, oleic acid was mixed with the 41-Å TOPO-capped nanoparticles at the QD molar fractions of 0.0, 0.1, 0.3, 0.5, 0.7, and 1.0. The respective π-A isotherms of the solutions are shown in Figure 7.10a. The average area of the two-component films ($A_{1,2}$) obeyed the followed linear equation: $A_{1,2} = X_1(A_1 - A_2) + A_2$, where X_1 is the QD mol fraction, A_1 is the area (2003 Å^2) per pure QD at a π of 25 mN/m, and A_2 is the area (28 Å^2) per pure oleic acid at a π of 25 mN/m. A variation of <5% or R^2 value of 0.9938 for linearity was obtained (Figure 7.10b). The collapse surface pressure at >38 mN/m of the two-component films is characteristic of the QDs and does not vary much with composition. Therefore, it can be concluded that the two components of the mixed monolayer films are immiscible and that there is minimum interdigitation among the TOPO caps on the QDs or with oleic acid molecules.

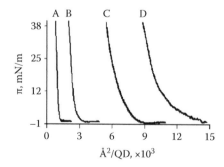

FIGURE 7.9 π-A isotherms of (A) 23 Å, (B) 41 Å, (C) 75 Å, and (D) 103 Å TOPO-capped CdSe QDs.

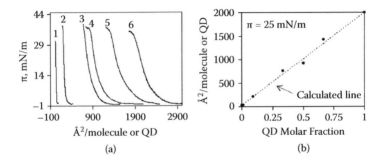

FIGURE 7.10 (a) The π-A isotherms of the 41 Å TOPO-capped CdSe QDs and oleic acid mixed monolayers at the QD molar fractions of [1] 0.0, [2] 0.1, [3] 0.3, [4] 0.5, [5] 0.7, and [6] 1.0. (b) Comparison between the calculated [dashed line] and experimental [points] average areas at 25 mN/m of the respective two-component films at the different QD molar fractions.

ODT-CAPPED

The limiting nanoparticle area was lower for the ODT-capped QDs than for the corresponding TOPO-capped nanoparticles. For example, the 23-Å QD has a value of 505 or 1086 Å²/QD when ODT- (Figure 7.11A) or TOPO-capped (Figure 7.11B), respectively. This suggests that replacement of the more bulky TOPO surface ligands with ODT molecules increases the packing density of the QDs through interdigitation of the longer ODT alkyl chains. However, this capacity for interdigitation facilitated 3D aggregate formation at the interface, which contributed to a smaller area/QD. This is explained in the next section devoted to the 2D self-assembly of the QDs.

The 23-Å ODT-capped CdSe QD solutions were "doped" with free ODT molecules at a 1:300 (QD:ODT) molar ratio. These free ODT molecules "dissolved" the QDs at the interface and reduced the formation of 3D aggregates; refer to the self-assembly section discussed later. The π-A isotherm of this mixed monolayer system is shown in Figure 7.11C, and its shape resembled that of the pure ODT monolayer

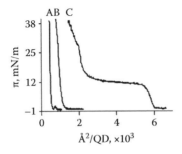

FIGURE 7.11 π-A isotherms of 23 Å CdSe QDs: (A) ODT-capped, (B) TOPO-capped, and (C) ODT-capped with 1:300 molar excess of free ODT.

(Figure 7.8A). The limiting nanoparticle area extrapolated from the π-A isotherm in Figure 7.11C is 6030 Å²/QD. If there is no alkyl chain interdigitation, the area/QD after "doping" with ODT should have been $[QD_{area} + (22.5 \times 300)]$ or $QD_{area} + 6750$ Å²/QD. In addition, the first collapse of the ODT monolayer film occurs at $\pi \approx 15$ mN/m, while that of the mixed monolayer with QDs occurs at $\pi \approx 10$ mN/m. This change in collapse surface pressure could be a result of miscibility and interdigitation of the two components at the interface.

The capacity of ODT to form a stable multilayer film at 1/3 its limiting molecular area was exploited to derive the limiting molecular area of a single 23-Å ODT-capped QD in the Langmuir film. The limiting nanoparticle area decreased from 6030 to 2474 Å²/QD from the monolayer to the "trilayer" film (Figure 7.11C). If this decrease of 3556 Å²/QD is divided by two, the area occupied by free ODT molecules in the "trilayer" mixed film is obtained, which is 1778 Å²/molecule. Subtracting this value from the total 2474 Å²/QD gives a limiting nanoparticle area of 696 Å²/QD for the ODT-capped QDs. This new value indicates that the number of 3D aggregates was reduced and that still surface ODT's alkyl chains on the QDs interdigitated.

Similar studies and calculations were not performed for the larger ODT-capped nanoparticles. "Doping" these QDs with free ODT molecules at the same 1:300 molar ratio did not help to break all of the aggregates. This could be partially attributed to the larger particle size and changes in the degree of alkyl chain interdigitation. For example, the surface ODT molecules should have less free space (more crowded) as they extend away from the surface of larger nanoparticles. This was confirmed with the TEM micrographs taken for the 103-Å QDs, as will be shown below.

SELF-ASSEMBLY OF THE CdSe QDs IN 2D: NATURE OF THE SURFACE CAP

LANGMUIR FILM (TOPO AND ODT)

At zero surface pressure, the QDs self-assembled at the air-water interface into 2D domains as shown by the epifluorescence microscope images (Figure 7.12A). Domain formation was more obvious as particle size increased. Upon compression, the domains organized into a more uniform film at a surface pressure of about 35 mN/m (Figure 7.12B). The nature of the surface cap on the QDs can also influence interparticle interactions and topography at the air-water interface. For example, the 23-Å TOPO-capped CdSe QDs presented a corrugated topography at the high surface pressure of 39 mN/m (Figure 7.12C). The film collapsed with further increase in surface pressure. This particular topography was not observed for the ODT-capped QDs of the same size even at higher surface pressures (Figure 7.12D).

The addition of 300 molar excess of free ODT molecules in solution "dissolved" the 23-Å ODT-capped QDs, which then presented more homogeneous topography as shown in Figure 7.12E. "Doping" the solutions of the larger ODT-capped QDs with free ODT molecules at the same molar ratio did not result in homogeneous films as above. Brighter domains were still visible at the air-water interface. Multilayer domains or 3D aggregates may contribute to the appearance of these brighter areas

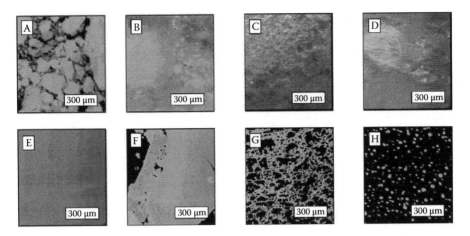

FIGURE 7.12 Epifluorescence images of the CdSe QDs taken directly at the air-water interface. (A) 103-Å TOPO-capped at 0 mN/m. (B) 103 Å TOPO-capped at 35 mN/ m. (C) 23-Å TOPO-capped at 39 mN/m. (D) 23-Å ODT-capped at 39 mN/m. (E) 23-Å ODT-capped + 300 molar excess free ODT at 35 mN/m. (F) 41-Å TOPO-CdSe at pH 10 and 50 mN/m. (G) 41-Å TOPO-capped mixed with oleic acid at a QD molar fraction of 1×10^{-3} at 25 mN/m. (H) 41-Å TOPO-capped mixed with oleic acid at a QD molar fraction of 1×10^{-4} at 25 mN/m. $\lambda_{ex} = UV$.

as a consequence of QD photoluminescence changes at the interface. For example, PL intensity of the QD Langmuir films was enhanced or underwent photochromism from red to green upon UV irradiation over time, as observed with the epifluorescence microscope. These changes in PL properties of the QDs may result from photooxidation of the nanocrystal surface or cap.[34,35] In addition, water molecules decreased the photoluminescence intensity of the TOPO-capped CdSe QDs. The effect of water molecules on the PL intensity of QDs in Langmuir films has been reported previously.[36] When the monolayer film was transferred to a quartz slide via the LB film deposition method, the PL from the QDs was easily detected on the dried film. However, an optical fiber attached to the spectrofluorimeter did not detect any PL emission from the CdSe QD monolayer when positioned directly above it at the air-water interface. This was a result of low PL intensity.

Some of the nanoparticles within 3D aggregates could be oriented more into the air and further away from the water molecules. This could result in a higher PL intensity in addition to the increased concentration of nanoparticles in these 3D domains. It is known that aggregation or high concentrations of QDs and fluorophores in general may result in lower or quenching of PL intensity of solutions because of inner filter effects, radiative and nonradiative transfer, excimer formation, or other. For example, the 23-Å TOPO-capped QD chloroform solutions decreased (6–75%) PL intensity and red-shifted (>10 nm) emission λ_{max} when the concentration level was increased from 7×10^{-6} to 30×10^{-6} M. However, it was assumed that photooxidation, interparticle interactions, or 3D self-assembly of the QDs at the interface enhanced PL intensity, and the aggregates appeared brighter. To support this assumption, QDs were spread onto a glycerol or basic (pH 10) subphase. The QDs had difficulties spreading through the interface and formed multilayer domains as illustrated in Figure 7.12F. These domains were brighter when imaged with the epifluorescence microscope, and their PL emission was detected with the optical fiber device.

MIXED LANGMUIR FILM (TOPO/FATTY ACID)

To further manipulate the 2D self-assembly of QDs at the air-water interface, mixed monolayer systems were also studied. The 41-Å TOPO-capped QDs were mixed with oleic acid at different molar fractions. Oleic acid was used because it should not replace the TOPO surface molecules on the QDs and may impart some organizational features because of its single double bond (18:1, *cis*-9). At the QD molar fraction of 1×10^{-3}, the 2D domains formed a totally new topography as shown in Figure 7.12G. The QDs self-assembled into domains, which formed a 2D structure such as those found in porous materials. At the lower QD molar fraction of 1×10^{-4}, separate domains between 10 and 41 μm in size were imaged as shown in Figure 7.12H.

These QD micro-domains moved collectively across the monolayer film even at high surface pressures. The conditions for imaging the topography were adjusted so that the images show the real domains as seen with the naked eye through the epifluorescence microscope measurements. The actual color of the domains may have varied slightly as a result. This observed "fluidity" of the mixed monolayer may be a result of the cis double bond on oleic acid. Unsaturated fatty acids have kinks and do not pack as well as saturated fatty acids. Therefore, an increase in the composition of unsaturated fatty acids results in increased lateral fluidity on a membrane or unstable/"fluid" monolayer film. When stearic acid, a saturated fatty acid, instead of oleic acid, was utilized to prepare the mixed monolayer films, static topographies were barely imaged. PL intensity was almost completely quenched. Imaging of the topography was difficult at the low QD molar fractions, which were needed to evaluate QD "fluidity" within the film.

Besides surface alkyl chain interdigitation and the closest possible packing arrangement of QDs, the degree of organization at the air-water interface may also be influenced by other factors. For example, surfactants such as TOPO and the crystalline structure of the CdSe nanoparticles could bring about intrinsic dipole moments.[33,38] Changes in surface cap and nanocrystal size may alter the direction and intensity of these polarities, which may be factors influencing the degree of nanoparticles organization at the interface.

To study the organization of the QDs at the air-water interface, TEM micrographs of LB (Langmuir-Blodgett) films were taken at the π of 20 mN/m. The QDs self-assembled into domains of variable architecture as shown in Figure 7.13A and Figure 7.13B. The nanoparticle organization within the 2D domains was influenced by the nature of the surface stabilizing molecule. TOPO-capped QDs had hexagonal-like close-packing arrangements as shown in Figure 7.13C. Deviations from the hexagonal packing of QDs could be attributed in part to the large particle size distribution of the samples. The ODT-capped nanocrystals were found to have areas in which the QDs were farther apart from each other (Figure 7.13D). A longer

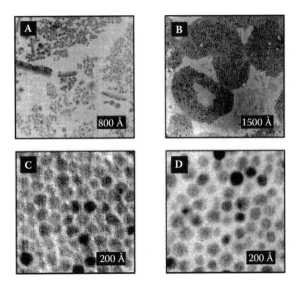

FIGURE 7.13 TEM micrographs of CdSe QD LB films deposited on carbon-coated copper grids at 20 mN/m. (A) 75-Å TOPO-capped. (B) 103-Å TOPO-capped. (C) 103-Å TOPO-capped. (D) 103-Å ODT-capped.

interparticle distance confirmed the presence of the longer ODT on the surface of the nanoparticles. Interdigitation was then more restricted at this large particle size. Some 3D domains within the LB films were detected as darker or indistinct areas.

MANIPULATION OF THE CdSe QDs LANGMUIR FILM: LANGMUIR-BLODGETT (LB) FILM

PHOTOLUMINESCENCE

QD Langmuir films were transferred onto hydrophilic or hydrophobic quartz slides utilizing the LB film deposition technique at different surface pressures. Deposition ratios higher than 0.8 were obtained for both substrates, indicating an effective film transfer. The ratio between the area decrease at the air-water interface, that is, the area of the monolayer transferred onto solid substrate, and the area of the solid substrate available for deposition, is defined as "deposition ratio." A good deposition ratio has values between 0.7 and 1, which is the ideal deposition case when all the solid substrate is covered with a monolayer. Emission from a single TOPO-capped QD LB monolayer film was detected. When deposited at higher surface pressures, the LB monolayer increased PL intensity (Figure 7.14). This results from a closer packing of the QD domains as shown in Figure 7.14. When imaged with the epifluorescence microscope, the topography of the QD LB films deposited at different surface pressures on the slides resembled that taken directly at the interface. Thus, the films were transferred nearly intact as assembled at the air-water interface.

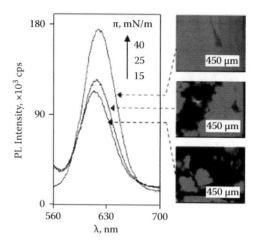

FIGURE 7.14 PL spectra and epifluorescence images (λ_{ex} = UV) of a 41-Å TOPO-capped CdSe QD LB monolayer film deposited on a hydrophilic quartz slide at different surface pressures.

NATURE OF THE SUBSTRATE

Photoluminescence intensity from the LB films on the hydrophilic quartz slide increased linearly (R^2 = 0.9115) with the number of QD monolayers (Figure 7.15a). This suggested that equal amounts of QDs were transferred after each deposition cycle, which resulted in a buildup of a homogeneous film. TOPO-capped QD deposited on the hydrophobic quartz substrate had a lower correlation coefficient for linearity, R^2 = 0.7748 (Figure 7.15b). However, the total PL intensity from the hydrophilic or hydrophobic LB films was within the same range after deposition of the 5

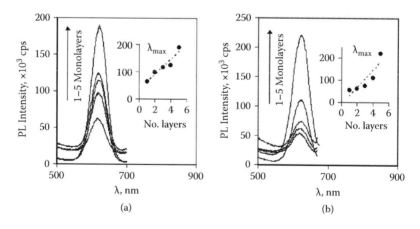

FIGURE 7.15 Increase in PL intensity of 41-Å TOPO-capped CdSe QDs LB films on (a) hydrophilic and (b) hydrophobic quartz slides as a function of the number of deposited monolayers. Deposition at 25 mN/m and λ_{ex} of 467 nm.

QD monolayers. Therefore, deviations from linearity could be partially attributed to variations in the deposition ratio, the distribution of the QD domains along the solid support, and the area in which PL intensity was recorded. The stability of the films was not evaluated herein.

One interesting phenomena is that using these QDs, no emission can be detected at the air-water interface. This is explained by the study of Cordero and others.[2] The authors used hexadecylamine-stabilized CdSe particles of 4.1 nm spread on the water surface. They have found that the water molecules were oxidizing the CdSe QDs from the interface, thus affecting the luminescence properties. Further, the Langmuir film of QDs was investigated having ZnS capping.

SYNTHESIS OF (CDSE)ZNS QUANTUM DOTS

The (CdSe)ZnS core-shell QDs capped with trioctylphosphine oxide (TOPO) ligand were prepared through a stepwise procedure as described elsewhere.[36] The precursor of (CdSe)ZnS core-shell QDs, nearly monodispersed TOPO-capped CdSe quantum dots, was synthesized by the method reported previously by Peng and Peng.[18] Cadmium oxide (CdO) and tetradecylphosphonic acid (TDPA) were used as the Cd precursor and the trioctylphosphine selenide as the Se precursor. The CdSe QDs were formed by pyrolysis of the Cd and Se precursors in a coordinating solvent, trioctylphosphine oxide (TOPO), at high temperature (250–270°C). The QDs were collected as powders by size-selective precipitation[16] with methanol, followed by drying under vacuum. The average size of the CdSe/TOPO QDs was determined by UV-Vis absorption spectroscopy of its chloroform solution.[25] An approximate particle concentration of CdSe/TOPO QD solutions can also be determined via UV-Vis spectroscopy.[37] By knowing the mass of CdSe/TOPO QDs used when preparing the solution and the total volume of the stock solution, the "molecular" weight of CdSe/TOPO quantum dots could be estimated.

ZnEt$_2$ and (TMS)$_2$S were used as Zn and S precursors for the ZnS cap. The amounts of Zn and S precursors needed to grow a ZnS shell of desired thickness for each CdSe sample were determined as follows. First, the average size of the CdSe core was estimated from UV-Vis data.[25] Next, the mole number of ZnS necessary to form a shell per mole of CdSe quantum dots was calculated on the basis of the volume of individual CdSe QDs and the desired thickness of the shell. It was assumed that the QD core and shell were spherical. The bulk density of ZnS was used in the calculations. Then, the amount of ZnS for one synthesis was calculated by knowing the mass of CdSe/TOPO QDs used and the "molecular" weight of CdSe/TOPO QDs estimated previously from UV-Vis measurements. Finally, the amounts of ZnEt$_2$ and (TMS)$_2$S were calculated from the amount of the ZnS needed in the synthesis. For example, if the CdSe core has an average size of 2.48 nm and a "molecular" weight of 32,000 Da, the amount of ZnS needed for a 1-nm shell for each single particle will be 1.599×10^{-19} g. Thus, by knowing the amount of CdSe QDs, the amounts of ZnEt$_2$ and (TMS)$_2$S were calculated. The common coating procedure was carried out by adding a ZnEt$_2$ and (TMS)$_2$S mixture solution dropwise into a coordination solution (TOPO/TOP as solvent) of CdSe QDs at high temperature under argon atmosphere.[36] Usually, the temperature during coating was lower than that used for growing the CdSe nanocrystals to avoid compromising the integrity of the native cores. The (CdSe)ZnS core-shell QDs were collected as a powder by precipitation, washed with methanol, and dried under vacuum.

HRTEM AND PHOTOPHYSICAL PROPERTIES

UV-Vis spectroscopy of QD solutions indicated an average CdSe core size or diameter of 2.9 nm.[25] The amounts of $ZnEt_2$ and $(TMS)_2S$ were then calculated for deposition of a 0.5-nm-thick ZnS shell on the QDs. Figure 7.16 shows the TEM micrograph and corresponding size distribution chart of (CdSe)ZnS/TOPO QDs. The calculated average size of (CdSe)ZnS/TOPO QDs was 3.7 nm. The QD size and the thickness of the TOPO surface cap helped to explain the limiting nanoparticle area of the QDs as derived from the surface pressure-area isotherm.

The UV-Vis absorption and photoluminescence spectra of (CdSe)ZnS/TOPO QD solutions are shown in Figure 7.17. The first electronic transition peak revealed a core size of 2.9 nm according to the literature.[25] The PL emission peak was found to be at 571 nm with a full width at half-maximum (fwhm) of 35 nm, which indicated a very small size distribution.

SURFACE PRESSURE- AND SURFACE POTENTIAL-AREA ISOTHERMS OF (CDSE)ZNS/TOPO QD MONOLAYERS

The three alkyl side chains of the surface TOPO molecules make the QDs soluble in many nonpolar solvents such as chloroform, hexane, and cyclohexane. Besides that, the large dipole moment derived from the phosphorus-oxygen bond[16,20] makes it possible for TOPO-capped QDs to form a stable monolayer at the air-water interface.

The surface pressure- and surface potential-area isotherms give the most important characteristics of the monolayer properties. The surface pressure-area isotherm

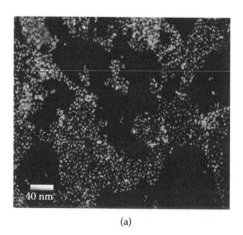

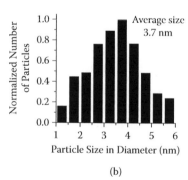

(a)   (b)

FIGURE 7.16 (a) TEM image and (b) size distribution chart of (CdSe)ZnS/TOPO quantum dots.

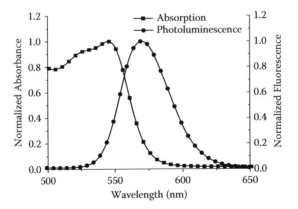

FIGURE 7.17 UV-Vis absorption and photoluminescence spectra of (CdSe)ZnS/TOPO quantum dots in $CHCl_3$ (4.4×10^{-6} M). $\lambda_{ex} = 350$ nm.

of (CdSe)ZnS/TOPO QDs is shown in Figure 7.18 and three distinct phases are well displayed. The film collapsed at a surface pressure of 45 mN·m^{-1}, indicating that the (CdSe)ZnS/TOPO QDs can form a stable Langmuir film at the air-water interface. This is similar to other amphiphilic molecules such as stearic acid, whose Langmuir film usually collapses at a surface pressure between 40 and 45 mN·m^{-1}. The limiting nanoparticle area of the QDs was obtained by extrapolating the linear part of the isotherm to zero surface pressure, 1500 Å^2.

As discussed before, the average particle diameter obtained from TEM measurement was 3.7 nm. The TOPO cap forms a monolayer at the surface of nanocrystals.[32]

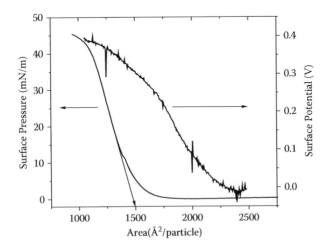

FIGURE 7.18 Surface pressure- and surface potential-area isotherms of (CdSe)ZnS/TOPO QDs.

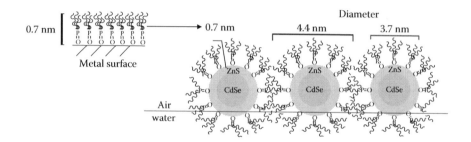

SCHEME 7.1 Size of QDs Obtained from TEM Micrographs and Surface Pressure-Area Isotherm. (a) Reported thickness of TOPO monolayer (0.7 nm) at metal surface. (b) Model of TOPO capped QDs. The TEM measure gave a diameter of bare QD of 3.7 nm because the TOPO is nonconductive; the diameter of TOPO-capped QDs was $(2 \times 0.7) + 3.7$ nm $= 5.1$ nm. (c) At the air-water interface, the TOPO moieties were compressed and resulted in a smaller diameter (4.4 nm) due to the interdigitation of the TOPO tails.

The thickness of a TOPO monolayer was reported to be 0.7 nm as measured on metal surfaces.[20] Then, the diameter of the TOPO-capped CdSe(ZnS) QDs should be $(2 \times 0.7) + 3.7$ or 5.1 nm. If assuming spherical QDs, an average diameter of 4.4 nm for a single QD was calculated from the limiting particle area of the QD Langmuir films. Obviously, at the air-water interface, the TOPO moieties were compressed and resulted in a smaller observed particle size due to interdigitation of the TOPO tails (Scheme 7.1).

The surface potential-area isotherm measures the dipole moment changes during compression of the monolayer at the air-water interface. This isotherm for the QDs is shown in Figure 7.18. When the average particle area was larger than 2250 Å², the QDs randomly existed at the air-water interface. As a result, the total surface potential contribution of QDs was zero. When the monolayer was compressed, the hydrophobic part of the QDs began to stretch out of the water and the change in the dipole moment resulted in a rapid increase of surface potential up to 220 mV, which corresponded to the lifting point of the surface pressure. When the surface area of the QDs in the monolayer was smaller than 1750 Å², the hydrophobic moieties of TOPO-capped QDs were lifted up into the air and the monolayer was in a liquid-condensed phase. From this point, a slight increase in the surface potential was observed although the change in surface pressure was still significant. This may be due to reorientation of the nanoparticle in order to maximize exposure of the hydrophobic TOPO moiety to the air. As the surface potential reached the maximum, the surface pressure-area isotherm showed the collapse of the monolayer. When the QDs were compressed at the air-water interface, the surface potential started to increase prior to an increase in the surface pressure. This is a common response when both curves are compared. The surface pressure increased considerably when the QDs started to interact among them. However, the film could be compressed with minimum particle-particle interactions as during the gas phase, resulting in a higher surface potential. This also explains why, at the solid phase of the isotherm, the surface potential curve changed its slope.

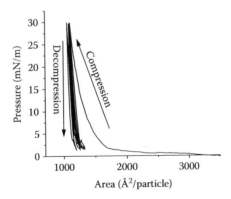

FIGURE 7.19 Multiple compression/decompression cycles of the (CdSe)ZnS/TOPO QD Langmuir film up to a surface pressure of 30 mN·m⁻¹.

STABILITY OF QD MONOLAYER

The stability of the QD monolayer at the air-water interface was examined by two different methods: compression/decompression cycle and kinetic measurements. Figure 7.19 shows the compression/decompression cycles of the QD monolayer up to a surface pressure of 30 mN·m⁻¹. A decrease of about 250 Å² in the apparent limiting nanoparticle area was observed after the first compression/decompression cycle. Then, a hysteresis behavior of the isotherm was observed for the successive six compression/decompression cycles that followed, which indicated long-term stability of the Langmuir film of the QDs.

Figure 7.20 shows the kinetic measurements of the QD Langmuir film at the air-water interface. The QD monolayer was compressed to a surface pressure of 30 mN·m⁻¹ and held constant at this value for a period of 40 min. The area per

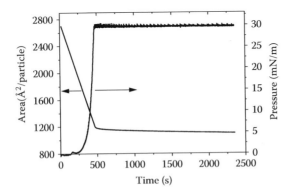

FIGURE 7.20 Changes in the surface area per QD at the air-water interface as a function of time and at a constant surface pressure of 30 mN·m⁻¹.

nanoparticle decreased to about 77 $Å^2 \cdot particle^{-1}$ over the 40-min period. This value corresponds to <6% of the limiting nanoparticle area of the QD monolayer. This slight decrease in the QD surface area is due to the rearrangement of particles to reduce the empty spaces in the monolayer or a partial aggregation of the QDs at a

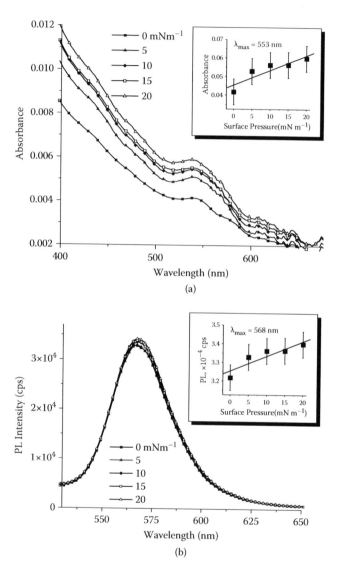

(a)

(b)

FIGURE 7.21 *In situ* (a) UV-Vis absorption and (b) photoluminescence spectra of the Langmuir film of the (CdSe)ZnS QDs at different surface pressures. The inset in panel (a) presents the absorbance at 553 nm as a function of the surface pressure. The inset in panel (b) presents the PL intensity of QDs at emission maximum (568 nm) as a function of the surface pressure. Linear fits of the data sets are also presented in both insets. The error bar represents the standard deviation of the data set.

surface pressure of 30 mN·m^{-1}. Therefore, the film kinetic data suggested the formation of a stable QD monolayer at the air-water interface.

In Situ UV-Vis Absorption and Photoluminescence Spectroscopies of (CdSe)ZnS QD Monolayers at the Air-Water Interface

The *in situ* UV-Vis absorption spectrum of the (CdSe)ZnS QD monolayer was measured at the air-water interface. Figure 7.21a shows the spectra of the (CdSe)ZnS QD monolayer at different surface pressures. The first electronic transition band was clearly observed. Although the absorbance at 533 nm showed an increased trend with increasing surface pressure of the monolayer, the linear fit of the absorption peak as a function of surface pressure gave poor linearity ($R = 0.90$), which implied that the QD nanoparticles were not uniformly distributed at the air-water interface.

The same phenomenon can be found in the *in situ* photoluminescence spectra of the QD Langmuir film (Figure 7.21b). In addition, the photoluminescence intensity of the QD Langmuir film showed only a slight increase with increasing surface pressure. Thus, in the following LB film preparation, we have used a surface pressure greater than 15 mN·m^{-1} for the deposition of the Langmuir film to ensure the homogeneity of the QDs in 2D.

Epifluorescence Microscopy of (CdSe)ZnS QD Langmuir Film at the Air-Water Interface

The interpretation made on the homogeneity of the QD Langmuir film can also be verified by topographical imaging of the QD Langmuir films. Epifluorescence microscopy was utilized in this study because it provides direct visual evidence of the topography of QD Langmuir films. Figure 7.22 shows the epifluorescence microscopic images of the QD monolayer at different surface pressures. The epifluorescence image of the QD Langmuir film at 0 mN·m^{-1} shows discrete domains of QDs (bright areas) and empty spaces (dark areas), which indicated that the highly hydrophobic TOPO-capped QDs form small domains when they are spread at the air-water interface. Upon compression of the QD Langmuir film, the domains began to interact and the monolayer became more homogeneous. When the surface pressure was >15 mN·m^{-1}, the QD Langmuir film was almost homogeneous as seen from the epifluorescence images (Figure 7.22E and Figure 7.22F).

As shown before, both the first absorption peak and PL intensity of the QD Langmuir film displayed poor linearity ($R = 0.90$ and 0.95, respectively) with increased surface pressure. In addition, the increasing trends of both absorbance and PL intensity are of little significance (Figure 7.21). As we can see, when the surface pressure was smaller than 15 mN·m^{-1} (Figure 7.22A–7.22C), the QD Langmuir film was composed of discrete domains. So it can be concluded that the observed results of the UV-Vis and PL measurements (Figure 7.21a and Figure 7.21b) were due to the heterogeneous topography of the QD Langmuir film at low surface pressures (0 to 15 mN·m^{-1}).

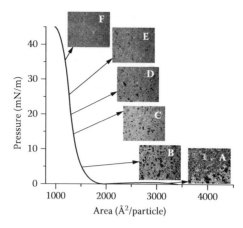

FIGURE 7.22 Epifluorescence microscopic images of the Langmuir film of (CdSe)ZnS/TOPO quantum dots at surface pressures of (A) 0, (B) 10, (C) 15, (D) 20, (E) 25, and (F) 35 $mN\cdot m^{-1}$. The image size is 895 $\mu m \times 713$ μm.

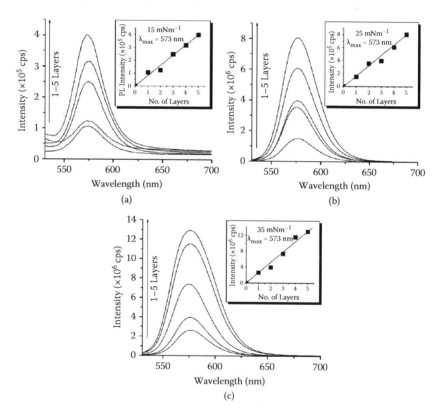

FIGURE 7.23 Photoluminescence spectra of LB films deposited at surface pressures of (a) 15, (b) 25, and (c) 35 $mN\cdot m^{-1}$. The insets show the PL intensity at emission maximum (573 nm) as a function of the number of layers.

PHOTOLUMINESCENCE STUDY OF LANGMUIR-BLODGETT FILMS OF (CdSe)ZnS/TOPO QDs

LB films were prepared at three surface pressures, namely, 15, 25, and 35 mN·m⁻¹. This choice was based on the epifluorescence measurements, which showed homogeneity in the topography of Langmuir films for surface pressure ≥15 mN·m⁻¹. The (CdSe)ZnS QD Langmuir film was deposited onto a chemically treated hydrophobic quartz substrate. The silanized quartz slide was suspended above the monolayer and gradually lowered vertically into the subphase at a deposition rate of 1.8 mm·min⁻¹. Because hydrophobic quartz slides were used, the dipping of the slides through the QD monolayers resulted in adsorption of hydrophobic moieties of the QDs, that is, TOPO. The calculated transfer ratio (barrier area swept/deposited area on the slide) was close to 1.0, indicating that the slide was completely coated by the QD monolayer. In this study, five layers were deposited at three different surface pressures. Figure 7.23 shows the photoluminescence spectra of LB films of (CdSe)ZnS QDs deposited at three different surface pressures. The photoluminescent emission maximum of the QDs at 573 nm for the LB film was identical within experimental error to the value of QDs in chloroform solution (λ_{max} = 571 nm). In all three cases, the PL intensity at 573 nm increased linearly with increasing number of layers, which indicated that homogeneous Langmuir films were deposited.

RECENT WORK

Lately, Pradhan and others[38] investigated the electronic conductivity in Langmuir films[39-42] using TOPO-protected CdSe QDs. Their experimental setup is based on the use of an interdigitated arrays (IDAs) electrode that is aligned vertically at the air-water interface for *in situ* voltammetric measurements at varied interparticle distances or different surface pressures. The authors found that upon photoirradiation with photon energies higher than that of the absorption threshold, the voltammetric currents increased substantially, with two of voltammetric peaks at positive potentials. Under photoirradiation, adsorption of ambient oxygen onto the particle surface leads to the efficient trapping of photogenerated electrons and hence an excess of holes in the particles [34,43-45] giving to the conductivity profiles a dynamic profile. The authors showed that the chemical stability of the CdSe QDs and the associated photogated charge transfer mechanism were manipulated by the trapping of photogenerated electrons by adsorbed oxygen and the process is reversible just like the case of photoetching. They were able to show that the voltammetric currents can be gated by photoirradiation. The difference between the Langmuir film behavior compared to the dropcast film is assigned to the presence of defects within the monolayer and thus leading to the effects mentioned above.

CONCLUSIONS

It is clear that the use of Langmuir films in controlling the interparticle distances and their application are getting more attention every day. The research on the surface engineering of quantum dots covers areas not studied in the past, whether the assembly, synthetic routine, presence or absence of ZnS capping. Properties like chemical and photophysical properties are studied with various methods, for example, UV-Vis and fluorescence spectroscopies. Assemblies of QDs are studied mainly as Langmuir-Blodgett films since this technique retains most of the Langmuir films' properties, thus giving the opportunity to investigate the films by AFM but mainly HRTEM. It was shown briefly that a wide range of properties can be investigated and photophysical properties depend on factors like capping, ligand, and irradiation. Without any doubt, the Langmuir film technique will gain popularity and more techniques will be adapted to allow *in situ* investigation of QDs monolayers.

ACKNOWLEDGMENTS

Dr. Leblanc appreciates financial support under the National Science Foundation (CH-0416095) and U.S. Army Research Office (DAAD19-03-1-0131).

REFERENCES

1. Chan, W.C.W. and Nieb, S.M. *Science* 1998, *281*, 2016.
2. Cordero, S.R.; Carson, P.J.; Estabrook, R.A.; Strouse, G.F. and Buratto S.K. *J. Phys. Chem. B* 2000, *104*, 12137.
3. Wu, X.Y.; Liu, H.J.; Liu, J.Q.; Haley, K.N.; Treadway, J.A.; Larson, J.P.; Ge, N.F.; Peale, F. and Bruchez, M.P. *Nat. Biotechnol.* 2003, *21*, 41.
4. Coe, S.; Woo, W. K.; Bawendi, M. and Bulovic, V. *Nature* 2002, *420*, 800.
5. Empedocles, S. and Bawendi, M. G. *Accounts Chem. Res.* 1999, *32*, 389.
6. Heath, J.R. *Science* 1995, *270*, 1315.
7. Nirmal, M.; Dabbousi, B.O.; Bawendi, M.G.; Macklin, J.J.; Trautman, J.K.; Harris, T.D. and Brus, L.E. *Nature* 1996, *383*, 802.
8. Bawendi, M.G.; Carroll, P.J.; Wilson, W.L. and Brus, L.E. *J. Chem. Phys.* 1992, *96*, 946.
9. Bawendi, M.G.; Steigerwald, M.L. and Brus, L.E. *Annu. Rev. Phys. Chem.* 1990, *41*, 477.
10. Nirmal, M. and Brus, L. *Accounts Chem. Res.* 1999, *32*, 407.
11. Katari, J.E.B.; Colvin, V.L. and Alivisatos, A.P. *J. Phys. Chem.* 1994, *98*, 4109.
12. Tittel, J.; Gohde, W.; Koberling, F.; Basche, T.; Kornowski, A.; Weller, H. and Eychmuller, A. *J. Phys. Chem. B,* 1997, *101*, 3013.
13. Mikulec, F.V.; Kuno, M.; Bennati, M.; Hall, D.A.; Griffin, R.G. and Bawendi, M.G. *J. Amer. Chem. Soc.* 2000, *122*, 2532.
14. Collier, C.P.; Vossmeyer, T. and Heath, J.R. *Annu. Rev. Phys. Chem.* 1998, *49*, 371.
15. Dabbousi, B.O.; Murray, C.B.; Rubner, M.F. and Bawendi, M.G. *Chem. Mater.* 1994, *6*, 216.
16. Murray, C.B.; Norris, D.J. and Bawendi, M.G. *J. Am. Chem. Soc.* 1993, *115*, 8706–8715.
17. Sui, G.; Orbulescu, J.; Ji, X.; Gattas-Asfura, K.M.; Leblanc, R.M. and Micic, M. *J. Cluster Sci.* 2003, *14*, 123.
18. Peng, Z.A. and Peng, X.G. *J. Amer. Chem. Soc.* 2001, *123*, 183.
19. Peng, X.G.; Schlamp, M.C.; Kadavanich, A.V. and Alivisatos, A.P. *J. Amer. Chem. Soc.* 1997, *119*, 7019.
20. Jiang, J.; Krauss, T.D. and Brus, L.E. *J. Phys. Chem. B.* 2000, *104*, 11936.
21. Alivisatos A.P. *J. Phys. Chem.* 1996, *100*, 13226.

22. Schmelz, O.; Mews, A.; Basche, T.; Herrmann, A. and Mullen, K. *Langmuir* 2001, *17*, 2861.
23. Striolo, A.; Ward, J.; Prausnitz, J.M.; Parak, W.J.; Zanchet, D.; Gerion, D.; Milliron, D. and Alivisatos, A.P. *J. Phys. Chem. B.* 2002, *106*, 5500.
24. Leatherdale, C.A.; Woo, W.-K.; Mikulec, F.V. and Bawendi, M.G. *J. Phys. Chem. B* 2002, *106*, 7619.
25. Yu, W.W.; Qu, L.; Guo, W. and Peng, X. *Chem. Mater.* 2003, *15*, 2854.
26. Yu, W.W.; Qu, L.; Guo, W. and Peng, X. *Chem. Mater.* 2004, *16*, 560.
27. Gattas-Asfura, K.M.; Constantine, C.A.; Lynn, M.J.; Thimann, D.A.; Ji, X. and Leblanc, R.M. *J. Am. Chem. Soc.* 2005, *127*, 14640.
28. Seker, F.; Meeker, K.; Kuech, T.F. and Ellis, A.B. *Chem. Rev.* 2000, *100*, 2505.
29. Wuister, S.F.; de Mello Donega, C. and Meijerink, A. *J. Phys. Chem. B* 2004, *108*, 17393.
30. Kalyuzhny, G. and Murray, R.W. *J. Phys. Chem. B* 2005, *109*, 7012.
31. Kuno, M.; Lee, J.K.; Dabbousi, B.O.; Mikulec, F.V. and Bawendi, M.G. *J. Chem. Phys.* 1997, *106*, 9869.
32. Taylor, J.; Kippeny, T. and Rosenthal, S.J. *J. Cluster Sci.* 2001, *12*, 571.
33. Kuzmenko, I.; Rapaport, H.; Kjaer, K.; Als-Nielsen, J.; Weissbuch, I.; Lahav, M. and Leiserowitz, L. *Chem. Rev.* 2001, *101*, 1659.
34. Aldana, J.; Wang, Y.A. and Peng, X. *J. Am. Chem. Soc.* 2001, *123*, 8844.
35. Myung, N.; Bae, Y. and Bard, A.J. *Nano Lett.* 2003, *3*, 747.
36. Hines, M.A. and Guyot-Sionnest, P.J. *J. Phys. Chem.* 1996, *100*, 468.
37. Harrison, M.T.; Kershaw, S.V.; Burt, M.G.; Rogach, A.L.; Kornowski, A.; Eychmuller, A. and Weller, H. *Pure Appl. Chem.* 2000, *72*, 295.
38. Pradhan, S.; Chen, S.; Wang, S.; Zou, J.; Kauzlarich, S.M. and Louie, A.Y. *Langmuir* 2006, *22*, 787 .
39. Chen, S.W. *Anal. Chim. Acta* 2003, *496*, 29.
40. Greene, I.A.; Wu, F.X.; Zhang, J.Z. and Chen, S.W. *J. Phys. Chem. B* 2003, *107*, 5733.
41. Yang, Y.Y.; Pradhan, S. and Chen, S.W. *J. Am. Chem. Soc.* 2004, *126*, 76.
42. Yang, Y.Y.; Chen, S.W.; Xue, Q.B.; Biris, A. and Zhao, W. *Electrochim. Acta* 2005, *50*, 3061.
43. Spanhel, L.; Haase, M.; Weller, H. and Henglein, A. *J. Am. Chem. Soc.* 1987, *109*, 5649.
44. Koberling, F.; Mews, A. and Basche, T. *Adv. Mater.* 2001, *13*, 672.
45. van Sark, W.G.J.H.M.; Frederix, P.L.T.M.; van den Heuvel, D.J.; Bol, A.A.; van Lingen, J.N.J.; Donega, C.D.; Gerritsen, H.C. and Meijerink, A. *J. Fluoresc.* 2002, *12*, 69.

8 Fundamental Forces in Powder Flow

N. Stevens, S. Tedeschi,
M. Djomlija, and B. Moudgil

CONTENTS

INTRODUCTION

The topic of powder flow behavior has been a subject of interest for many centuries. Early research on powder properties dates back to Reynolds, Prandlt, Coulomb, and Mohr (1). The foundation laid by these pioneers has allowed the development of the principles by which powders are understood today.

Powder flow properties have significant influence on many unit operations in modern industry including filtration, grinding, mixing, and agglomeration (2–5). These unit operations are commonly used for the food, pharmaceutical, paint and coatings, plastic, agriculture, mining, ceramic, chemical, printing, electronic, paper, textile, and wood industries. With such a vast number of processes involving bulk solids there is a need for a fundamental understanding of the behavior of such materials. It has been estimated that industries processing bulk solids total one trillion dollars per year in gross sales in the United States alone and operate at only 63% capacity. By comparison, industries that rely on fluid transport processes operate at 84% of design capacity (6).

Economic setbacks for industry when concerning powder flow usually result from one of two extremes. This can be illustrated if we consider powder discharge from a hopper. The first extreme is if the powder is too cohesive, resulting in an arching or rat-holing phenomenon. The other extreme is if the powder is too free-flowing, it is possible to flood a unit operation. These phenomena make accurate powder characterization essential. Granular materials behave differently from any other form of matter—solids, liquids, or gases. They tend to exhibit complex behaviors sometimes resembling a solid and other times a liquid or gas (7). Thus the characterization of granular materials presents a challenge unlike any other.

To characterize and understand a powder, it will first be defined. A powder is a two-phase matrix of solid particles, in intimate contact with neighboring particles, surrounded by a gas phase, typically, but not limited to, air. Cohesion between particles is understood to be the sum of attractive forces. The sum of these forces, acting in the bulk phase, defines what is commonly referred to as the powder's flow properties. The flowability, or strength of a powder, is typically characterized by measuring its response to shear forces under an applied normal load.

In order to understand bulk powder properties and how they may influence a particular process, it is necessary to have knowledge of what comprises cohesive forces and how this contributes to the overall behavior of a powder. This chapter will review the fundamental theory of powder flow and its characterization, the underlying forces relevant to particle cohesion, their relationship to bulk properties, and state-of-the-art research to understand and control powder behavior.

POWDER PROPERTIES

With roughly one-half of industrial products and three-quarters of the raw materials in the chemical industry alone being composed of granular materials it is important to obtain an in-depth understanding of powder handling (8). Granular material is defined as any material regardless of its shape and size that is composed of individual solid particles. Despite its importance, powders, or granular materials, are not as well understood as compared to gases and liquids. This has dire consequences. Merrow (6) investigated 39 solid-processing plants on their startup times and compared them to startup times of plants processing gases and liquids and found that solid-processing plants required roughly on average seven times more time. Major contributors to the slow startup are the tendency of solids to plug, stick, and flow erratically.

To develop a more thorough understanding of powders, one must approach the problem from a macroscopic and microscopic point of view. At a macroscopic level, powder behavior is understood from average behavior of the bulk and the powder is treated as a continuum. The microscopic properties of the individual particle that comprise the bulk is what governs the bulk behavior and, therefore, must also be considered in order to develop an understanding of powder flow. Unlike individual molecules of a gas, it cannot be assumed that each powder particle is the same. Variation in shape, size, and the multitude of possible orientations of particles in the bulk affect the bulk properties. However, experiments do show to some extent that the microscopic and macroscopic properties are related.

POWDER FLOW

Powders are neither solid, liquid, nor gas; however, they have traits of each associated with them. Powders are not a solid, although they have strength and can withstand some deformation. They can flow under certain circumstances much like a liquid and can be compressed to a certain degree like a gas. Powders are in a class of their own. Powders may flow due to stress that is mechanically applied or flow initiated under its own weight; however, the principles of its flow are distinctly different from those of a liquid. If the stresses are not sufficiently large enough, flow will not commence. Due to individual particle interlocking, frictional forces arise and give powders shear strength which enables them to from piles, a trait not associated with liquids. Cohesion is the mutual attraction of the particle of the solid for one another and is measured on the bulk scale as the amount of shear stress necessary to cause yielding at conditions of zero normal pressure. A number of factors contribute to cohesion, the predominant factor in a normal dry condition being van der Waals forces along with electrostatic forces and friction forces induced by the orientation and roughness of the particles. Under humid conditions, the adsorption of water vapor leads to formation of liquid bridges and in some instances caking. When sufficient stress is applied and flow is initiated, it will not depend on the rate of flow as in the case of fluids. Controlling flow rates of powders is difficult in practice.

STATE OF STRESS

In order to establish flow in a powder, the strength of the powder must be overcome. The stress exerted on the powder must be greater than the yield stress. Stress, being a continuum concept, is considered along certain planes within the bulk as opposed to individual particle's interactions with one another. Local stresses exist within the mass of the powder, which transmits an average stress across these planes. Considering an element of volume within the continuum and considering all forces acting on it, surface and body, if this volume is reduced to zero so that only a point remains, the body forces would go to zero and only surface forces would remain. At this point, the surface forces are in equilibrium with one another and this is known as the state of stress of the material for the point considered. The total state of stress at a point in the bulk is represented by nine components, three normal stresses and three shear stresses, shown in Figure 8.1.

The state of stress can also be represented by three invariants known as the principal stresses (8). This state of stress is identical to those of the stresses above; however, the orientation of the axes have changed. Along the principal axes only normal stresses reside; shear stresses are zero, shown in Figure 8.2.

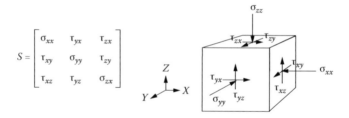

FIGURE 8.1 Stresses acting on a unit of volume in Cartesian coordinates.

The formation of shear planes within the bulk of the material is primarily influenced by the major (σ_1) and minor (σ_3) principal stresses. The role the intermediate (σ_2) principal stress plays is not well understood and is considered to be insignificant and, therefore, often neglected (8).

MOHR CIRCLE

When powders are consolidated they undergo a change in strength. The stress during consolidation is referred to as the consolidation stress. The consolidation stress provides the strength of the powder when exposed to these conditions, the history if you will. A two-dimensional representation of the state of stress can be illustrated by a circle in normal and shear stress space. All points on the Mohr circle are equivalently equal. The largest circle furthest to the right represents the consolidation state of the powder and all subsequent circles to the left represents the state of stress of the powder at lower normal loads that leads to incipient failure.

Powders are assumed to obey the linear Coulomb yield criteria described by equation (8.1).

$$\tau = c + \sigma \tan \phi \tag{8.1}$$

Here τ is the stress at which the powder begins to deform plastically when confined to a normal stress σ. For cohesive powders, the Coulomb yield criteria, also known as the yield locus, is in actuality slightly curved but is approximated by a straight line by plotting it inclined from the normal axis at an angle of ϕ, the angle of internal

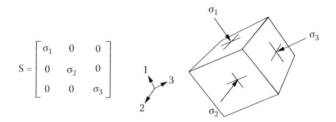

FIGURE 8.2 Stresses acting on a unit volume along principal coordinates.

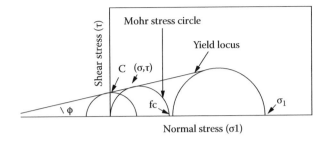

FIGURE 8.3 Mohr circle diagram [33].

friction, and intersecting the shear stress at the value of C the cohesion (8). Stress values below this line are not sufficient to initiate flow. The yield locus will be tangent to Mohr circles with stresses high enough to instigate shear. There exists a family of yield loci, one for each consolidation stress.

A measure of the strength of a powder is determined by its unconfined yield strength, f_c, which is defined as the major principal stress acting on the bulk necessary to cause failure at a free surface (9). This can be seen in Figure 8.3 as the Mohr circle that intersects the origin. There is a unique strength value for every consolidation stress σ_1. The unconfined yield strength is related to material properties through the following equation:

$$f_c = \frac{2\cos\phi}{1-\sin\phi}C \tag{8.2}$$

Plotting unconfined yield stress versus consolidation stress, also known as the flow function, indicates a material tendency to either flow or form a stable free surface. Material with a relatively large f_c value compared to consolidation stress, σ_1, will flow poorly (4).

DIRECT SHEAR TESTERS

There are a number of different direct shear testers that have been developed to characterize powder flow. Common commercial testers include the Jenike shear tester, Peschel ring shear tester, Schulze shear tester, and Johanson Indicizer. They all work on the same basic principles differing only in geometry, application of shear force, and amount of material used for testing. One of the more common shear testers is the Schulze cell and this will be discussed in further detail later. A schematic of the Schulze cell is shown in Figure 8.4.

The Schulze cell consists of an annular cell, lid, and two lever arms. For a typical test, material is loaded into the cell and the lid is held in place by two lever arms, which are connected to load sensors. A consolidation load is placed on the lid, consolidation the material beneath it. The base of the cell is rotated as the lid is held in place by the lever arms and the load sensor measures the shear stress. Once a steady state is reached, the rotation is stopped and shear stress versus distance sheared is

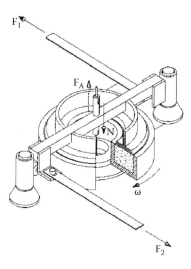

FIGURE 8.4 Cross-sectional schematic of the Schulze shear cell. The annular region of the cell is filled with powder with the two load arms connected to the stationary top. The bottom of the cell is rotated to shear the sample consolidated by applying different normal loads [63]. Reprinted from *Powder Handling and Processing*, 8(3) Schulze, D., Flowability and time consolidation measurements using a ring shear tester, Copyright (1996), with permission from Trans Tech Publications.

recorded. This is known as pre-shear and establishes the history of the powder tested. This point is identified on the consolidation Mohr circle shown on Figure 8.5. One set of experiments is carried out at this consolidation load by reducing the load with one lighter then the consolidation load and failing the sample. To develop a complete flow function, this process is repeated for several loads while in between the sample is pre-sheared with the consolidation load to return the sample to the initial condition.

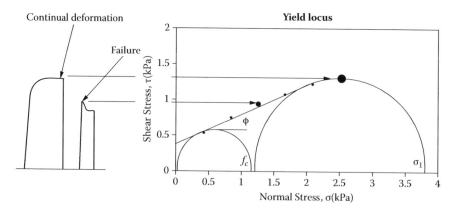

FIGURE 8.5 A typical result from consolidation and failure of a powder sample. The corresponding points on the Mohr circle are shown to illustrate the procedure of generating the yield locus [33].

An understanding of powder behavior starts with consideration of ensemble powder properties and techniques to measure them. In order to fully develop the knowledge base required to predict powder flow behavior, the interparticle forces relevant to powders and their contribution to bulk powder properties must also be understood. The next section will discuss the fundamental interparticle forces and the understanding, to date, of their correlation to bulk powder properties.

INTERPARTICLE FORCES AND THE RELATIONSHIP TO POWDER PROPERTIES

Particle interactions in a powder consist of the sum of both the attractive (or cohesive) forces holding the particles together and the repulsive forces acting to separate them. The forces common to powder systems include attraction by van der Waals forces, cohesion by capillary forces in the presence of liquid bridges, and repulsion by electrostatic forces (10). In addition, the powder will also experience friction when shear forces ample to initial flow are applied and particles in contact must move past each other. An illustration depicting these forces can be seen in Figure 8.6. Here F_{ad} is the force of adhesion due to van der Waals and capillary bridge forces, F_{es} is the electrostatic force, F_F and F_S are the friction and shear forces, respectively, acting at the particle level, and F and S are the overall normal and shear forces acting on the ensemble of particles. Each of these forces will now be examined in more detail in order to develop an understanding of their interactions and how these forces relate back to dry powder systems.

VAN DER WAALS FORCES

The most common interparticle forces for powders arise from the electromagnetic interaction, columbic attraction, or dipoles created by the separation of the positive and negative portions of atoms or molecules (11,12). These dipoles arise from a number of mechanisms, including the creation of fluctuating induced dipoles by vibrational motion, permanent dipoles, or polar molecules. There are three main contributions, collectively called van der Waals forces, which include Keesom interactions, or dipole-dipole interactions; Debye interactions, or dipole-induced dipole interactions; and London dispersion interactions, or instantaneous, fluctuating dipole-induced dipole interactions. An illustration of these forces can be seen in

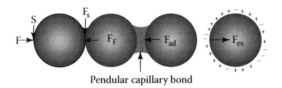

Pendular capillary bond

FIGURE 8.6 Illustration of important interparticle forces in powders and aerosols. Here F_{ad} is the force of adhesion due to van der Waals and capillary bridge forces, F_{es} is the electrostatic force, F_F and F_S are the friction and shear forces respectively acting at the particle level, and F and S are the overall normal and shear forces acting on the ensemble of particles [64].

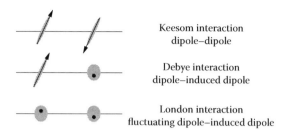

Keesom interaction
dipole–dipole

Debye interaction
dipole–induced dipole

London interaction
fluctuating dipole–induced dipole

FIGURE 8.7 A depiction of the three interactions that are collectively known as van der Waals forces. Van der Waals forces include: Keesom interactions, or dipole-dipole interactions; Debye interactions, or dipole-induced dipole interactions; and London dispersion interactions, or instantaneous, fluctuating dipole-induced dipole interactions [64].

Figure 8.7. London dispersion forces are considered to be the predominant part of van der Waals forces between particles (13,14).

The theory behind these forces was derived from the Ideal Gas Law, which predicts the behavior of monatomic and simple diatomic gases. To explain the deviation of more complex molecules from the Ideal Gas Law, van der Waals suggested a modification to the Ideal Gas Law to include intermolecular interactions. This equation is known as the van der Waals equation of state (14,15).

$$\left(P - \frac{\alpha}{V^2}\right)(V - \beta) = RT \tag{8.3}$$

Here P is pressure, V is volume, R is the universal gas constant, T is temperature, and α and β are constants specific to a particular gas. The constant β describes the finite volume of the molecules comprising the gas, and the constant α takes into account the attractive forces between the molecules. A paired potential form of the van der Waals equation for interaction included, for the first time, both an attractive and repulsive term. This is known as the Lennard-Jones potential (16):

$$W_{a/a} = \frac{C}{r^6} + \frac{B}{r^{12}} \tag{8.4}$$

Here the net potential energy between two atoms, $W_{a/a}$, separated by a distance, r, is described by the competition of an attractive London dispersion force (17), C, related to the synchronization of instantaneous dipoles created when the energy fields of neighboring atoms overlap (18), and a Born repulsion term, B, arising from the overlap of electron clouds.

MICROSCOPIC APPROACH TO VAN DER WAALS FORCES

In 1937, Hamaker expanded the concept of the van der Waals forces from single atoms and molecules to all atoms in solid bodies (13,19). He assumed that each atom in one body interacts with all atoms in another body and added these forces by pairwise summation. Bradley (20), Derjarguin (18), and de Boer (21) all followed

Hamaker to improve the understanding of the forces acting between bodies. Their approach considered the interaction energy, W, of two spheres of equal size, calculated by equation (8.5). If the radius of the particles is greater than the separation distance, the results can be simplified to the more common equation for the van der Waals energy of interaction between two spheres shown in equation (8.6).

$$W_{sph/sph} = \frac{A}{6}\left[\frac{2R^2}{H(4R+H)} + \frac{2R^2}{(2R+H)^2} + \ln\left(1 - \frac{4R^2}{(2R+H)^2}\right)\right] \quad (8.5)$$

$$W_{sph/sph} = \frac{AD}{12H} \quad (8.6)$$

Here A is the Hamaker constant, R is the particle radius, H is the separation distance between the particle surfaces, and D is the particle diameter. The significance of this work is that the van der Waals forces can now be considered by the geometry of the body and an interaction term that is dependent only on the material properties, not shape.

MACROSCOPIC APPROACH TO VAN DER WAALS FORCES

The microscopic approach used by Bradley, Derjarguin, and de Boer has one drawback in that it does not take into account the effect of the interaction of dipoles within one body on other dipoles within the same body. In an attempt to include these interactions, Lifshitz (22) modified the Hamaker theory, ignoring the atoms completely, and assuming each material as its own body. Lifshitz developed a calculation for van der Waals forces using the difference in dielectric properties. The final expression Lifshitz derived for the Hamaker constant is shown in equation (8.7).

$$A_{132} = \frac{3}{2}k_B T \sum_{n=0}^{\infty}\left[\frac{\varepsilon_1(iv_n) - \varepsilon_3(iv_n)}{\varepsilon_1(iv_n) + \varepsilon_3(iv_n)}\right]\left[\frac{\varepsilon_2(iv_n) - \varepsilon_3(iv_n)}{\varepsilon_2(iv_n) + \varepsilon_3(iv_n)}\right] \quad (8.7)$$

Here the subscripts 1, 2, and 3 refer to material 1 and 2 interacting in medium 3; k_B is Boltzmann's constant, T is the temperature; and $\varepsilon(v)$ is the dielectric permittivity of the material at a frequency v. There is some complexity involved in employing this equation as it is quite difficult to measure the dielectric response of materials over the wide range of frequencies required for the calculation. Work has been done by Ninham and Parsegian (23) and Hough and White (24) to simplify the measurement to specific characteristic frequencies in the ultraviolet (UV) range. The result is the Tabor-Winterton approximation (25):

$$A_{132} = \frac{3}{4}k_B T\left(\frac{\varepsilon_1 - \varepsilon_3}{\varepsilon_1 + \varepsilon_3}\right)\left(\frac{\varepsilon_1 - \varepsilon_3}{\varepsilon_1 + \varepsilon_3}\right)$$

$$+ \frac{3hv_e}{8\sqrt{2}}\frac{\left(n_1^2 - n_3^2\right)\left(n_2^2 - n_3^2\right)}{\left(n_1^2 + n_3^2\right)^{1/2}\left(n_2^2 + n_3^2\right)^{1/2}\left\{\left(n_1^2 + n_3^2\right)^{1/2} + \left(n_2^2 + n_3^2\right)^{1/2}\right\}} \quad (8.8)$$

Here n_i is the refractive index of the respective material and h is Plank's constant. It is important to note that where the medium and the solids have the same index of refraction ($n_1 = n_2$ and $\varepsilon_1 = \varepsilon_2$), the Hamaker constant, A_{132}, becomes zero. This method is known as index matching and can be used to minimize or eliminate the effect of the Hamaker constant.

Thus far, discussion has focused on the relationships for interaction using interaction energies and has ignored interaction forces. For molecules, this is reasonable, but as larger bodies are studied, force is of more interest. To overcome the challenges presented by the geometry of different surfaces, Derjaguin developed an approximation to relate the force of an interaction between a body of some geometry and the energy of interaction per unit area of two parallel flat surfaces (18). The relationship between the interaction force for bodies of different geometries and the interaction of planar surfaces can then be extracted, making it possible to compare the interaction force of bodies of different sizes and geometries. For this reason, interaction force measurements are often reported as force normalized by radius.

SURFACE ENERGY APPROACH TO ADHESION

In the previous section, the interaction between surfaces is determined as the surfaces are at some distance apart. However, when the surfaces are in contact, as is the case for powder in the bulk state, a different approach can also be considered using the surface energies of the surfaces described by the work of adhesion. Work of adhesion is defined as the energy needed to separate two surfaces from contact to an infinite separation distance and takes into consideration the surface energy of the bodies in contact as seen in equation (8.9) (26).

$$W_a = \gamma_{12} + \gamma_{23} \qquad (8.9)$$

Here W_a is the work of adhesion, γ_{12} is the surface energy of surface 1 in medium 2, and γ_{23} is the surface energy of surface 3 in medium 2. Since it is experimentally impractical to measure solid-gas and liquid-solid surface energies, it is acceptable to combine the work of adhesion with Young's equation. Young (27) described the contact angle of a liquid drop resting at equilibrium on a smooth surface as a balance between solid-liquid, liquid-vapor, and solid-vapor interfacial energies as shown in Figure 8.8.

The mathematical expression of the interfacial energy balance was later expressed in equation form by Dupré (28):

$$\cos\theta = (\gamma_{23} - \gamma_{13})/\gamma_{12} \qquad (8.10)$$

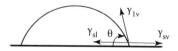

FIGURE 8.8 A liquid drop on an ideal solid surface at equilibrium [65].

Here γ_{12} is the surface energy at the liquid-vapor interface, γ_{23} is the surface energy at the solid-vapor interface, γ_{13} is the surface energy at the solid-liquid interface, and θ is the contact angle between γ_{12} and γ_{13} through the liquid phase. The relationship between work of adhesion and Young's equation for contact angles is shown below:

$$W_a = \gamma_{12}(1 + \cos\theta) \tag{8.11}$$

From this equation, the work of adhesion can be calculated by measuring the contact angle of a liquid, with a known surface energy, on the material of interest. By modifying the surface energies of the bodies through coatings and chemical surface modification, the work needed to separate them can also be modified. Therefore, by controlling the surface energy of a solid surface, it is possible to modify the cohesive forces in a particle bed.

CAPILLARY FORCES

Another important source of cohesion arises from capillary forces, which arise from the presence of pendular liquid bonds between two particles. The presence of liquid is known to have a significant effect on the mechanical properties, transport, and dispersion of powder (29). The addition of liquid binders to bulk solids has been employed to increase the cohesive properties of the bulk material in attempts to decrease the dustiness of a bulk material, enhance agglomeration processes, increase green strength during tablet and mold production, and prevent segregation tendencies (29). Condensation of water vapor, present as relative humidity in the air, is alone sufficient to create liquid bridges (30,31). Since most powder processing and transport occurs in ambient conditions, and in the presence of some degree of atmospheric humidity, an understanding of the effect of capillary forces on powder properties is essential. The different regimes of liquid distribution in a powder sample are as shown in Figure 8.9. Capillary forces increase the cohesion between the particles at points of contact. For this reason, the degree of powder packing is an important factor of the system (32). High solids fraction powders have more particle contacts, leading to an increased contribution of adhesive capillary forces, ultimately increasing the powder strength. On the other hand, low solids fraction powders have fewer

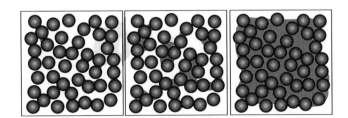

FIGURE 8.9 Regimes of liquid distribution in a powder sample. Left—Pendular state, where individual liquid bridges exist between particles; Center—Funicular state, two or more liquid bridges on adjacent particles start to coalesce with each other; Right—Capillary state, all the interstices between the particles are completely filled with liquid [66].

interparticle contacts and therefore less adhesion, leading to higher powder flowability. The equation for the attractive forces due to the presence of liquid bridges between two spheres is given in equation (8.12):

$$F_{cap} = -2\pi R \gamma_{lv} \cos\theta \qquad (8.12)$$

Here R is the radius of the capillary, γ_{lv} is the surface tension of the liquid phase, θ is the contact angle of the liquid at the particle surface, and the sign of the equation represents the force direction, which is cohesive in nature. The adhesion force due to the presence of a liquid bridge is calculated by the Laplace equation, equation (8.13),

$$\Delta P = \gamma \left(\frac{1}{r_1} - \frac{1}{r_2} \right) \qquad (8.13)$$

Here ΔP is the pressure difference across the liquid interface, and r_1 and r_2 are the radii of the particles. Decreasing particle size will therefore have a large effect on the adhesion force when liquid bridges are present.

The forces acting in cohesive powders in the presence of liquid bridges have been shown to be dependent on the geometry and dynamics of the transport process. The nature of the liquid also plays a key role in determining the boundary condition for defining the process of shearing of a cohesive powder as the flow initiates. In the presence of a volatile liquid like water in the pendular bridges, the liquid-vapor equilibrium is reached quickly and hence the shearing process would involve a constant capillary radius condition. On the other hand, nonvolatile liquids such as oil are limited by a constant volume condition.

In a study by Esayanur (33), theoretical formulae have been developed for estimating the capillary force in the presence of a nonvolatile liquid bridge for the sphere/plate and the sphere/sphere geometry. The theoretical estimates were validated by direct AFM measurement of the force/distance profile between surfaces in the presence of an oil liquid bridge. These expressions for capillary forces were used to develop a model to predict the unconfined yield strength of cohesive powders. This will be discussed in detail later in the chapter.

ROUGHNESS

Thus far, interparticle forces for ideal, smooth surfaces have been considered; however, real surfaces typically have some degree of inherent roughness. On the atomic level, surfaces typically exhibit roughness due to the size of atoms and their spacing within the crystal structure. These perturbations on the particle surface will influence the effective interparticle spacing, the consequences of which are multifold since van der Waals forces are sensitive to the area of the surfaces in contact and separation distance between particles. A visualization of the effect of surface roughness can be seen in Figure 8.10. Increasing roughness decreases the contact area between the surfaces as well as increases the separation distance of the average surface plane. The classical model for the effect of roughness on van der Waals adhesion

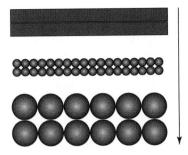

FIGURE 8.10 Change in contact area with increasing roughness [64]. The illustration shows the decrease in contact area that occurs with increasing surface roughness and is attributed to the difference in interparticle forces between smooth and rough surfaces. Rumpf [19,20] initially derived the equation for the contribution of surface asperities to the decrease in van der Waals forces. This work was later expanded by Rabinovich et al. to more closely define surface roughness [39,40] and include the effect of roughness on capillary forces [41,43, 44,67].

was created by Rumpf (2,5) and shown in equation (8.14):

$$F_{ad} = \frac{A}{6H_0^2}\left[\frac{rR}{r+R} + \frac{R}{\left(1+\dfrac{r}{H_0}\right)^2}\right] \tag{8.14}$$

Here A is the Hamaker constant, H_0 is the interparticle separation distance, R is the radius of the particle, and r is the radius of the asperity on the surface, and an illustration of which is found in Figure 8.11. This model, however, significantly

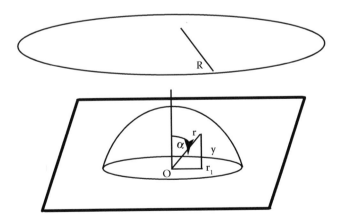

FIGURE 8.11 Asperity as described by the Rumpf model for the interaction of an adhering particle with a rough surface. A hemispherical asperity of radius r is shown and the origin at the surface interacting along the ordinate axis with a spherical adhering particle.

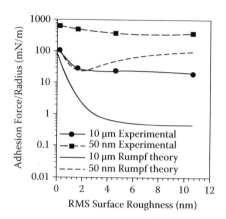

FIGURE 8.12 Experimentally measured normalized adhesion force between the glass sphere or AFM tip and the model substrates as a function of increasing nanoscale roughness. The theoretical adhesion calculated with the modified Rumpf model (Eq. 14) with for silica/silicon and silica/titanium, for silicon nitride/silicon and silicon nitride/titanium, $H_0 = 0.3$ nm, and $R_{sphere} = 10$ mm or $R_{tip} = 50$ nm underestimates the measured adhesion force by more than an order of magnitude [40]. Reprinted from *Journal of Colloid and Interface Science, 232*(1) Rabinovich *et. al.*, Capillary Forces Between Nanoscale Rough Surfaces (II), Copyright (2000), with permission from Elsevier.

underestimates the adhesion force when compared to experimental findings shown in Figure 8.12. The model has since been modified by others, including Greenwood and Williamson (34), Fuller and Tabor (35), Czarnecki and Dabros (36), and Xie (37), and their work summarized and expanded by Adler (38). A comparison of different

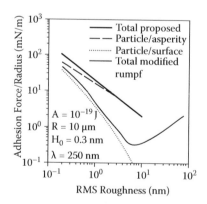

FIGURE 8.13 Comparison of adhesion models. Prediction of the total, contact, and noncontact force of adhesion. Also plotted for comparison is the predicted adhesion force from the modified Rumpf model, equation (8.15), using the same parameters except for peak-to peak distance, which is not accounted for in the model [39]. Reprinted from *Journal of Colloid and Interface Science, 232*(1) Rabinovich *et. al.*, Capillary Forces Between Nanoscale Rough Surfaces (I), Copyright (2000), with permission from Elsevier.

adhesion models can be seen in Figure 8.13. Research by Rabinovich and coworkers (38–40) have further refined Rumpf's model to more accurately account for the effect of roughness on adhesive forces.

$$F_{ad} = \frac{AR}{6H_0^2} \left[\frac{1}{\left(1 + \frac{32Rk_1 rms}{\lambda^2}\right)} + \frac{1}{\left(1 + \frac{k_1 rms}{H_0}\right)^2} \right] \tag{8.15}$$

Here k_1, the maximum peak-to-peak height between asperities, rms is the root mean squared roughness of the surface, and λ is the average peak-to-peak distance between asperities. The first term accounts for the particle contact with the asperity surface and the second term represents the interaction between the particle and the average surface plane. Rabinovich and others showed that surface roughness on the nanometer scale (<2 nm) greatly increases the force of adhesion, and as the scale of the roughness increases, there is a subsequent decrease in adhesion forces. This equation was also expanded to include the second-order roughness on the asperity in the form of equation (8.16). Experimental results are shown in Figure 8.14.

$$F_{ad} = \frac{AR}{6H_0^2} \left[\frac{1}{1 + 58R \cdot rms_2/\lambda_2^2} \right] + \frac{AR}{6H_0^2} \left[\frac{1}{(1 + 58R \cdot rms_1/\lambda_1^2)(1 + 1.82 \cdot rms_2/H_0)^2} \right] \tag{8.16}$$

Rabinovich and coworkers have also investigated the effect of nanoscale surface roughness on the critical humidity required for the onset of capillary forces in many systems (38,41–44). The presence of surface roughness on a nanoscale has been shown to have a profound effect on the critical humidity required to induce capillary adhesion. This can be seen in Figure 8.15, a graph of the adhesion force measured by atomic force microscopy of silica surfaces of varying roughness over a range of relative humidity values. Increasing roughness increased the critical relative humidity

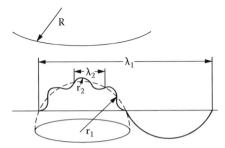

FIGURE 8.14 Asperity schematic for surfaces with two roughness scales. Schematic illustration of the geometric model used to calculate adhesion force in the proposed model. The surface is proposed to consist of an array of spherical asperities and troughs with the origin of the asperities positioned below the average surface [40]. Reprinted from *Journal of Colloid and Interface Science, 232*(1) Rabinovich *et. al.*, Capillary Forces Between Nanoscale Rough Surfaces (II), Copyright (2000), with permission from Elsevier.

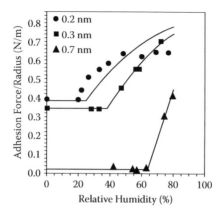

FIGURE 8.15 Adhesion profiles for silica. Force of adhesion as a function of relative humidity for silica surfaces of increasing roughness. Circles represent the oxidized silicon wafer (0.2 nm measured *rms* roughness), squares represent the PE-CVD substrate (0.3 nm measured *rms* roughness), and triangles represent the etched PE-CVD surface (0.7 nm measured *rms* roughness). Also shown are theoretical predictions for both the dry adhesion and capillary adhesion regimes [43]. Reprinted from *Advances in Colloid and Interface Science, 96*(1-3) Rabinovich *et. al.*, Capillary Forces Surfaces with Nanoscale Roughness, Copyright (2002), with permission from Elsevier.

required for the onset of capillary adhesion as well as decreased the overall adhesion force. Surface roughness will decrease the minimum amount of water needed to form liquid bridges at particle contact points since the capillary can form on the asperities as shown in Figure 8.16a. However, the volume of liquid present is small and the onset of capillary adhesion shifts to higher relative humidity levels such that the liquid bridges form between the two particle surfaces. This is shown in Figure 8.16b. The effect is seen regardless of substrate, as shown in Figure 8.17. Their study also

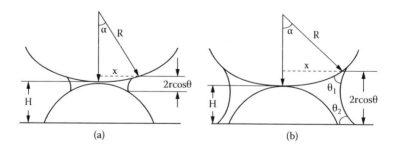

FIGURE 8.16 Particle/asperity capillaries. Schematic illustration of the meniscus formed in a capillary created by the adhesion of an ideally smooth particle to an asperity on a flat substrate (a) for conditions in which the relative humidity is low, and (b) for which the relative humidity is high, [43]. Reprinted from *Advances in Colloid and Interface Science, 96*(1–3) Rabinovich *et. al.*, Capillary Forces Surfaces with Nanoscale Roughness, Copyright (2002), with permission from Elsevier.

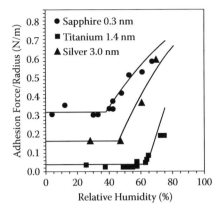

FIGURE 8.17 Force of adhesion as a function of relative humidity for substrates of sapphire, measured *rms* roughness of 0.3, titanium, measured *rms* roughness of 1.4, and silver, measured *rms* roughness 3.0 nm. Solid lines are theoretical predictions for both dry adhesion and capillary adhesion regimes [43]. Reprinted from *Advances in Colloid and Interface Science,* 96(1-3) Rabinovich *et. al.*, Capillary Forces Surfaces with Nanoscale Roughness, Copyright (2002), with permission from Elsevier.

validated the use of equation (8.16) to predict the force of adhesion as a function of particle size, humidity, and roughness for practical systems.

ELECTROSTATIC FORCES

Electrostatic forces are also considered important for powders. Static electricity arises when charges on the surface of a material are not free to move to neutralize each other or do so slowly (45). These charges remain on the surface where they can interact with their surroundings. Particles with a similar net surface charge, greater in magnitude than the attractive forces, will experience columbic repulsion. The general equation for the electrostatic repulsive force is shown below:

$$F_e = \frac{q_1 q_2}{4\pi e_0 r^2} \tag{8.17}$$

Here q_1 and q_2 are the charge on each particle, e_0 is the unit charge of an electron, and r is the radius of the particle.

Particles continually gain, lose, and transfer charge between themselves and the surrounding environment. There are four basic ways by which charge can be imparted to the particle. The first method is by direct ionization of the particles themselves. The second is by ionization of particles by electromagnetic radiation. Another route for charging is by static electrification, which occurs when particles acquire charge from the surrounding environment or other particles. The final way is through collisions with ions or ion clusters.

The amount of charge a particle can obtain by any of these methods is material dependent. The method is also ineffective above 50% relative humidity since the presence of water will nullify the charge (45,46). The presence of water does, however, lead to cohesion due to capillary forces, as discussed above.

Now that an understanding of the fundamental forces underlying adhesion between particles has been developed, their significance to powders will now be presented.

EXTENDING INTERPARTICLE FORCES TO BULK PROPERTIES

Most of the current design strategies and characterization techniques for industrial powders are based on a continuum approach similar to the field of soil mechanics. Solutions to most engineering problems due to powder flow have been handled on a macroscale, focusing on the quickest solution to resume the function of a piece of equipment or process. Due to the reduction in size of powders to the nanoscale, the surface properties and interparticle forces play a significant role in determining the flow characteristics. Powder characterization using ensemble techniques, such as direct shear testing, are no longer adequate due to an insufficient understanding of the changes in powder properties with respect to operating variables. This, along with the drive toward the production of multifunctional powders and an increase in complexity of synthesis processes, has led to the need for additional understanding of particle-scale interactions in powders.

Work by Esayanur (33) focused on developing an interparticle force-based framework to predict the unconfined yield strength of cohesive powders, leading to a correlation between the two scales of measurement. To understand the role of interparticle forces in determining the strength of cohesive powders, the particle scale interactions were characterized using atomic force microscopy (AFM), contact angle, surface tension, and coefficient of friction. The bulk-scale properties such as unconfined yield strength, packing structure, and size of the shear zone were also investigated. Esayanur demonstrated the effect of liquid addition by measuring the

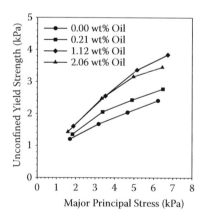

FIGURE 8.18 The dependence of unconfined yield strength on major principal stress for quartz. The experiments were conducted in the Schulze cell as a function of the amount of oil added [33].

change in unconfined yield strength of a fine quartz powder using a Schulze cell technique. An increase in bulk strength of up to 150% was found with the addition of only 1 to 2% oil by weight, as shown in Figure 8.18. These results were then compared to direct measurements using AFM; and with the use of the theoretical formulae developed through that work, a model was developed to predict the unconfined yield strength of cohesive powders.

The PERC model was developed for a system of monodisperse randomly packed, smooth spherical particles. The model was then extended to include the effect of surface roughness (K) and friction (μ), which are first-order contributions to the unconfined yield strength. Apart from the contribution due to capillary forces, the effects of van der Waals in the presence of surface roughness and frictional forces were included to refine the model and extend its applicability to real powders (33). As a result, equation (8.18) was developed.

$$f_c = \frac{4\gamma \cos\theta \, k_r n_l p (Cn_c \rho_{sol})^{\frac{1}{2}}}{\left(k_h^{\frac{3}{2}}\right) R \cos\phi (3\rho_{liq})^{\frac{1}{2}}} + \frac{\left(k_r^{\frac{3}{2}}\right)\mu nn_l p}{R \cos\phi}\left(\frac{A_H}{12\pi H^2 K} + 2\gamma\cos\theta\right) \quad (8.18)$$

Here A_H is the Hamaker constant, H is the distance of closest approach, K is the roughness factor, and μ is the friction coefficient. The first term on the right side of equation (8.18) represents the contribution of capillary forces and the second term represents the adhesion force in the dry state. The validity of this expression was then verified by the comparison of its results to those obtained by direct measurement for a dry system (Figure 8.19) and for a system with capillary forces present (Figure 8.20).

The development of equation (8.18) has enabled the link between macroscopic and microscopic powder scales. Therefore, powder flow can be described with respect to

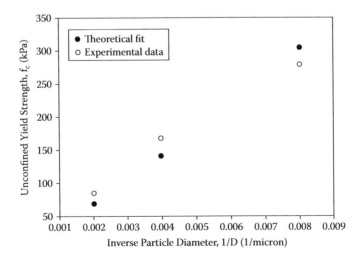

FIGURE 8.19 Unconfined yield strength (f_c) as a function of the inverse particle diameter (D) for dry powder [33].

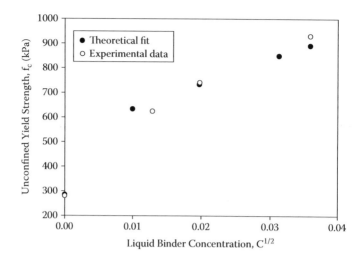

FIGURE 8.20 Unconfined yield strength (f_c) as a function of the square root of liquid binder concentration ($C^{1/2}$) [33].

particle characteristics and interparticle forces. Under static conditions, such as during powder storage, changing conditions can lead to an increase in powder strength, resulting in unit operation complications. Through modeling, the powder is treated as a solid body, using fracture mechanics theory as an approach to determine the stresses required for the recommencement of powder flow. The following section will focus on the conditions leading to a gain in powder strength and recent research in this area.

MODELING OF BULK-SCALE PROPERTIES

Bulk solids such as food powders, detergents, pharmaceuticals, feedstock, fertilizers, and inorganic salts often gain strength during storage. This phenomenon is known as caking and can be defined as the process by which free-flowing materials are transformed into lumps and agglomerates due to changes in atmospheric or process conditions. Traditional interparticle forces such as the formation of pendular liquid bridges, van der Waals, or electrostatic forces can cause an increase in bulk cohesive; strength, however, none of these phenomena alone cause caking. Caking occurs when the surface of the particle is modified over time to create interparticle bonds due to the formation of solid bridges between similar or dissimilar materials. Thus, the presence of traditional interparticle forces may initiate the caking process, but the advent of caking requires some mechanism to change these forces into solid bridges.

Several mechanisms to explain the formation of solid liquid bridges have been explained in literature. These mechanisms include crystallization, sintering, chemical bonding, partial melting, and glass transition (3,47–50). Much research has been directed at understanding the mechanisms of caking as well as the material properties and process conditions necessary to induce a caking event. Pietsch (50,51) and Specht (52) observed that cake strength increases with an increase in moisture content. The effect of relative humidity has only recently received attention in the research of caking (53,54) using the Schulze shear cell. It can be seen in Figure 8.17 that the slope of the yield locus increases with an increase in moisture content. This indicates that moisture is adsorbed by the particles at higher relative humidity. The moisture sorption causes the increase in shear stress, which in turn increases the unconfined yield strength.

Empirical results have also demonstrated that the caking strength may increase as moisture is removed from the particle bed (49,52,55). This is attributed to the forming of solid interparticle bridges, thus increasing the mean particle size. Therefore, powder flow becomes more difficult as the resulting cohesion increases. The effect of particle size on the strength of powders has been widely researched. Rumpf was the first to theorize that the particle size influences the strength of a powder (3) and this has been demonstrated empirically (52). The increase in strength is due to the increased number of interparticle contacts. As more interparticle contacts are formed, the sum of the interparticle forces increases and therefore the cohesive powder strength also increases.

Rumpf (3) was the first to propose a theory for the tensile strength of agglomerates and many of the present models are based on his work. He developed expressions for the strength of agglomerates with various types of interparticle bonds. One mechanism for agglomeration is the formation of liquid bridges. Rumpf suggests that the tensile strength of an agglomerate is proportional to the inverse of the particle diameter squared as in equation (8.17).

$$\sigma_t = \frac{k_o(1-\varepsilon)}{\pi d_p^2} H \tag{8.19}$$

In equation (8.19), σ is the tensile strength of the agglomerate, ε is the porosity of the cake, d_p is the particle diameter, k_o is the coordination number (number of contacts per particle), and H is the strength of the interparticle bond. The flaw in this equation is based on the unrealistic assumption that the bridges fail simultaneously and that all the particles are of equal size. Rumpf also proposed a theory for agglomeration due to the formation of solid crystalline bridges. This equation is based on moisture content of the material and the concentration of material in the bridge and shown in equation (8.18).

$$\sigma_t \approx y_k q \frac{\rho}{\rho_k}(1-\varepsilon)\sigma_k \tag{8.20}$$

In equation (8.20), y_k is the concentration of the dissolved species k, q is the moisture content of the particles before caking, ρ is the density of the particle, ρ_k is the density of k in the crystal bridge, and σ_k is the strength of a crystal bridge. Other researchers have used Rumpf's model to verify experimental data for various materials. Pietsch (50) applies Rumpf's model to investigate the influence of drying rate on the tensile strength of pellets bound by salt bridges. Tanaka (56) further developed Rumpf's model by incorporating the structure of the agglomerate. He included the effects of heat and mass transfer on the formation of the solid bridges. Tanaka used a model of contacting spheres with pendular water as shown Figure 8.21.

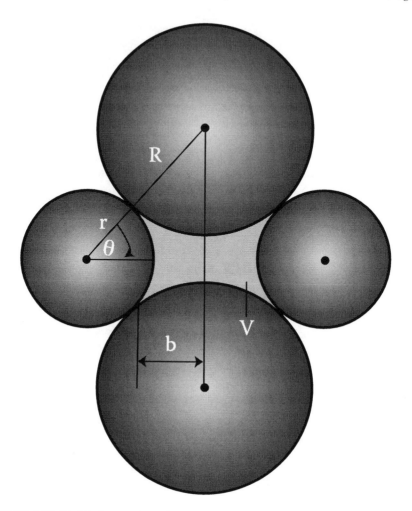

FIGURE 8.21 Model of contacting spheres with pendular water used to calculate the volume and width of the bridge [56].

The particles in the agglomerate are assumed to be monosized. A fictitious sphere of radius r approximates the curvature of the bridge. The volume of the bridge and the narrowest width of the bridge are derived as a function of R. Hence the bridge volume and narrowest width are related by the R according to equation (8.21).

$$\frac{b}{R} = 0.82\left(V/R^3\right)^{0.25} \tag{8.21}$$

Here V is the volume of the bridge, b is the narrowest width of the bridge, and R is the particle radius. Equation (8.21) implies that the cake fails at the narrowest width of the bridge, the neck. The volume V between two particles is a function of the total volume of a single particle V_t and the number of contact points per particle. Tanaka used the relationship between the volume and the width of the solid bridge combined with Rumpf's model to form equation (8.22) for the tensile strength of powders,

$$\sigma_t = 0.17 \frac{(1-\varepsilon)}{\varepsilon} \left(\frac{8\varepsilon}{3(1-\varepsilon_o)} C_e(q/100)X^{1/(1-X)} \right)^{1/2} \sigma_k \qquad (8.22)$$

σ_z is the tensile strength of the recrystallized solid bridge, C_e is the equilibrium concentration, q is the initial moisture content, ε_c is the porosity of the recrystallized bridge, and X is a lumped parameter, which is function of the temperature and humidity.

The principles of fracture mechanics have been applied to the field of particle technology as an alternative approach to Rumpf's theory in determining the strength of agglomerates. A theory has been developed to explain the failure of solids caused by flaws or imperfections in the structure of the solid and the elastic and plastic deformation of the material. The current models proposed by Kendall (57) and Adams (58) are based on Linear Elastic Fracture Mechanics (LEFM). The failure of very brittle materials can be described and analyzed using LEFM (59).

LEFM is based on an energy balance in which the strain energy released at the crack tip provides the driving force to create new surfaces (60). In order for fracture to occur, the rate at which energy is released in the solid must be equal to or greater than the cleavage resistance R_c. Thus the elastic energy release rate of the crack G must be equal to the crack resistance R_c before crack propagation and fracture can occur. The Griffith criterion for fracture is given by applying Hooke's law ($\sigma = \varepsilon E$). Equation (8.23) is obtained.

$$R_c = \beta^2 \pi \sigma \varepsilon \alpha \qquad (8.23)$$

An alternative approach to the Griffith fracture criterion, developed by Irwin, is investigating the stress field at the tip of the crack (61). The stress intensity factor K defines the behavior of the crack under an applied load. This approach is simply the study of stress and strain fields near the tip of a crack in an elastic solid. Since the stresses are elastic, they must be proportional to the stress (or applied load) σ. Irwin proposed equation (8.24) to describe the stress field near the tip of a crack.

$$K_c = \beta \sigma f (\pi a)^{1/2} \qquad (8.24)$$

Here β is a dimensionless geometric parameter, a is the crack length, and σ is the applied load. The crack will grow when K reaches a critical value, K_c. At the time of fracture, the stresses in the crack tip are equal to those in the elastic solid and K_c becomes the toughness of the material. K_c is a measure for the crack resistance of a material and is called the plane strain fracture toughness (62). It can be shown that the elastic energy release rate is equivalent to the stress intensity factor (59).

$$G = \frac{K^2}{E} \qquad (8.25)$$

With that said, most real materials exhibit plastic deformation during failure, so for this condition, Elastic-Plastic Fracture Mechanics (EPFM) must be applied.

EPFM incorporates the plastic deformation at the crack tip during fracture that is neglected in LEFM. For LEFM the stress-strain curve is linear and it follows Hooke's law; however, for EPFM, the stress-strain curve is nonlinear. The stress-strain curve is approximated by the Ramberg-Osgood equation and when expressed in terms of fracture energy required to break a caking bond, the fracture stress, σ_f, is expressed as shown in equation (8.26):

$$\sigma_f = \left(\frac{FJ_R}{Ha} \right)^{1/(n+1)} \tag{8.26}$$

Here n is the strain hardening exponent, F is the plastic modulus, J_R is the plastic fracture energy, H is a geometric factor, and a is the crack length. Specht took this one step further and considered the probability that not every particle would form an interparticle bond using Discrete Element modeling and tomography. The term "granularity" is used to describe the spacing between the major forces contributing to the cake strength. This parameter gives an indication of the size of the particle clusters between each contact force, that is, the number of particles not associated with a major contact force. A decrease in the granularity implies that there exists a greater number of major force chains. The force chain images are analyzed using image analysis to determine the relative spacing between major force chains within the shear zone (68).

A relationship between the consolidation stress, moisture content, and the mean granularity of a caked powder bed can be described by equation (8.27):

$$f_c = \left(\sigma_{f,tot} \frac{1+\sin\phi}{1-\sin\phi} \right) \left(\frac{d_f}{d_{fo}} \right)^3 \tag{8.27}$$

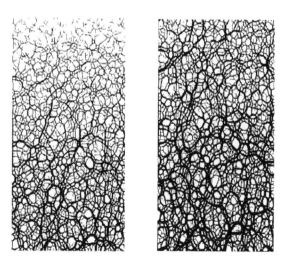

FIGURE 8.22 Force chain structure in a DEM simulated powder bed with a high applied load (right) and a low applied load (left) [68].

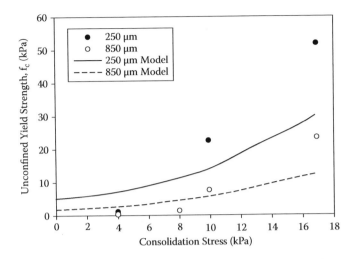

FIGURE 8.23 Unconfined yield strength as a function of the consolidation stress compared to the model of equation 8.27 [52].

Here f_c is the unconfined yield stress, d_f is the mean granularity after caking, and d_{f_0} is mean granularity prior to caking. Specht's model was shown to predict the trend of data; however, the magnitude was not exact, due to the use of spherical particles in DEM simulations. Using nonspherical particles in the DEM simulations would allow a greater number of contacts, creating a greater number of force chains in the model. Although the exact magnitude of the data is not predicted, this model provides a good base for the incorporation of strength as a function of consolidation stress as shown in Figure 8.23.

CONCLUDING REMARKS

Understanding powder flow is imperative to overcome powder handling complexities in modern industry. By establishing the contribution of microscopic and macroscopic forces to powder strength, the knowledge base required to control powder flow can be developed. Although powder flow theory is still incomplete, the research presented in this chapter has led to vast improvements in powder handling. In order to realize this goal, the contribution of additional complexities associated with the physical and chemical properties of real particles must be further considered.

ACKNOWLEDGMENTS

The authors acknowledge the financial support of the Particle Engineering Research Center (PERC) at University of Florida, the United States Army, Edgewood Chemical and Biological Center, and the Industrial Partners of the PERC for support of this research. Any opinions, findings, and conclusions or recommendations expressed in this material are those of the author(s) and do not necessarily reflect those of the United States Army or the Industrial Partners of the PERC.

REFERENCES

1. Ennis, B.J., Solids handling: Bulk solids flow characterization, in *Perry's Chemical Engineering Handbook*, R.H. Perry, Editor. 2007. p. 8.
2. Rumpf, H., Die Wissenschaft des Agglomerierens. *Chem.-Ing.-Tech.*, 1974. 46: p. 1–11.
3. Rumpf, H., Grundlagen und methoden des granulierens. *Chemie-Ing.-Techn.*, 1958. 30(3): p. 144–158.
4. Rumpf, H., *Particle Technology*. 1990, London: Chapman Hall.
5. Rumpf, H., K. Sommer, and K. Steier, Mechanismen der Haftkraftverstärkung bei der Partikelhaftung durch plastisches Verformen, Sintern und viskoelastisches Flieben. *Chem. Ing. Tech.*, 1976. 48: p. 300–307.
6. Merrow, E.W., Estimating the startup times for solids processing plants. *Chem. Eng.*, 1988. 95(15): p. 89–92.
7. Jenike, A.W., *Storage and Flow of Solids*. 1964, University of Utah: Salt Lake City.
8. Rietema, K., Powders, what are they. *Powder Technol.*, 1984. 37(Jan): p. 5–23.
9. Inoya, B.J., *Powder Technology*. 1984, New York: Hemisphere Publishing Corp.
10. Zimon, A.D., *Adhesion of Dust and Powder*. 1969, New York: Plenum Press.
11. Callister, W.D., *Materials Science and Engineering, An Introduction*. 1997, New York: Wiley & Sons, Inc.
12. London, F., Eisenschitz, E., The relation between the van der Weals forces and the homeopolar valence forces. *Zeits. f. Physik*, 1930. 60(491).
13. Hamaker, H.C., The London van der Waals attraction between spherical particles. *Physica IV*, 1937. 10(1048).
14. van der Waals, J.D., The equation of state for gases and liquids, in *Nobel Lecture*. 1910.
15. van der Waals, J.D., *Over de Continui'teit van de Gas- En Vloeistoftoestand (On the Continuity of the Gaseous and Liquid States)*. 1873, Leiden Univeristy, the Netherlands.
16. Lennard-Jones, J.E.D., B.M., Cohesion at a crystal surface. *Trans. Faraday Soc.*, 1928. 24(92).
17. London, F., The general theory of molecular forces. *Trans. Faraday Soc.*, 1937. 33(8).
18. Derjaguin, B.V., *Kolloid Zeitschrift*, 1934. 69(155).
19. Hamaker, H.C., *Physica*, 1937. 4(1058).
20. Bradley, R.S., The cohesive force between solid surfaces and the surface energy of solids. *Philos. Mag.*, 1932. 13(86): p. 853–862.
21. de Boer, J.H., *Trans. Faraday Soc.*, 1936. 32: p. 10–18.
22. Lifshitz, E.M., The theory of molecular attractive forces between solids. *Sov. Phys. JETP*, 1956. 2(73).
23. Ninham, B.W. and V.A. Parsegian, van der Waals forces—Special characteristics in lipid-water systems and a general method of calculation based on Lifshitz Theory. *Biophys. J.*, 1970. 10(646).
24. Hough, D.B. and L.R. White, The calculation of Hamaker constants from Lifshitz Theory with applications to Weddting phenomena. *Adv.Colloid and Interf. Sci.*, 1980. 14(1): p. 3–41.
25. Tabor, D. and H. Winterton, Direct measurement of normal and retarded van der Waals forces. *Proc. Roy. Soc. Lond. Ser. a—Math. Phys. Sci.*, 1969. 312(1511).

26. Israelachvili, J., *Intermolecular & Surface Forces*. 2nd Edition. 1992, San Diego: Academic Press, Inc.

27. Young, T., An essay for the cohesion of fluids. *Philos. Trans. Roy. Soc. Lond.*, 1805. 95: p. 65–87.

28. Dupré, A., in *Theorie Mechanique de la Chaleur*. 1869, Gauthier-Villars: Paris.

29. Reed, J.S., *Principles of Ceramic Processing*. 2nd ed. 1995, New York: John Wiley & Sons.

30. Coelho, M.C., Upper humidity limit of the pendular state. *Powder Techn.*, 1979. 23(2): p. 203–07.

31. Coelho, M.C. and N. Harnby, Effect of humidity on form of water-retention in a powder. *Powder Techn.*, 1978. 20(2): p. 197–200.

32. Coelho, M.C. and N. Harnby, Moisture bonding in powders. *Powder Techn.*, 1978. 20(2): p. 201–205.

33. Esayanur, M.S., Interparticle force based methodology for prediction of cohesive powder flow properties, in *Materials Science and Engineering*. 2004, University of Florida: Gainesville.

34. Greenwood, J.A. and J.B.P. Williamson, Contact of nominally flat surfaces. *Proc. Roy. Soc. Lond., Ser. A: Math., Phys. Eng. Sci.*, 1966. 295(1442): p. 300–319.

35. Fuller, K.N.G. and D. Tabor, The effect of surface roughness on the adhesion of elastic solids. *Proc. Roy. Soc. Lond. A*, 1975. 345: p. 327–342.

36. Czarnecki, J. and T. Dabros, Attenuation of the van der Waals attraction energy in the particle semi-infinite medium system due to the roughness of the particle surface. *J. Colloid Interf. Sci.*, 1980. 78(1): p. 25–30.

37. Xie, H.Y., The role of interparticle forces in the fluidization of fine particles. *Powder Techn.*, 1997. 94(2): p. 99–108.

38. Adler, J., *Interaction of Non-Ideal Surfaces in Particulate Systems*. 2001, University of Florida: Gainesville.

39. Rabinovich, Y.I., et al., Adhesion between nanoscale rough surfaces—I. Role of asperity geometry. *J. Colloid Interf. Sci.*, 2000. 232(1): p. 10–16.

40. Rabinovich, Y.I., et al., Adhesion between nanoscale rough surfaces—II. Measurement and comparison with theory. *J. Colloid Interf. Sci.*, 2000. 232(1): p. 17–24.

41. Ata, A., Y.I. Rabinovich, and R.K. Singh, Role of surface roughness in capillary adhesion. *J. Adhesion Sci. Techn.*, 2002. 16(4): p. 337–346.

42. Nguyen, A.V., et al., Attraction between hydrophobic surfaces studied by atomic force microscopy. *Int. J. Min. Process.*, 2003. 72(1–4): p. 215–225.

43. Rabinovich, Y.I., et al., Capillary forces between surfaces with nanoscale roughness. *Adv. Colloid Interf. Sci.*, 2002. 96(1–3): p. 213–230.

44. Rabinovich, Y.I., M.S. Esayanur, and B.M. Moudgil, Capillary forces between two spheres with a fixed volume liquid bridge: Theory and experiment. *Langmuir*, 2005. 21(24): p. 10992–10997.

45. Reist, P.C., *Aerosol Science and Technology*. Second ed. 1993, New York: McGraw-Hill, Inc. 379.

46. Ren, J., et al., Electrostatic dispersion of fine particles in the air. *Powder Techn.*, 2001. 120(3): p. 187.

47. Aguilera, J., J. del Valle, and M. Karel, Caking phenomena in amorphous food powders. *Trends Food Sci. Techn.*, 1995. 6: p. 149–155.

48. Hancock, B.C. and S.L. Shamblin, Water vapour sorption by pharmaceutical sugars. *Pharm. Sci. Techn. Today*, 1998. 1(8): p. 345–351.

49. Johanson, J.R. and B.O. Paul, Eliminating caking problems. *Chem. Process.*, 1996. 59(8): p. 71–75.

50. Pietsch, W.B., Strength of agglomerates bound by salt bridges. *Can. J. Chem. Eng.*, 1969. 47(4): p. 403.

51. Pietsch, W.B., Adhesion and agglomeration of solids during storage flow and handling—A survey. *J. Eng. Ind.*, 1969. 91(2): p. 435–438.
52. Specht, D.W., Caking of granular materials: An experimental and theoretical study, in *Particle Engineering Research Center*. 2006, University of Florida: Gainesville. p. 141.
53. Leaper, M.C., et al., Measuring the tensile strength of caked sugar produced from humidity cycling. *Proc. Inst. Mech. Eng. Part E—J. Process Mech. Eng.*, 2003. 217(E1): p. 41–47.
54. Teunou, E. and J.J. Fitzpatrick, Effect of relative humidity and temperature on food powder flowability. *J. Food Eng.*, 1999. 42(2): p. 109–116.
55. Cleaver, J.A.S., et al., Moisture-induced caking of boric acid powder. *Powder Techn.*, 2004. 146(1–2): p. 93–101.
56. Tanaka, T., Evaluating caking strength of powders. *Ind. Eng. Chem. Prod. Res. Dev.*, 1978. 17(3): p. 241–246.
57. Kendall, K., Agglomerate strength. *Powder Metall.*, 1988. 31(1): p. 28–31.
58. Adams, M.J., The strength of particulate solids. *J. Powder Bulk Solids Techn.*, 1985. 9(4): p. 15–20.
59. Broek, D., *The Practical Use of Fracture Mechanics*. 1988, Dordrecht, the Netherlands: Kluwer Academic Publishers.
60. Griffith, A.A., The phenomena of rupture and flow in solids. *Phil. Trans. Roy. Soc. Lond. A*, 1921. 221: p. 163–167.
61. Irwin, G.R., Fracture I. *Handb. Phys.*, 1957. IV: p. 558–590.
62. Broek, D., *Elementary Engineering Fracture Mechanics*. 1974, Leyden, the Netherlands: Noordhoff International Publishing.
63. Schulze, D., Flowability and time consolidation measurements using a ring shear tester. *Powder Handling & Processing*, 1996. 8(3): p. 221–226.
64. Tedeschi, S.T., Improving aerosol dispersion through a fundamental understanding of interparticle forces, in *Materials Science and Engineering*. 2007, University of Florida: Gainesville. p. 177.
65. Stevens, N.I., Contact angle measurements on particulate systems, in *Ian Wark Research Institute*. 2005, University of South Australia.
66. Flemmer, C.L., On the regime boundaries of moisture in granular materials. *Powder Techn.*, 1991. 66(2): p. 191–194.
67. Yoon, R.-H., D.H. Flinn, and Y.I. Rabinovich, Hydrophobic interactions between dissimilar surfaces. *J. Colloid Interf. Sc.*, 1997. 185(2): p. 363–370.
68. Bilgili, E., et al., Stress inhomogeneity in powder specimens tested in the jenike shear cell: myth or fact? *Part. Part. Syst. Characterization*, 2004. 21(4): p. 293–302.

9 Characterization of Pharmaceutical Aerosols and Dry Powder Inhalers for Pulmonary Drug Delivery

M.S. Coates, P. Tang, H.-K. Chan,
D.F. Fletcher, and J.A. Raper

CONTENTS

INTRODUCTION

The growing use of the inhalation route for the delivery of new types of drugs, such as proteins for both local and systemic effects (1,2), and the need to manufacture and deliver fine particles to the targeted parts of the respiratory tract have led to more interest in the characterization of not only the size of the aerosol particles, but also their morphology and how they are dispersed by inhaler devices. In recent years, these devices are most commonly dry powder inhalers (3).

The influence of particle size on the clinical outcome of aerosols has been extensively reported (4,5). In order for aerosol particles to be transported through the respiratory system, their aerodynamic diameter must be less than 5 μm, with the optimum size in the range 2–3 μm (5). Traditionally in the pharmaceutical industry, the measurement of aerodynamic diameter of these aerosol particles has been achieved by the use of cascade impactors (6) and multistage liquid impingers (7). However, these techniques are time-consuming and yield only time-averaged particle size distributions. Laser diffraction, while not providing a direct measure of aerodynamic diameter, has gained favor by researchers and industry due to ease of use and instantaneous measurement (8). Time of flight analysis (9) gives aerodynamic size but is not amenable to *in situ* analysis due to the low concentration requirements of the instruments. As the size analysis of pharmaceutical aerosols is well understood, we will concentrate here on the particle morphology as represented by the fractal dimension and dispersion capabilities of dry powder inhalers.

FRACTAL DIMENSION CHARACTERIZATION

A fractal object, as defined by Mandelbrot, has a dimension D greater than the geometric or physical dimension (0 for a set of disconnected points, 1 for a curve, 2 for a surface, and 3 for a solid volume) but less than or equal to the embedding dimension in an enclosed space (the embedding Euclidean space dimension is usually 3) (10). For objects to be fractal, ideally they have to be self-similar, independently of the length scale. This means that if part of the object is cut out and this part is magnified, the resulting object will look exactly the same as the original one. Although an ideal fractal object should theoretically be homogeneous, it is essential to examine the length scale at which the fractality takes place. Some investigators have allocated separate values of surface fractal dimension to a protein surface since self-similarity of the protein surface differs with length scale (11–13). Although self-similarity is occasionally only valid for a limited range, it may still be reasonable to use fractal concepts carefully with the full knowledge that the model is approximate (14).

Fractal dimension can be related to the mass or surface of an object. Surface fractal dimension, D_S, has been widely used to characterize particle surface roughness (15–17). Characterization of particle surface is important because surface roughness is recognized to affect physicochemical properties of pharmaceutical products (18) and is therefore a crucial factor in the manufacturing process and product performance. Furthermore, surfaces of most materials, amorphous and crystalline, porous and nonporous, are reported to be fractal on the molecular scale (19). D_S varies from 2 for a perfectly smooth surface to 3 for a very rough surface. If the surface of an object is fractal, its perimeter (P) varies with the yardstick (β) as shown below (20–22):

$$P \propto \beta^{2-D_S} \tag{9.1}$$

In order to get an accurate value for D_S, the resolution of β has to be fine enough so that each detail of irregularity can be resolved. Various techniques are available to obtain D_S, such as image analysis, light scattering, gas adsorption, mercury intrusion porosimetry, and one using atomic force microscopy. These methods have been reviewed in detail by Tang, Chan, and Raper (23). Recently, scanning tunneling microscope (24), power spectrum and R/S method (25), triangulation method (26), and nonequilibrium techniques (27) were used to obtain D_S. Here, we will focus on gas adsorption and light scattering because these methods are most appropriate to characterize aerosol inhalant particles.

GAS ADSORPTION

Theory

There are three different gas adsorption methods available to obtain surface fractal dimension. The first method uses several adsorbates having molecules of different cross-sectional area (28). The surface fractal dimension is then determined from equation (9.2) (29):

$$S \propto \sigma^{-(D_S-2)/2} \tag{9.2}$$

where σ is the adsorbate molecular cross-sectional area.

The surface area of the molecules is usually predicted using the correlations derived by McClellan and Harnsberger, which were developed assuming that the same surface area was obtained by adsorption of different adsorbates (30). This, however, is not true if the particle surface is rough. Adsorbate molecules having larger cross-sectional area cannot access the finer structure of the surface and hence the surface area will be underestimated. For this reason, the determination of D_S from the first method might not be accurate.

In the second method, instead of using different adsorbates, particles are fractionated into narrow particle size distribution and the Brunauer-Emmett-Teller (BET) surface area is measured for each size fraction (31). D_S can then be determined from equation (9.3) (29):

$$S \propto D_p{}^{D_S-3} \tag{9.3}$$

where S is the specific surface area (m^2/g) and D_p is the mean particle diameter after fractionation. While this method is effective, it is tedious to use and therefore a third method was developed.

The third method is more favorable than the first two methods since it only requires a single adsorption isotherm to be measured. D_S is deducible from the isotherm utilizing a theory developed by Frenkel (32), Halsey (33), and Hill (34), which was extended to fractal surfaces by Pfeifer and others (35). At the early stages of adsorption (low P/P_O), the effect of surface tension is negligible and the interactions between adsorbate molecules and the particles are mainly due to van der Waals forces. In this situation, D_S can be determined as follows:

$$\ln\left(\frac{V}{V_m}\right) = C + \left(\frac{D_S - 3}{3}\right) \ln\left[\ln\left(\frac{P_O}{P}\right)\right]$$

(9.4)

where V is the volume of gas adsorbate at an equilibrium pressure P, V_m the volume of gas for a monolayer adsorption, P the adsorption equilibrium pressure of a gas, P_O the saturation pressure of the gas at the given temperature, and C the pre-exponential factor.

At higher P/P_O, the surface tension (or capillary condensation) effect becomes more pronounced and then the following applies:

$$\ln\left(\frac{V}{V_m}\right) = C + (D_S - 3) \ln\left[\ln\left(\frac{P_O}{P}\right)\right]$$

(9.5)

The van der Waals forces and the effect of surface tension represent limiting cases; however, in general, the adsorption forces are a mixture of the two (36). Ismail and Pfeifer used δ to determine which effect dominates (36,37):

$$\delta = 3\left[1 + \left(\frac{D_S - 3}{3}\right)\right] - 2$$

(9.6)

$\delta \geq 0$ indicates that capillary condensation is negligible.

Limitation of the Extended Frenkel-Halsey-Hill (FHH) Theory

The appropriate pressure range and therefore the range of thickness of the adsorbed layer coverage to obtain the correct D_S was first established by Tang and colleagues (38). In order to determine the number of adsorbed layers (n) (37), it is first necessary to calculate the volume of monolayer coverage (V_m) from the BET equation. From the BET equation,

$$\frac{P}{V(P_o - P)} = \frac{1}{V_m \beta} + \frac{(\beta - 1)P}{V_m \beta P_o},$$

V_m is determined from the slope and intercept of a plot of

$$\frac{P}{V(P_o - P)}$$

versus P/P_O (28), where β is a constant related to the energy of adsorption in the first adsorbed layer. To calculate the number of adsorbed layers, an iterative process was required (Figure 9.1).

The adsorption isotherms of wrinkled bovine serum albumin (BSA) particles, prepared by spray drying (39), show two linear regions (Figure 9.2a) while those from

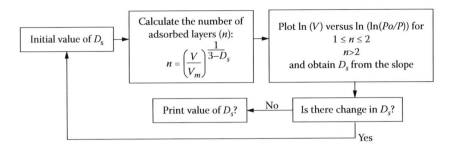

FIGURE 9.1 Iterative process to determine surface fractal dimension from N_2 adsorption isotherms.

smooth BSA show more than 2 linear regions (Figure 9.3a). The iterative process was used to calculate the number of adsorbed layers (Figure 9.1) and the appropriate adsorption isotherms were reconstructed (Figure 9.2b & Figure 9.2c, Figure 3b–d). For smooth BSA, the correct value of D_S was generated from monolayer coverage (Figure 9.3b). The value of D_S (2.12 + 0.04) agrees with that obtained by light scattering (2.10 + 0.04) (38). Electron micrographs also show that the particles are smooth and therefore D_S should be close to 2. However, D_S was higher as the surface coverage increased. This is because the adsorbed gas molecules no longer probed the

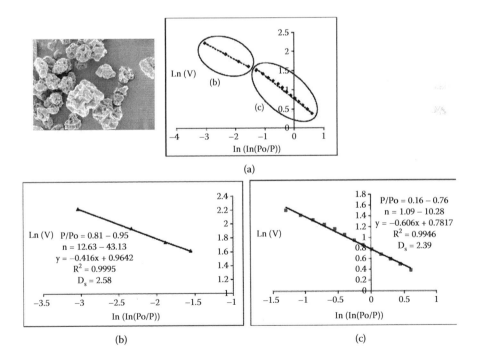

FIGURE 9.2 (a) N_2 adsorption isotherms for rough BSA particles (shown in the photo) for numbers of adsorbed layers (n) 1.09-43.13; (b) for $n = 12.63$-43.13; (c) for $n = 1.09$-10.28.

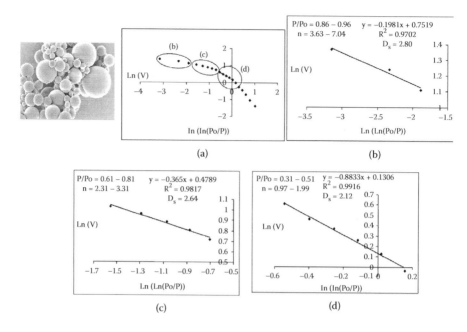

FIGURE 9.3 (a) N_2 adsorption isotherms for smooth BSA particles (shown in the photo) for numbers of adsorbed layers (n) 0.97-7.04; (b) for n = 0.97-1.99; (c) for n = 3.63-7.04; (d) for n = 2.31-3.31. Note that 5 points in (a) were excluded from fractal plot (d) because monolayer coverage was not reached at those points.

single particle surface but instead probed the outline of the particle clumps, indicating an apparent increase of D_S. Particle roughness reduces the contact area between particles and hence this effect became evident at much higher coverage (pressure) for rough particles. For $n = 1.09 - 10.28$, D_S (2.39) agrees with that obtained by light scattering (2.39 ± 0.05). Hence, the adsorbed gas molecules did not probe the particles as clumps until the multilayer built up significantly, enough to fill in all the wrinkles. This shows that accurate determination of D_S from adsorption isotherms using the extended FHH theory depends on the appropriate choice of coverage region. Application of this theory to obtaining D_S is thus limited by the number of adsorbed layers on the particles. Recently, it was found that for silica materials while the D_S obtained from N_2 adsorption differed from that obtained by small angle x-ray scattering (SAXS), D_S from butane adsorption agrees with SAXS (40). This observation cannot be explained with the theory we proposed earlier (38). It was reported that N_2 may show selective adsorption on the surface of hydroxylated silicas while butane is unlikely to show this selective adsorption (41). When this phenomenon takes place, the extended FHH theory is no longer valid for the determination of D_S.

LIGHT SCATTERING

The quickest and simplest method to obtain D_S is by light scattering, which utilizes the Rayleigh-Gans-Debye (RGD) scattering theory. This theory, however, has some limitations since it is only valid when the light wavelength, λ, is much larger than that of the scatterers (i.e., $\lambda > 20 \times$ radius of particle) and the scatterers have similar refractive indices to the surrounding medium (i.e., $|m - 1| \ll 1$, where m is the complex refractive index of the particle relative to that of the surrounding medium (42)). While this theory ignores the effect of multiple scattering (which would change the scattering intensity), it has been shown experimentally and theoretically that it does not change the evaluated fractal dimension of the aggregates (43).

If an object is uniform and has a distinct surface, scattering from the surface will reveal the roughness. The scattering intensity, I, can then be related to the D_S as

$$I \propto \left[\frac{4\eta\pi}{\lambda} \sin\left(\frac{\theta}{2}\right) \right]^{-6+D_S} \tag{9.7}$$

where η is the refractive index of suspending fluid and θ is the scattering angle.

Scattering momentum, q ($q = (4\eta\pi/\lambda) \sin(\theta/2)$), represents the magnification scale. Higher q values indicate higher magnification. To determine D_S, the magnification has to be high enough so that the particle surface structure can be resolved, as illustrated in Figure 9.4 (44).

D_S is obtainable from equation (9.7) only when $q > D_P^{-1}$ (Porod region in Figure 9.4) where the detailed structure of the particle surface can be resolved. When $D_A^{-1} < q < D_P^{-1}$, the fractal region gives the mass fractal dimension, which represents the packing density of the aggregates.

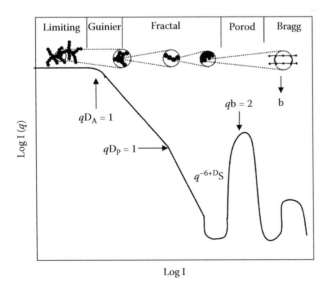

FIGURE 9.4 Scattering curve for colloidal aggregate (D_P: primary particle diameter; D_A: aggregate diameter, D_S: surface fractal dimension, b: bond length).

Bovine Serum Albumin (BSA) Particles with Different Surface Roughness

Recent work by the authors has revealed that BSA particles of differing surface roughness, shown in Figure 9.2 and Figure 9.3, yield a significant difference in aerosol performance (45). Wrinkled particles showed better dispersion coupled with higher fine particle fraction (FPF; fraction of particles less than 5 μm in diameter) in the aerosol cloud. Using a Rotahaler® (GlaxoSmithKline) and a Dinkihaler® (Aventis), the FPF of wrinkled particles are approximately 10–20% higher than the FPF generated from the dispersion of smooth particles. Furthermore, the aerosol performance of the wrinkled particles showed less dependence on the inhaler choice and air suction flow. The smooth and wrinkled particles had similar physical properties: moisture content, true density, particle size distribution, amorphous state, and aerodynamic diameters. Therefore, it was postulated that the high FPF of wrinkled particles occurs as a result of the surface asperities increasing the interparticulate distance, and reducing the contact area between the particles, hence decreasing the powder cohesiveness.

The determination of the aerodynamic diameter (D_A) was done via the "spherical envelope density" (ρ_{SH}) method where the powder was subjected to tapping in glass tubes (45):

$$D_A = d_V \sqrt{\rho_{SE}} \quad \rho_{SH} = \frac{\text{mass of wrinkled particle used for tapping}}{\text{void space} + \text{true volume of wrinkled particles}} \quad (9.8)$$

$$\text{void space} = \text{tapped volume of wrinkled particles} - \quad (9.9)$$
$$\text{tapped volume of smooth particles}$$

d_V is the equivalent spherical particle diameter obtained by laser diffraction (Malvern Mastersizer S, Worcestershire, U.K.). This method has two flaws. First, it does not take into account the reduction of drag coefficient due to the surface wrinkleness of the particles. Second, interlocking between particles, exacerbated by polydispersity, will reduce the void space calculated in equation (9.9). To rectify this problem, we developed a method to calculate aerodynamic diameter (D_A) taking into account the drag coefficient (46), which is explained in the next subsection.

Computation of the Drag Coefficient (C_D) to Predict the Aerodynamic Diameter (D_A) of Particles with Rough Surfaces

Particle sizing equipment that measures aerodynamic diameter is readily available but there are biases associated with the use of these instruments (47). Particularly, complete dispersion of micron-sized dry particulate solids may be difficult to achieve due to strong cohesive forces. Tang, Chan, and Raper (46) have developed a predictive calculation, which serves to calculate D_A iteratively as follows:

$$U_1 \rightarrow Re = \frac{U\rho_L d_V}{\mu_L} \rightarrow C_D \rightarrow U_2 = \sqrt{\frac{2[V_P(\rho_P - \rho_L)g]}{A_A \rho_L C_D}} \rightarrow U_2 - U_1 = 0 \xrightarrow{Yes} D_A = \sqrt{\frac{18U\mu_L}{g}}$$
$$\downarrow No$$

where Re is particle Reynolds number; μ_L and ρ_L, viscosity and density of suspending fluid, respectively; U, particle terminal settling velocity; V_P and ρ_P, volume and

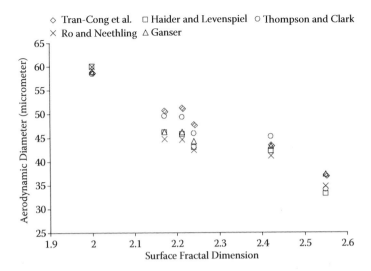

FIGURE 9.5 Plot of aerodynamic diameter, calculated using different C_Ds as a function of surface fractal dimension. A_A and V_P were set to be 300 μm^2 and 4500 μm^3, respectively.

density of the particle, respectively; g, gravitational acceleration; A_A, projected area of the particle normal to the settling direction.

Chhabra, Agarwal, and Sinha (48) compared literature reported C_D expressions and found that C_Ds calculated by Ganser (49), and Haider and Levenspiel (50) are the most accurate, with an average error of 16.3%, and these C_Ds were therefore employed in our work. Furthermore, C_Ds derived by Ro and Neethling (51); Tran-Cong, Gay, and Michaelides (52); and Thompson and Clark (53) were employed because of their simplicity and ease of use, which make them more appealing and attractive. A set of model fractal objects having D_S varying from 2.00 to 2.55 and uniform volume was constructed. The D_S was determined from Richardson's plot (log perimeter versus log yardstick) (21). Using the predictive model, the D_A decreases with increasing D_S due to the increased drag forces exerted by the surrounding fluid (Figure 9.5).

VALIDATION OF THE AERODYNAMIC DIAMETER USING THE PREDICTIVE MODEL

To confirm the accuracy of the predictive model, bovine serum albumin particles with different degrees of surface corrugation were prepared via spray drying (39) (Figure 9.6). The D_S of these powders were determined using light scattering, with

(a)   (b)   (c)   (d)

FIGURE 9.6 Scanning electron microscope (SEM) photos of BSA particles (a) $D_S = 2.06$; (b) $D_S = 2.18$; (c) $D_S = 2.35$; (d) $D_S = 2.41$.

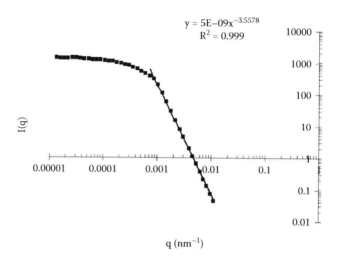

FIGURE 9.7 Scattering curves obtained from light scattering of BSA particles with $D_S = 2.44$.

a typical scattering curve shown in Figure 9.7. SEM photos (Figure 9.6) confirm the increasing degree of corrugation which reflected in higher D_S values measured.

The BSA particles had similar physical characteristics, that is, size, polydispersity, moisture content, and amorphous nature. The D_As of these particles were calculated and compared with values measured using a Particle Size Distribution Analyzer (PSD) 3603 (TSI, Shoreview, U.S.A.), which measures the time taken for a particle to accelerate between two laser beams and relates this to the aerodynamic diameter. The particle size distribution determined by the PSD3603 shows the presence of large aggregates. This, combined with the absence of particles >10 μm in both the laser diffraction results and electron microscope photos (Figure 9.6), indicates that complete dispersion could not be achieved using the PSD3603's dry powder disperser. This has been recognized by the instrument manufacturer in their instructions for the measurement of aerodynamic diameters of standard spherical aerosol particles and hence these large aggregates have to be eliminated to give accurate results in the instrument calculation. The D_As measured by the PSD3603 were therefore recalculated after eliminating these big aggregates. For smooth BSA particles ($D_S = 2.06$), even after removal of large agglomerates, there is approximately a 20% difference between measured and calculated D_As (Figure 9.8). As smooth particles have more contact area than wrinkled ones, they are more cohesive and hence dispersion into individual particles is more difficult. The comparison between measured and computed D_As for wrinkled BSA (Figure 9.8) shows that

1. D_As computed by the Ro and Neethling C_D correlation agree for particles with D_S 2.41 and 2.35. The maximum difference for particles with D_S 2.18 is 26%.
2. Using the Haider and Levenspiel, and Thompson and Clark correlations, D_A computed matches the measured value only for the particles with D_S 2.18.

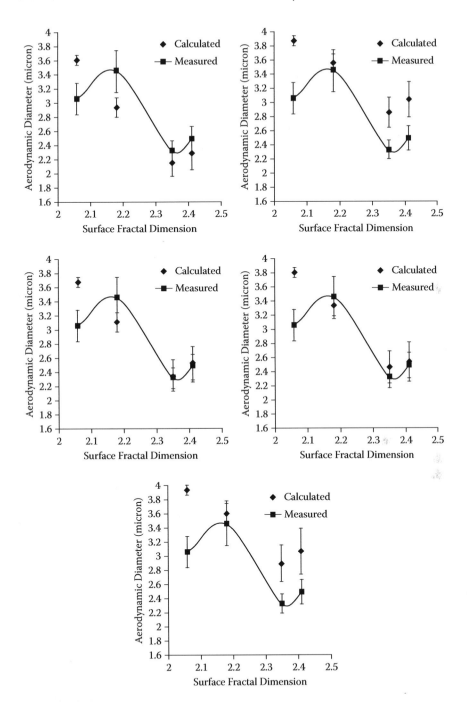

FIGURE 9.8 Aerodynamic diameters measured by Particle Size Distribution Analyser (PSD)3603 and computed from the model utilizing the drag coefficient from (a) Ro and Neethling; (b) Haider and Levenspiel; (c) Ganser; (d) Tran Cong et al; (e) Thompson and Clark.

Particles with other D_S values have maximum differences of approximately 30% and 20% calculated from C_D by Haider and Levenspiel (50), and Thompson and Clark (53), respectively.

3. D_As calculated using the C_D correlations by Ganser (49) and Tran-Cong et al. (52) agree with the measured values.

D_As calculated from the C_Ds employed here follow the expected trend of decreasing value with increasing surface roughness. Comparison with the measured values from the PSD3603 shows that Ganser's and Tran-Cong et al.'s correlation predicts D_A most accurately, followed by Ro and Neethling's, Haider and Levenspiel's, and Thompson and Clark's. This calculation technique, therefore, serves to readily predict D_A in situations when direct experimental measurement is inaccurate or insufficient due to equipment limitations or difficulty in dispersing powders into individual particles in the aerosol.

COMPUTATIONAL MODELING OF DRY POWDER INHALERS

Dry powder inhalers are designed to convert a dry powder drug formulation into a respirable aerosol cloud for inhalation (54). First introduced onto the market in the 1960s, they were primarily developed to satisfy the need to deliver large drug doses (2), but became increasingly popular at the time CFC replacement became an environmental issue (55). Today there are a large number of commercially available dry powder inhalers, which vary in dispersion efficiency due to their design characteristics, with an even larger number of devices presently in development (56).

Despite the large number of inventions of dry powder inhalers, studies detailing the fundamental mechanisms of agglomerate dispersion are scarce. This consequent lack of detailed knowledge is a major hindrance in improving the performance and reproducibility of dry powder delivery systems. The deagglomeration of drug particles to form a fine respirable aerosol cloud in dry powder inhalers is thought to be achieved by three major mechanisms: (a) particle interaction with the shear flow and turbulence, (b) particle-device impaction, and (c) particle-particle impaction (56). As each of the mechanisms is controlled by the air flowfield generated in the device, a detailed knowledge of the flowfield generated in dry powder inhalers and how the flowfield interacts with the drug agglomerates to produce respirable aerosol clouds is essential to develop an understanding of their performance.

Computational Fluid Dynamics (CFD) has found wide application in the study of pharmaceutical unit operations for drug manufacturing, for example, flow in mixing vessels (57,58) and spray dryer design (59,60). Recently, CFD has been used to model the airflow in propellant metered-dose inhalers (61) and holding chambers (62,63). Although CFD modeling is also available for dry powder inhalers (56,64,65), few studies in the open literature have utilized CFD to enhance understanding of the factors affecting the performance of dry powder inhaler systems.

In recent years, Coates and others (66–70) have developed CFD models to simulate and examine the flowfield generated in the Aerolizer (a commercially available dry powder inhaler, shown in Figure 9.9). The computational models developed were used in conjunction with experimental aerosol characterization techniques to create

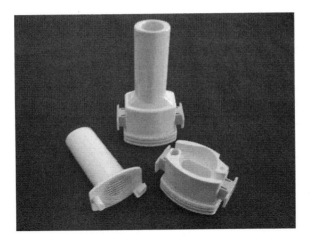

FIGURE 9.9 Photograph of the Aerolizer® (produced by Plastiape S.p.A. for Novartis).

a generic tool to study powder deagglomeration. Through the correct application of this tool, significant information has been gained on the factors affecting agglomerate break-up in dry powder inhalers, with specific focus on the effects of device design and operating conditions.

COMPUTATIONAL METHODOLOGY

The nature of the air flowfield generated within the inhaler was simulated by solving the Reynolds Averaged Navier Stokes (RANS) equations together with the SST (Shear Stress Transport) turbulence model (71) and automatic wall functions using the commercial CFD code ANSYS CFX (www.ansys.com/cfx), as previously described (66). Solving the RANS equations in conjunction with a suitable turbulence model is the most appropriate approach to solving turbulent flows in complex geometries (and is generally adopted for practical engineering problems) as it provides a good approximation of the flowfield without the need for excessive computational requirements (71). Figure 9.10 shows a simplified flowchart of the general step-by-step procedures employed when performing CFD simulations.

In addition to simulating the airflow generated in the inhaler, the CFD models developed were also capable of modeling the flow of single, unagglomerated particles within the device. Lagrangian particle tracking was performed as a postprocessing operation, in which the fate of a large number of mannitol particles were tracked through the fluid after release from the capsule region and subjected to drag forces and turbulent dispersion. This was modeled using the approach of Gosman and Ioannides (73). By setting the different walls within the device to have a zero coefficient of restitution, it was possible to determine the frequency and location of wall impactions. Modifications to the computational code were made to enable the speed of all particle collisions with the different walls of the device to be determined.

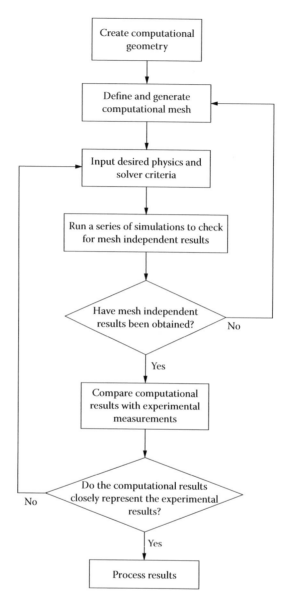

FIGURE 9.10 Simplified flowchart of the step-by-step procedures employed when performing CFD simulations.

The computational model developed was validated using Laser Doppler Velocimetry (LDV) techniques by measuring axial and tangential velocities of a large number of measurement points across the exit of the inhaler mouthpiece, which were then compared directly with the corresponding CFD results. Good agreement was observed between the computational and experimental results over a range of device designs

FIGURE 9.11 Visual comparison between the particle tracks obtained from the computational model and those captured from the high-speed photography. Note that (a) shows the physical geometry of the inhaler whilst (b) displays the computational geometry, comprised only of the surfaces in contact with the fluid.

and flow rates, demonstrating that the computational model could accurately simulate the device flowfield (66). In order to provide a visual comparison between the CFD and experimental results, high-speed photography was performed to visualize the motion of the drug particles within transparent Aerolizer devices. Figure 9.11a shows the motion the drug particles take when dispersed by the Aerolizer. Comparing this figure with the particle tracks obtained when the computational models were used to simulate the dispersion of 10,000 drug particles (Figure 9.11b) provides additional evidence of good agreement between the experimental and computational results, and shows that the CFD model closely represents the flow of particles within the inhaler.

Aerosol Characterization Methodology

The dispersion performance of the inhaler was determined using a spray-dried mannitol powder (particle size $d_{50} = 3.2$ μm, span $((d_{90}-d_{10})/d_{50}) = 1.3$) and a four-stage (plus filter) liquid impinger. The multi-stage liquid impinger (MSLI) is the recommended apparatus for determining the aerodynamic size distribution of dry powder aerosols as the wetted stages prevent solid particles from bouncing between stages (74). Based on the well-known Stokes number equation (75), the MSLI is designed so that each stage of the impinger reduces in cut-off diameter, allowing particles of different size to be collected on the stages due to their inertial properties. The resulting mass of powder deposited on each stage can then be determined (using HPLC analysis) to obtain the size distribution of the collected aerosol. The dispersion performance of the inhaler was characterized by determining the fine particle fraction of the emitted aerosol cloud, defined as the mass fraction of particles smaller than 5 μm, referenced against either the total mass of powder loaded into (FPF$_{Loaded}$) or the total mass of powder emitted (FPF$_{Em}$) from the device.

The Effect of Design on the Performance of a Dry Powder Inhaler

The design of a dry powder inhaler is vital to control the size range of drug particles emitted from the device (76,77). All passive dry powder inhalers on the market today are designed with three common design features, namely a mouthpiece, air inlets, and a powder storage/dispensing system. Additional design features, such as grids and rotating balls, can also be added to the device to provide further particle breakup. However, there are currently few detailed studies in the open literature examining how inhaler design contributes to the overall inhaler performance. Recent studies performed by Coates et al. (66) have significantly enhanced this knowledge, leading to the following findings.

Effect of Grid Structure

A large number of dry powder inhalers are designed with an inhaler grid, primarily included to prevent the drug-filled capsule from exiting the device during inhalation. However, the inhaler grid is also included to generate high turbulence levels within the device and to enhance dispersion by inducing powder deagglomeration through powder impaction with the grid structure. Voss and Finlay (78) have shown that turbulence plays a definite, although not necessarily the dominant, role in fine particle dispersion and that mechanical impaction of the powder on a grid is not an effective breakup mechanism. The presence of the grid provides the opportunity for increased or reduced turbulence to affect the inhaler performance.

Coates and colleagues (66) found that the structure of the inhaler grid plays a significant role in the overall performance of the device, as well as on the amount of powder retained in the inhaler mouthpiece, without significantly affecting the inhaler dispersion performance. Examining the flow of particles experienced in the Aerolizer with three different grid structures (as illustrated in Figure 9.12), it was found that the "full" inhaler grid acts to convert the high-velocity tangential airflow entering the device into a low-velocity, predominantly axial airflow exiting the mouthpiece (Figure 9.13a). As the grid voidage was increased, the effect of the grid on reducing the level of swirl in the device reduced, with a predominantly tangential flow occurring in the mouthpiece for the "grid 2" case (Figure 9.13c). This increased the tangential flow of particles in the inhaler mouthpiece, which increased the degree of particle-mouthpiece contact and consequently the amount of powder retained in the mouthpiece. For the full grid case, 17% of the total amount of powder loaded in the capsule was retained in the mouthpiece after dispersion, which increased to 25 and 34% for the grid 1 and grid 2 cases, respectively (Table 9.1). This led to a reduction in the FPF_{Loaded}, reducing the overall performance of the inhaler (Table 9.1). In each case, the original Aerolizer grid performed better than the higher voidage modified grids.

Furthermore, the structure of the inhaler grid had a strong effect on the intensity of grid turbulence, with a reduction in the magnitude of the turbulence generated immediately upstream of the grid observed as the voidage of the grid was increased. In addition, the number of high-speed particle-grid impactions decreased with increasing grid voidage. Therefore, increasing the voidage of the grid reduced the deagglomeration potential of the flowfield generated in the device, due to lower grid

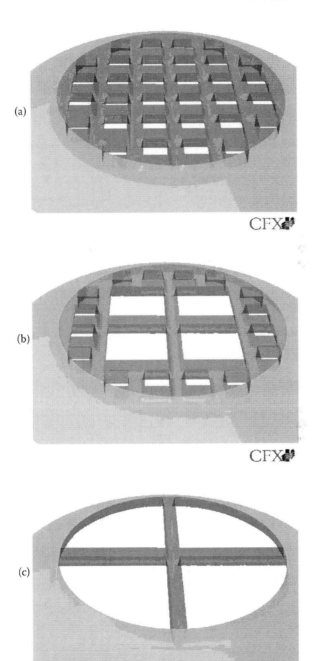

FIGURE 9.12 Schematic of the grids used to study the effect of grid structure on the inhaler performance. (a) Full case (b) Grid 1 case (c) Grid 2 case.

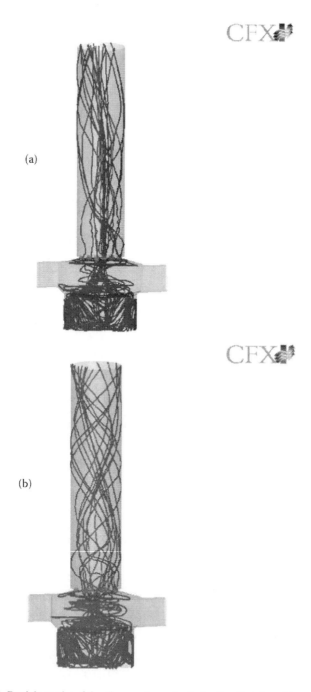

FIGURE 9.13 Particle tracks of the dispersed powder indicating that the grid structure affects the tangential component of the flowfield in the mouthpiece. (a) Full grid case (b) Grid 1 case (c) Grid 2 case.

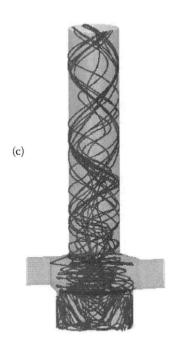

(c)

FIGURE 9.13 (Continued)

TABLE 9.1

Summary of the FPF$_{Loaded}$, FPF$_{Emitted}$, Mouthpiece Retention and Throat Produced from the Aerolizer for the Different Design Cases Studied

	FPF$_{Loaded}$ (%wt.)	FPF$_{Emitted}$ (%wt.)	Mouthpiece Retention (%wt.)	Throat Deposition (%wt.)
Grid Structure				
Full grid	47.6 (2.2)	62.4 (1.4)	17.2 (0.5)	4.9 (0.5)
Grid 1 case	39.2 (2.0)	59.0 (2.1)	24.9 (0.1)	7.1 (0.7)
Grid 2 case	35.2 (1.9)	61.3 (2.0)	33.9 (1.2)	5.8 (1.5)
Mouthpiece Length				
Full length	46.7 (2.2)	62.4 (1.4)	17.2 (2.0)	4.9 (0.5)
¾ length	45.4 (1.6)	61.7 (1.7)	20.8 (1.0)	6.7 (0.1)
½ length	46.3 (1.5)	61.2 (1.0)	17.7 (1.3)	6.2 (1.2)
Capsule Effects				
Powder outside capsule	35.5 (2.2)	46.8 (2.1)	21.7 (2.3)	13.9 (2.4)
No capsule present	40.4 (2.2)	51.9 (0.6)	21.0 (1.4)	12.6 (0.4)
Powder inside capsule	46.7 (2.2)	62.4 (1.9)	18.5 (1.9)	5.7 (1.3)

turbulence and fewer particle-grid impactions. However, as the voidage of the grid was increased, the number of drug particle impactions with the mouthpiece increased, increasing the deagglomeration potential of the device flowfield. The observation of no significant difference in the inhaler dispersion performance suggested that the increased deagglomeration potential (due to particle-mouthpiece impactions) was balanced out by the reduced deagglomeration potential (due to higher grid turbulence and particle-grid impactions). This gave rise to the similarity in inhaler dispersion performance observed experimentally (Table 9.1).

Effect of Mouthpiece Length

The length of an inhaler mouthpiece controls the level of flow development through the mouthpiece, which may have a significant effect on the nature of the flowfield exiting the device. Undeveloped flow can contain regions of high velocity that can enhance throat impaction upon inhalation. It is believed that more development of the flow through the mouthpiece leads to a more uniform flow profile at the device exit. This uniform profile reduces the regions of high velocity, potentially reducing throat deposition and improving the overall inhaler performance.

The length of the Aerolizer mouthpiece was found to play no significant role in the overall inhaler performance. No significant difference in the performance of the inhaler or amount of throat impaction was observed as the length of the mouthpiece was reduced to two-thirds and one-third of the original size (Table 9.1). However, a slight reduction in the amount of powder retained in the device after dispersion was observed as the mouthpiece length was reduced (Table 9.1), indicating that a shorter mouthpiece will slightly improve the overall performance of the inhaler by increasing the amount of powder emitted from the device.

Effect of Mouthpiece Shape

The effect of the mouthpiece shape was studied in order to determine its effect on drug delivery and device performance. Modified mouthpieces that had either circular or oval shapes with an increased cross-sectional area were constructed and studied both experimentally and computationally. Increasing the mouthpiece flow area was found to reduce the axial velocity and thus reduce throat deposition, a factor that could be important in certain applications. Despite the significant reduction in mouthpiece deposition, there was no difference in the FPF_{Loaded} values between the different designs, suggesting mouthpiece redesign would not affect device performance.

Role of the Rotating Capsule

A large number of the dry powder inhalers on the market use a capsule to store and dispense the drug formulation. Upon inhalation, the flowfield generated within these dry powder inhalers acts to rotate and/or vibrate the capsule to eject the powder contained in the capsule through the capsule holes into the surrounding flowfield. It is believed that agglomerates could break up through a number of capsule-induced deagglomeration mechanisms:

1. Powder agglomerates could impact with the internal walls of the capsule, prior to ejection, as it rotates or vibrates.
2. Forcing powder agglomerates through the small capsule holes could cause large agglomerates to break up, preventing slugs of powder from exiting the capsule.
3. High-speed impactions with the surrounding walls of the device could occur as the particles are ejected from the capsule.
4. The moving capsule could act as a rotor to deagglomerate ejected particles through mechanical impaction. Some of these mechanisms may dominate or interact to affect particle dispersion in the inhaler.

The presence, but not the size, of a capsule was found to have a significant effect on the inhaler performance. No significant difference in the overall inhaler performance was observed as the size of the capsule used to disperse the drug powder (sizes 3, 4, and 5) was varied. When powder was loaded directly into the device, onto the surface adjacent to the spinning capsule, the presence of the capsule was found to actually reduce the performance of the inhaler by reducing the overall turbulence levels generated in the device. Figure 9.14 shows the levels of turbulence experienced in the Aerolizer with and without the presence of the rotating capsule, demonstrating that the highly turbulent core generated in the Aerolizer without the capsule present is eliminated by the presence of the capsule. This reduced the overall levels of turbulence generated in the device by more than 65%, reducing the overall inhaler performance (from 41% without the capsule to 36% with the capsule present) (Table 9.1).

However, when powder was loaded inside the capsule, additional "initial capsule deagglomeration" was provided by the capsule (produced by forcing powder through the capsule holes) that could far outweigh the performance reduction due to reduced turbulence levels. The overall performance of the inhaler increased to 47% when powder was loaded into the capsule, demonstrating that the capsule provides additional strong mechanisms of powder deagglomeration (Table 9.1). The results further showed that capsule-particle impaction is a weak mechanism for deagglomeration. The potential role of the capsule acting as a rotor to cause deagglomeration by mechanical impaction has an insignificant effect on the overall performance of the inhaler.

Device Operating Conditions

The major disadvantage of passive dry powder inhalers is their performance dependence on the patient's inspiratory airflow. As all patients breathe differently, they generate different inhalation profiles, which can strongly affect inhaler dispersion performance and lead to large inter-patient dose variability. This potentially reduces the suitability of dry powder inhalers to diseases where precise dosing is essential. Hence, it is vital to understand how the operating conditions of dry powder inhalers affect the overall performance of the device, and subsequently the amount of lung deposition.

Influence of Airflow Rate

Inspiratory airflow through a dry powder inhaler controls both the turbulence levels generated in the device and the intensity of particle impactions, which are pivotal

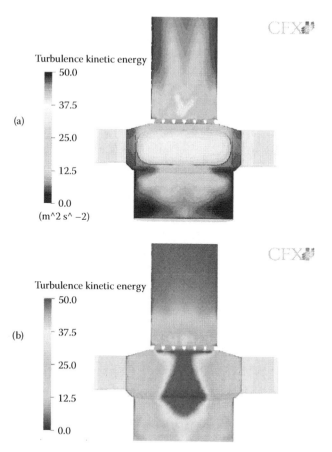

FIGURE 9.14 Nature of the flowfield generated in the Aerolizer® with and without the presence of the rotating capsule. (a) Capsule present (b) No capsule present.

to the inhaler dispersion performance. Numerous empirical studies have shown that inhaler performance is strongly affected by the inspiratory flow rate (77,79,80), each concluding that the performance of a dry powder inhaler generally increases with flow rate due to an increase in the deagglomerating forces generated in the device. However, Coates and others (67) hypothesized that critical flow turbulence levels and particle impaction velocities must exist at which the inhaler dispersion performance is maximized; that is, above these critical levels, dispersion would not improve further. Despite the lack of this important information, vital for optimal inhaler design, no such studies have attempted to determine the typical levels of turbulence and particle impaction velocities required to maximize the dispersion performance of dry powder inhaler systems. While it is difficult to obtain quantitative measures of these critical conditions, it is possible to determine qualitatively whether they exist.

Increasing the device airflow rate initially improved the dispersion performance of the Aerolizer, up to a critical level, where the inhaler dispersion performance no longer improved with increased flow. It was found that a critical flow rate of

65 1 min^{-1} was required to maximize the Aerolizer dispersion performance with the mannitol powder used. Further increases in the device flow rate did not lead to improved inhaler dispersion performance. This indicates that the deagglomerating forces required to maximize the dispersion performance of the Aerolizer/mannitol system had been generated in the device at 65 1 min^{-1} and the CFD models could then be used to quantify these values. However, in order to do this, it was required to understand which turbulent mechanisms are dominant in determining particle breakup.

The turbulence kinetic energy, a commonly used parameter to assess turbulence intensity, is a measure of the absolute turbulence level generated in the device, whereas the integral scale strain rate (ISSR: defined as the turbulence eddy dissipation rate divided by the turbulence kinetic energy) is a measure of the velocity gradient across the integral scale eddies (the most energetic occurring in a turbulent flow (81)) and is hence a more appropriate parameter to study agglomerate breakup. Studying the ISSR generated in the Aerolizer over a wide range of flow rates, it was found that an increasing trend in the ISSR was observed with flow rate. At the critical device flow rate of 65 1 min^{-1}, a volume-averaged ISSR of 5400 s^{-1} was generated in the inhaler. Further increases in the integral scale strain rate did not improve dispersion. Similarly, at the critical device flow rate, particle impaction with the grid and inhaler base occurred at average velocities of 19.0 and 12.7 m s^{-1}, respectively. The computational model developed was unable to determine which deagglomeration parameter (i.e., integral scale strain rates or particle impaction velocity) had the most significant effect on the inhaler dispersion performance.

This study was performed to quantify integral scale strain rates and particle-device impaction velocities that occurred at a flow rate where the Aerolizer dispersion performance was maximized. However, it is unclear whether these are the critical turbulence levels and particle impaction velocities required to maximize the performance of other dry powder inhalers; hence the validity of the findings needed to be confirmed. In order to test the findings, the study methodology was applied to the Aerolizer with different air inlet sizes (namely the full air inlet and designs consisting of two-thirds and one-third the original air inlet size) (Figure 9.15). Reducing the air inlet size increases the velocity of the entrained airflow, which increases the turbulence levels generated in the device. Therefore, at the same device flow rate, a different turbulence distribution is generated in the device for each air inlet size, which can be used to determine the effect that a varying turbulence distribution has on the inhaler dispersion performance.

The experimental results showed that the size of the inhaler air inlet had a varying effect on the inhaler performance at different flow rates. The inhaler air inlet size controls the levels of turbulence and particle impaction velocities generated in the device, as well as the flow development rate and device emptying times. At low flow rates (30 and 45 1 min^{-1}), reducing the air inlet size increased the inhaler dispersion performance by increasing the turbulence levels and particle impaction velocities generated in the device above the critical levels. These results demonstrated that the optimum dispersion performance of a dry powder inhaler can be predicted if details of the device flowfield are known. Reducing the size of the air inlet also reduced the

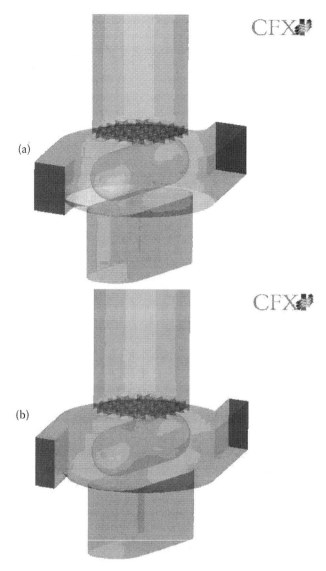

FIGURE 9.15 Schematic of the device designs used to study the effect of the air inlet size on inhaler performance. (a) Full air inlet case (b) 2/3 air inlet case (c) 1/3 air inlet case.

time taken for powder to empty the device, but increased the time required for flow to be fully developed within the device. At higher device flow rates (60 1 min^{-1}), reducing the size of the air inlet reduced the inhaler dispersion performance because a large amount of powder was released from the device when both the turbulence levels and particle impaction velocities were below their fully developed values. This result highlights the importance of minimizing the amount of powder released from the device prior to full flow development.

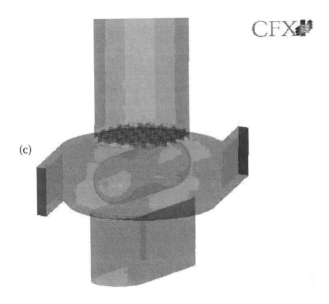

(c)

FIGURE 9.15 (Continued)

Inspiratory flow rate also controls the velocity of airflow exiting the device, which affects the amount of powder deposition in the throat. Increased throat deposition reduces the overall inhaler performance (referring to the ability of the device to disperse drug agglomerates, taking into account capsule, device, and throat retention). Consequently, an optimal flow rate will exist at which the overall inhaler performance is maximized. When examining the overall performance of the inhaler, powder retention in the capsule, device and throat need to be considered. The overall performance of the Aerolizer was also maximized at a flow rate of 65 l min^{-1}, as throat deposition and capsule retention were relatively low while the inhaler dispersion performance was optimized.

FUTURE FOCUS

As the aforementioned results have demonstrated, computational fluid dynamics is well-suited for solving the air flowfield generated in complex geometries. However, there are limited models available in existing CFD codes to examine agglomerate breakup. Discrete element method (DEM) modeling, which uses microscopic force balances to model the forces acting between individual particles in agglomerates, can be used to develop models of agglomerate breakup due to interaction with the turbulent flowfield and impaction with solid walls (82–84). These models can be coupled with CFD to examine powder deagglomeration on a particle scale. Several DEM codes are now commercially available that permit CFD coupling either on-line or as a postprocessing operation. Despite limitations in the number of particles that can be simulated, successful development of a CFD-DEM model would provide a powerful tool to study agglomerate dispersion in dry powder inhalers and could potentially eliminate the need for experimental aerosol characterization.

CONCLUSIONS

The work described in this review encompasses a range of characterization studies aimed at better describing the behavior of aerosol particles delivered to the respiratory system by dry powder inhalers. Characterization of the particle morphology in the form of surface fractal dimension and the ability to predict aerodynamic diameter from fractal dimension leads to a better understanding of the potential clinical aspects of therapeutic aerosol drug delivery. The CFD characterization of the performance of a typical dry powder inhaler complements the particle characterization to further our understanding of the dispersion process through the delivery device. Both aspects are required if we are to optimize the use of inhalation drug therapy for both current and future uses.

REFERENCES

1. Cipolla, D.C., Clark, A.R., Chan, H-K., Gonda, I., and Shire, S.J., Assessment of aerosol delivery systems for recombinant human desoxyribonuclease. *S.T.P. Pharma Sci*, 1994. **4**: p. 50–62.
2. Chan, H.-K., Clark, A., Gonda, I., Mumenthaler, M., and Hsu, C., Spray dried powder and powder blends of recombinant human deoxyribonuclease rhDNase for aerosol delivery. *Pharm Res*, 1997. **14**: p. 431–437.
3. Clark, A.R., Medical aerosol inhalers: Past, present and future. *Aerosol Sci Technol*, 1995. **22**: p. 374–391.
4. Edwards, D.A., Ben-Jebria, A., and Langer, R., Recent advances in pulmonary drug delivery using large porous inhaled particles. *J Appl Physiol*, 1998. **84**: p. 379–385.
5. Clay, M.M., Pavia, D., and Clarke, W., Effect of aerosol particle size on bronchodilation with nebulized turbutaline in asthmatic subjects. *Thorax*, 1986. **41**: p. 364–368.
6. de Boer, A.H., Gjaltema, D., Hagedoorn, P., and Frijlink, H.W., Characterization of inhalation aerosols: a critical evaluation of cascade impactor analysis and laser diffraction technique. *Int J Pharm*, 2002. **249**(1–2): p. 219–231.
7. Mitchell, J.P., and Nagel, M.W., Cascade impactors for the size characterization of aerosols from medical inhalers: their uses and limitations. *J Aerosol Med*, 2003. **16**(4): p. 341–377.
8. Baron, P.A., and Willeke, K., *Aerosol Measurement—Principles, Techniques, and Applications*. 2nd ed. 2001, New York: John Wiley & Sons.
9. Mitchell, J.P., Nagel, M.W., Wiersema,K.J., and Doyle, C.C., Aerodynamic particle size analysis of aerosols from pressurized metered-dose inhalers: comparison of Andersen 8-stage cascade impactor, next generation pharmaceutical impactor, and model 3321 aerodynamic particle sizer aerosol spectrometer. *AAPS Pharm Sci Technol*, 2003. **4**(4): p. 1–9.
10. Mandelbrot, B.B., *The Fractal Geometry of Nature*. 1983, New York: W.H. Freeman and Company.
11. Aqvist, J., and Tapia, O., Surface fractality as a guide for studying protein-protein interactions. *J Mol Graphics*, 1987. **5**(1): p. 30–34.
12. Farin, D., and Avnir, D., The fractal nature of molecule-surface chemical activities and physical interactions in porous materials, in *Characerisation of Porous Solids*, K.K. Unger, Rouquerol, J., Sing, K.S.W., Kral, H., Editor. 1988, Elsevier: Amsterdam. p. 421–432.
13. Lewis, M., and Rees, D.C., Fractal surface of proteins. *Science*, 1985. **230**: p. 1163–1165.

14. Baish, J.W., and Jain, R.K., Fractals and cancer. *Cancer Res,* 2000. **60**: p. 3683–3688.

15. Tatlier, M., and Kiwi-Minsher, L., Catalytic activity of FeZSM-5 zeolites in benzene hydroxylation by N2O: the role of geometry characterized by fractal dimensions. *Catal Commun,* 2005. **6**(11): p. 731–736.

16. Sant, S., Nadeau, V., and Hildgen, P., Effect of porosity on the release kinetics of propafenone-loaded PEG-g-PLA nanoparticles. *J Control Release,* 2005. **107**(2): p. 203–214.

17. Gomez-Serrano, V., Cuerda-Correa, E.M., Fernandez-Gonzalez, M.C., Alexandre-Franco, M.F., and Macias-Garcia, A., Preparation of activated carbons from walnut wood: a study of microporosity and fractal dimension. *Smart Mater Struct,* 2005. **14**(2): p. 363–368.

18. Ramadan, M.A., and Tawashi, R., Effects of surface geometry and morphic features on the flow characteristics of microspheres suspensions. *J Pharm Sci,* 1990. **79**(10): p. 929–932.

19. Farin, D., and Avnir, D., Reactive fractal surfaces. *J Phys Chem-US,* 1987. **91**(22): p. 5517–5521.

20. Farin, D., and Avnir, D., Use of fractal geometry to determine effects of surface morphology on drug dissolution. *J Pharm Sci,* 1992. **81**: p. 54–57.

21. Kaye, B.H., Specification of the ruggedness and/or texture of a fine particle profile by its fractal dimension. *Powder Technol,* 1978. **21**: p. 1–16.

22. Witten, T.A.J., and Sander, L.M., Diffusion-limited aggregation: a kinetic critical phenomenon. *Phys Rev Lett,* 1981. **47**: p. 1400–1403.

23. Tang, P., Chan, H-K., and Raper, J., Fractal geometry in pharmaceutical and biological applications-a review. *Encyclopedia of Pharmaceutical Technology,* 2004. DOI: 10.1081/E-EPT-120020310: p. 1–16.

24. Gorobei, N.N., Luk'yanenko, A.S., and Chmel, A.E., Self-similar evolution of the surface morphology of a stressed amorphous alloy foil. *J Exp Theor Phys,* 2005. **101**(3): p. 468–471.

25. Fu, T., Shen, Y.G., Zhou, Z.F., and Li, K.Y., Surface morphology of sputter deposited W-Si-N composite coatings characterized by atomic force microscopy. *Mat Sci Eng B-Solid,* 2005. **123**(2): p. 158–162.

26. Go, J.-Y., Pyun, S-I., and Cho, S-I., An experimental study on cell-impedance-controlled lithium transport through $Li_{1-\delta}CoO_2$ film electrode with fractal surface by analyses of potentiostatic current transient and linear sweep voltammogram. *Electrochim Acta,* 2005. **50**(27): p. 5435–5443.

27. Potapov, A.A., Bulavkin, V.V., German, V.A., and Vyacheslavova, O.F., Fractal signature methods for profiling of processed surfaces. *Tech Phys,* 2005. **50**(5): p. 560–575.

28. Suzuki, T., and Yano, T., Fractal surface structure of food materials recognized by different molecules. *Agric Biol Chem,* 1991. **55**(4): p. 967–971.

29. Pfeifer, P., and Avnir, D., Chemistry in noninteger dimensions between two and three. I. Fractal theory of heterogeneous surfaces. *J Chem Phys,* 1983. **79**(7): p. 3558–3565.

30. McClellan, A.L., and Harnsberger, H.F., Cross-sectional areas of molecules adsorbed on solid surfaces. *J Colloid Interf Sci,* 1967. **23**: p. 577–599.

31. Suzuki, T., and Yano, T., Fractal structure analysis of some food materials. *Agri Biol Chem,* 1990. **54**(12): p. 3131–3135.

32. Frenkel, J., *Kinetic Theory of Liquids.* 1946, Oxford: Clarendon.

33. Halsey, G.D., Physical adsorption on non-uniform surfaces. *J Chem Phys,* 1948. **16**: p. 931–937.

34. Hill, T.L., Theory of physical adsorption. *Adv Catal,* 1952. **4**: p. 211–258.

35. Pfeifer, P., Wu, Y.J., Cole, M., and Krim, J., Multilayer adsorption on a fractally rough surface. *Phys Rev Lett*, 1989. **62**: p. 1997.

36. Wu, M.K., The roughness of aerosol particles: surface fractal dimension measured using nitrogen adsorption. *Aerosol Sci Tech*, 1996. **25**: p. 392–398.

37. Ismail, I.M.K., and Pfeifer, P., Fractal analysis and surface roughness of nonporous carbon fibres and carbon backs. *Langmuir*, 1994. **10**(5): p. 1532–1538.

38. Tang, P., Chew, N.Y.K., Chan, H-K., and Raper, J., Limitation of determination of surface fractal dimension using N2 adsorption isotherms and modified Frenkel-Halsey-Hill theory. *Langmuir*, 2003. **7**: p. 2632–2638.

39. Chew, N.Y.K., Tang, P., Chan, H-K., and Raper, J.A., How much particle surface corrugation is sufficient to improve aerosol performance of powders? *Pharm Res*, 2005. **22**(1): p. 148–152.

40. Watt-Smith, M.J., Edler, K.J., and Rigby, S.P., An experimental study on gas adsorption on fractal surfaces. *Langmuir*, 2005. **21**: p. 2281–2292.

41. Gregg, S.J., and Sing, K.S.W., *Adsorption, Surface Area and Porosity*. 2nd ed. 1982, London: Academic Press.

42. Amal, R., Fractal structure and kinetics of aggregating colloidal hematite, in *Chemical Engineering*. 1991, New South Wales University: Sydney.

43. Chen, Z., Sheng, P., Weitz, D.A., Lindsay, H.M., Lin, M.Y., and Meakin, P., Optical properties of aggregate clusters. *Phys Rev B*, 1988. **37**(10): p. 5232–5235.

44. Schaefer, D.W., and Hurd, A.J., Growth and structure of combustion aerosols. *Aerosol Sci Tech*, 1990. **12**: p. 876–891.

45. Chew, N.Y.K., and Chan, H.K., Use of solid corrugated particles to enhance powder aerosol performance. *Pharm Res*, 2001. **18**(11): p. 1570–1577.

46. Tang, P., Chan, H-K., and Raper, J.A., Prediction of aerodynamic diameter of particles with rough surfaces. *Powder Technol*, 2004. **147**: p. 64–78.

47. Mitchell, J.P., and Nagel, M.W., Time-of-flight aerodynamic particle size analysers: their use and limitations for the evaluation of medical aerosols. *J Aerosol Med*, 1999. **12**(4): p. 217–240.

48. Chhabra, R.P., Agarwal,L., and Sinha, N.K., Drag on non-spherical particles: an evaluation of available methods. *Powder Tech*, 1999. **101**: p. 288–295.

49. Ganser, G.H., A rational approach to drag prediction of spherical and non-spherical particles. *Powder Technol*, 1993. **77**: p. 143.

50. Haider, A., and Levenspiel, O., Drag coefficient and terminal velocity of spherical and non-spherical particles. *Powder Technol*, 1989. **58**: p. 63–70.

51. Ro, K.S., and Neethling, J.B., Terminal settling characteristics of bioparticles. *Res J WPCF*, 1990. **62**: p. 901–906.

52. Tran-Cong, S., Gay, M., and Michaelides, E., Drag coefficients of irregularly shaped particles. *Powder Technol*, 2004. **139**: p. 21–32.

53. Thompson, T.L., and Clark, N.N., A holistic approach to particle drag prediction. *Powder Technol*, 1991. **67**(1): p. 57–66.

54. Dunbar, C.A., Hickey, A.J., and Holzner, P., Dispersion and characterization of pharmaceutical dry powder aerosols. *KONA*, 1998. **16**: p. 7–44.

55. Byron, P.R., Drug delivery devices: issues in drug development. *Proceedings of the American Thoracic Society*, 2004. **1**: p. 321–328.

56. Schuler, C., Bakshi, A., Tuttle, D., Smith, A., Paboojian, S., Snyder, H., Rasmussen, D., and Clark A. Inhale's dry-powder pulmonary drug delivery system: challenges to current modeling of gas-solid flows. In *FEDSM99: 3rd ASME/JSME Joint Fluids Engineering conference & 1999 ASME Fluids Engineering summer meeting*. 1999.

57. Micale, G., Montante, G., Grisafi, F., Brucato, A., and Godfrey, J., CFD simulation of particle distribution in stirred vessels. *Chem Eng Res Des*, 2000. **78**(3): p. 435–444.

58. Aubin, J., Fletcher, D.F., and Xuereb, C., Modelling turbulent flow in stirred tanks with CFD: the influence of the modelling approach, turbulence model and numerical scheme. *Exper Therm Fluid Sci*, 2004. **28**: p. 431–445.

59. Oakley, D.E., Scale-up of spray dryers with the aid of computational fluid dynamics. *Drying Tech*, 1994. **12**: p. 217–233.

60. Harvie, D.J.E., Langrish, T.A.G., and Fletcher, D.F., A computational fluid dynamics study of a tall-form spray dryer. *Trans I Chem E C*, 2002. **80**(3): p. 163–175.

61. Versteeg, H.K., Hargrave, G., Harrington, L., Shrubb, I., and Hodson, D. The use of computational fluid dynamics (CFD) to predict pMDI air flow and aerosol plume formation. In *Respiratory Drug Delivery VII, vol. I*. 2002. Raleigh, North Carolina: Serentec Press Inc.

62. Prabhakarpandian, B., Sundaram, S., and Makhijani, V.B. Evaluation of dose delivery using MDI-spacer combinations. In *Respiratory Drug Delivery VIII, vol. I*. 2002. Tuscon, Arizona: Serentec Press Inc.

63. Matida, E.A., Rimkus, M., Grgic, B., Lange, C.F., and Finlay, W.H., New add-on spacer design concept for dry-powder inhalers. J Aerosol Sci, 2004. **35**: p. 823–833.

64. Ligotke, M.W. Development and characterization of a dry powder inhaler. In *Respiratory Drug Delivery VIII, vol. 1*. 2002. Tuscon, Arizona: Serentec Press Inc.

65. *Computer-Optimised Pulmonary Delivery in Humans of Inhaled Therapies (COPHIT)*. www-mitton.ansys.com

66. Coates, M.S., Chan, H-K, Fletcher, D.F., and Raper, J.A., Effect of design on the performance of a dry powder inhaler using computational fluid dynamics. Part 1: Grid structure and mouthpiece length. *J Pharm Sci*, 2004. **93**(11): p. 2863–2876.

67. Coates, M.S., Chan, H-K, Fletcher, D.F., and Raper, J.A., The influence of air flow on the performance of a dry powder inhaler using computational and experimental analyses. *Pharm Res*, 2005. **22**(9): p. 1445–1453.

68. Coates, M.S., Chan, H-K, Fletcher, D.F., and Raper, J.A., The role of capsule on the performance of a dry powder inhaler using computational and experimental analyses. *Pharm Res*, 2005. **22**(6): p. 923–932.

69. Coates, M.S., Chan, H-K, Fletcher, D.F., and Raper, J.A., Effect of design on the performance of a dry powder inhaler using computational fluid dynamics. Part 2: Air inlet size. *J Pharm Sci*, 2006. **95**(6): p. 1382–1392.

70. Coates, M.S., Chan, H-K, Fletcher, D.F., and Chiou, H., Influence of mouthpiece geometry on the aerosol delivery performance of a dry powder inhaler. *Pharm Res*, 2007. **24**(8): p. 1450–1456.

71. Menter, F.R., Two-equation eddy-viscosity models for engineering applications. *AIAA J*, 1994. **32**(8): p. 269–289.

72. Ferziger, J.H., and Perić, M., *Computational Methods for Fluid Dynamics*. 1996, Berlin-Heidelberg: Springer-Verlag.

73. Gosman, A.D., and Ioannides, E., Aspects of computer simulation of liquid fuelled combustors. *AIAA J*, 1981. **7**(6): p. 482–490.

74. British Pharmacopeia, *Appendix XII, Aerodynamic assessment of fine particles-fine particle dose and particle size distribution, Apparatus C*. 2001.

75. Clift, R., Grace, J.R., and Weber, M.E., *Bubbles, Drops and Particles*. 1978, New York: Academic Press Inc., Ltd.

76. Steckel, H., and Müller, B.W., In vitro evaluation of dry powder inhalers I: drug deposition of commonly used devices. *Int J Pharm*, 1997. **154**: p. 19–29.

77. Chew, N.Y.K., and Chan, H.K., Influence of Particle Size, Air Flow, and Inhaler Device on the Dispersion of Mannitol Powders as Aerosols. *Pharm Res*, 1999. **16**(7): p. 1098–1103.

78. Voss, A., and Finlay, W.H., Deagglomeration of dry powder pharmaceutical aerosols. *Int J Pharm*, 2002. **248**: p. 39–50.

79.   Chew, N.Y.K., and Chan, H-K, *In vitro* aerosol performance and dose uniformity between the Foradile Aeroliser and the Oxis Turbuhaler. *J Aerosol Med,* 2001. **14**: p. 495–501.

80.   de Boer, A.H., Winter, H.M.I., and Lerk, C.F., Inhalation characterisitics and their effects on *in-vitro* drug delivery from dry powder inhalers. Part 1. Inhalation characteristics, work of breathing and volunteers' preference in dependence of the inhaler resistance. *Int J Pharm,* 1996. **130**: p. 231–244.

81.   Finlay, W.H., *The Mechanics of Inhaled Pharmaceutical Aerosols.* 2001, London: Academic Press Inc., Ltd.

82.   Yang, R.Y., Zou, R.P., and Yu, A.B., Computer simulation of the packing of fine particles. *Phys Rev E,* 2000. **62**(3): p. 3900–3908.

83.   Subero, J., Ning, Z., Ghadiri, M., and Thornton, C., Effect of interface energy on the impact strength of agglomerates. *Powder Technol,* 1999. **105**: p. 66–73.

84.   Moreno, R., Ghadiri, M., and Antony, S.J., Effect of the impact angle on the breakage of agglomerates: a numerical study using DEM. *Powder Technol,* 2003. **130**: p. 132–137.

10 Imaging of Particle Size and Concentration in Heterogeneous Scattering Media Using Multispectral Diffuse Optical Tomography

C. Li and H. Jiang

CONTENTS

INTRODUCTION

When incident light propagates through a particle, it is scattered or absorbed. In optical spectroscopy studies, it has been shown that scattering spectra are correlated with tissue morphology (cellular size and concentration) (Beauvoit et al., 1993; Cerussi, 2002; Mourant et al., 1998; Perelman et al., 1998). These experimental studies have suggested that both nuclei and mitochondria contribute to tissue scattering significantly. On the other hand, it is well known in pathology that tumor cells/nuclei are considerably enlarged relative to normal ones (Cotran, Robbins, and Kumar, 1994). Thus significant clinical value would result from the exploration of scattering spectra. The purpose of this chapter is to review optical spectroscopy and imaging techniques for measuring particle size and concentration. The review is focused on an emerging tomographic technique called diffuse optical tomography (DOT), while optical spectroscopic techniques are described.

Several assumptions are generally made in these optical techniques for particle sizing: particles are spherical; no particle-particle interactions exist; and photon diffusion equation is valid, which means that light scattering is much larger than the absorption.

This chapter is organized as follows: various optical spectroscopic techniques for particle sizing are first reviewed; multispectral diffuse optical tomography for particle sizing is then detailed; discussion and conclusions are finally made.

PARTICLE SIZE MEASUREMENTS WITH OPTICAL SPECTROSCOPIC TECHNIQUES

When light propagates through mammalian tissues, the scatterers are cell organelles of various sizes, such as nucleus and mitochondria, which have higher refractive index than the surrounding cytoplasm (Brunsting and Mullaney, 1974). Perelman and colleagues have shown that light singly backscattered from an epithelial layer of tissue such as the skin has a wavelength-dependent periodic pattern (Perelman et al., 1998). They found that the periodicity of the pattern increased with nuclear size and the amplitude of the periodic signal was related to the density of nucleus. After analyzing the periodic pattern, the nucleus size and density could be extracted. However, the periodic pattern was overwhelmed by the diffuse background or by the multiple scattered light since the backscattered light is only a small portion of the scattered light. One way to overcome the problem is to use a model to mathematically describe the single backscattered light. But this method has to be remodeled for different tissues under investigation, which is inconvenient (Perelman et al., 1998). Another robust approach, also proposed by Backman et al., is to use polarized light to differentiate the single scattered light from multiple scattered light background (Backman et al., 1999). It was reported that the initially polarized light lost its polarization after scattered propagation in turbid media such as biological tissues (Anderson, 1991; Demos and Alfano, 1996; Yoo and Alfano, 1989). In contrast, the single backscattered light kept its polarization (Demos, Alfano, and Dmons, 1997). After subtracting the unpolarized light, the polarized component of the backscattered light from the epithelial layer of the tissue was obtained. An alternative approach to differentiate the backscattering from background is to utilize a probe geometry that optimizes the detection of single scattered light (Amelink et al., 2003). It was demonstrated that the single optical fiber approach was highly sensitive to the light backscattered from layered superficial tissues.

Cells have complicated structures and the organelles inside a cell vary in size from 0.1 to 10 microns. All the organelles contribute to the scattering, which complicates the cell-scattering phenomena. With fiber-optic, polarized elastic-scattering spectroscopy techniques, Mourant and colleagues estimated that the average scatterer radius in tissue was from 0.5 to 1.0 μm, which is much smaller than the nucleus (Mourant, Johnson, and Freyer, 2001). Using polarized light spectroscopy, the particle size distribution in mammal cells was measured and the results suggested that small particles (possibly the mitochondria) contribute most to the scattering. However, other subcellular structures, such as the nucleoli and the nucleus, may also contribute significantly (Bartlett et al., 2004). Gurjar and others demonstrated that the spectrum of the single backscattering component was capable of providing the cell nuclei size distribution, which means that single backscattering was dominated by the nucleus (Gurjar et al., 2001).

PARTICLE SIZE MEASUREMENTS WITH MULTISPECTRAL DIFFUSE OPTICAL TOMOGRAPHY

Currently, two noninvasive optical imaging methods are used to measure particle size distribution. One is light scattering spectroscopy (LSS) (Gurjar et al., 2001), in which polarized light was delivered to the epithelial tissue and the single backscattering light that kept the polarization was analyzed to extract the morphological features and the refractive index of the scatterers. This method is limited primarily to superficial surface imaging. The other one is a tomographic imaging method based on multispectral diffuse optical tomography (Li and Jiang, 2004b), in which the tomographic scattering images of tissue at multiple wavelengths were obtained with diffuse optical tomography and the scattering spectra was then used to extract the scatterer's size information with a Mie theory-based reconstruction method. This tomographic method is reviewed in detail in the following sections.

OPTICAL DIFFUSION THEORY

The Boltzmann transport equation describes incoherent photon propagation through highly scattering media such as tissue. The equation in time domain is written as

$$\left(\frac{1}{c}\frac{\partial}{\partial t}+\hat{s}\bullet\nabla+\mu_{tr}(\hat{r})\right)\phi(\hat{r},\hat{s},t)=\mu_{s}(r)\int_{4\pi}\Theta(\hat{s},\hat{s}')\,\phi(\hat{r},\hat{s}',t)d\hat{s}+q(\hat{r},\hat{s},t) \quad (10.1)$$

where $\phi(r,\hat{s}',t)$ is the radiance (W/(cm^2 sr)) at position $\hat{r}$, at time t, propagating along the unit vector $\hat{s}$. $\mu_{tr}(\hat{r})=\mu_{a}(\hat{r})+\mu_{s}(\hat{r})$ is the transport cross section at position $\hat{r}$. $\mu_{a}(\hat{r})$ and $\mu_{s}(\hat{r})$, the absorption and scattering coefficients, are the inverse of the absorption and scattering mean free path, respectively. c is the speed of light in the medium. The function $\Theta(\hat{s},\hat{s}')$ is the probability density function over all solid angles of the change in photon propagation direction from $\hat{s}$ to $\hat{s}'$ due to an elastic scattering event, which satisfies the condition: $\int_{4\pi}\Theta(\hat{s},\hat{s}')d\hat{s}=1$. $q(\hat{r},\hat{s},t)$ is the photon power generated at position $\hat{r}$ along direction $\hat{s}$. This equation reflects the energy conservation in the medium. However, the equation often must be simplified to be mathematically manageable. One way to simplify it is to expand the equation with spherical harmonics and truncate the series at the Nth term, namely, P_N approximation. Thus, the quantities in equation (10.1) can be expressed as

$$\phi(\hat{r},\hat{s},t) = \sum_{L}^{\infty} \sum_{m=-L}^{L} \left(\frac{2L+1}{4\pi} \right)^{\frac{1}{2}} \psi_{L,m}(\hat{r},t) Y_{L,m}(\hat{s}) \tag{10.2}$$

$$q(\hat{r},\hat{s},t) = \sum_{L}^{\infty} \sum_{m=-L}^{L} \left(\frac{2L+1}{4\pi} \right)^{\frac{1}{2}} q_{L,m}(\hat{r},t) Y_{L,m}(\hat{s}) \tag{10.3}$$

$$\Theta(\hat{s},\hat{s}') = \sum_{L}^{\infty} \left(\frac{2L+1}{4\pi} \right)^{\frac{1}{2}} \Theta_{L} P_{L}(\cos\theta) \tag{10.4}$$

where

$$\left(\frac{2L+1}{4\pi} \right)^{\frac{1}{2}} \tag{10.5}$$

is the normalization factor, $Y_{L,m}$ the spherical harmonic of order L at degree m, and P_L the Legendre polynomial of order L.

The P_1 approximation is obtained when $N = 1$ from the P_N approximation. After simplification, the following equations are obtained (Arridge, inverse problem, 1999):

$$\left(\frac{1}{c} \frac{\partial}{\partial t} + \mu_{tr}(\hat{r}) \right) \Phi(\hat{r},t) + \nabla \cdot \hat{J}(\hat{r},t) = \mu_s(\hat{r})\Phi(\hat{r},t) + \Theta_0 q_{0,0}(\hat{r},t) \tag{10.6}$$

$$\left(\frac{1}{c} \frac{\partial}{\partial t} + \mu_{tr}(\hat{r}) \right) \hat{J}(\hat{r},t) + \frac{1}{3}\nabla\Phi(\hat{r},t) = \Theta_1 \mu_s(\hat{r})\hat{J}(\hat{r},t) + \hat{q}_1(\hat{r},t) \tag{10.7}$$

where $\Phi(\hat{r},t) = \psi_{0,0}(\hat{r},t)$ is the photon fluence, and

$$\hat{J}(\hat{r},t) = \left[\frac{1}{\sqrt{2}}\left(\psi_{1,-1}(\hat{r},t) - \psi_{1,1}(\hat{r},t) \right), \frac{1}{i\sqrt{2}}\left(\psi_{1,-1}(\hat{r},t) + \psi_{1,1}(\hat{r},t) \right), \psi_{1,0}(\hat{r},t) \right]^{T} \tag{10.8}$$

is the photon flux.

The P_1 approximation can be further simplified by making the following assumptions:

$$\frac{\partial \hat{J}}{\partial t} = 0, \ \hat{q}_1 = 0 \tag{10.9}$$

The approximation, namely diffuse approximation, is usually justified only if the scattering coefficient is much larger than the absorption coefficient in order to satisfy the first assumption,

$$\frac{\partial \hat{J}}{\partial t} = 0.$$

(10.10)

The second assumption, $\hat{q}_1 = 0$, means that the photon source is isotropic. The diffuse approximation leads to the following equation, namely the diffusion equation, in the time domain:

$$-\nabla \cdot D(\hat{r})\nabla\Phi(\hat{r},t) + \mu_a \Phi(\hat{r},t) + \frac{1}{c}\frac{\partial\Phi(\hat{r},t)}{\partial t} = q_0(\hat{r},t),$$

(10.11)

or in the frequency domain:

$$-\nabla \cdot D(\hat{r})\nabla\Phi(\hat{r},\omega) + \mu_a \Phi(\hat{r},\omega) + \frac{i\omega}{c}\Phi(\hat{r},\omega) = q_0(\hat{r},\omega),$$

(10.12)

or in the continue wave domain:

$$-\nabla \cdot D(\hat{r})\nabla\Phi(\hat{r}) + \mu_a \Phi(\hat{r}) = q_0(\hat{r})$$

(10.13)

where

$$D(\hat{r}) = \frac{1}{3(\mu_a + \mu'_s)}$$

(10.14)

is the diffusion coefficient, $\mu'_s = (1 - \Theta_1)\mu_s$ the reduced scattering coefficient, and $q_0(\hat{r}) = q_{0,0}(\hat{r})$ the isotropic source.

In infinite homogeneous media, the solution to the diffusion equation can be obtained through the Green function method (Patterson, Chance, and Wilson, 1989). However, for realistic finite homo- or heterogeneous media such as tissue, the boundary effects/conditions must be accounted for.

Diffuse Optical Tomography

Reconstruction Algorithm

There are two procedures involved in DOT image reconstruction. The first one is the forward solution procedure, in which the distribution of light in the medium of interest is predicted. The second one, the inverse solution procedure, is used to iteratively update the initially guessed optical property distribution of the medium through an optimization method such as Newton method by minimizing the squared difference between the data computed from the diffusion equation and measured around the surface of the medium.

Forward Solution Procedure

Since an analytical solution to the Boltzmann transport or the diffusion equation is not available in a realistic situation (Boas et al., 1994), numerical methods must be used for most cases. The finite element method (FEM) is a natural choice because it can be used to solve the diffusion equation in inhomogeneous media with an arbitrary geometry, although other numerical methods such as the finite difference method (Hielscher et al., 2004), finite volume method (Ren et al., 2004), and boundary element method (Ripoll and Ntziachristoz, 2003) have been used in DOT. The FEM applications in DOT have been discussed in detail elsewhere (Arridge, 1993; Paulsen and Jiang, 1995). Here we follow Paulsen and Jiang (1995) and briefly describe the FEM method in the CW DOT.

Using the finite element discretization, the steady-state photon diffuse equation coupled with the type III boundary conditions, $-D\nabla\Phi \bullet \hat{n} = \alpha\Phi$, can be transformed into the following matrix form

$$[A]\{\Phi\} = \{b\}, \tag{10.15}$$

where α is the BC coefficient related to the internal reflection at the boundary; the elements of matrix $[A]$ are $a_{ij} = \langle -D\nabla\phi_j \bullet \nabla\phi_i - \mu_a\phi_j\phi_i \rangle$, where $\langle \rangle$ indicates integration over the problem domain; $\{b\}$ is

$$-\langle S\phi_i \rangle + \alpha \sum_{j=1}^{M} \Phi_j \int \phi_i\phi_j ds; \tag{10.16}$$

$S = S_0\delta(r - r_0)$ where S_0 is the source strength and $\delta(r - r_0)$ is the Dirac delta function for a source at r_0; M is the number of boundary nodes; and ϕ_i and ϕ_j are locally spatially varying Lagrangian basis functions at nodes i and j, respectively. The vector $\{\Phi\} = [\Phi_1, \Phi_2, ..., \Phi_N]$ is the photon density.

Four critical parameters (the BC coefficient α, the source strength S_0, and the initial guess of D and μ_a) should be accurately determined by a preprocessing data optimization scheme for the forward computation. We assume all the source intensities are the same and equal to S_0. The pre-processing data optimization scheme for determining the four initial parameters is discussed in detail elsewhere (Iftimia and Jiang, 2000). Briefly, X-square errors,

$$X^2 = \sum_{i=1}^{M1} \left[\Phi_i^{(m)} - \tilde{\Phi}_i^{(c)} \right]^2, \tag{10.17}$$

are minimized as the function of the above four parameters, where M1 is the number of boundary measurements (e.g., $M1 = 64 \times 64$), $\Phi_i^{(m)}$ is the measured photon density from a given experimental heterogeneous medium, and $\tilde{\Phi}_i^{(c)}$ is the computed photon density from a numerical simulation of a homogeneous medium with the same geometry as the experimental medium. Given reasonable ranges for the four parameters, the minimum X^2 corresponds to the best initial guess of the four parameters.

Inverse Solution Procedure

An example of the inverse solution procedure is given here based on Taylor expansion or Newton method. We assume that the computed and/or the measured Φ are analytic functions of D and μ_a, and that D and μ_a are independent parameters. Φ then can be Taylor expanded about an assumed (D, μ_a) distribution, which is a perturbation away from some other distribution, $(\tilde{D}, \tilde{\mu}_a)$, and the expansion is expressed as

$$\Phi(\tilde{D}, \tilde{\mu}_a) = \Phi(D, \mu_a) + \frac{\partial \Phi}{\partial D} \Delta D + \frac{\partial \Phi}{\partial \mu_a} \Delta \mu_a + \cdots \tag{10.18}$$

where $\Delta D = \tilde{D} - D$ and $\Delta \mu_a = \tilde{\mu}_a - \mu_a$. If the assumed optical property distribution is close to the true one, the high-order items in the expansion can be neglected and we obtain

$$J \Delta \chi = \Psi^o - \Psi^c \tag{10.19}$$

where

$$J = \begin{bmatrix} \dfrac{\partial \Psi_1}{\partial D_1} & \dfrac{\partial \Psi_1}{\partial D_2} & \cdots & \dfrac{\partial \Psi_1}{\partial D_K} & \dfrac{\partial \Psi_1}{\partial \mu_{a1}} & \dfrac{\partial \Psi_1}{\partial \mu_{a2}} & \cdots & \dfrac{\partial \Psi_1}{\partial \mu_{aL}} \\[2ex] \dfrac{\partial \Psi_2}{\partial D_1} & \dfrac{\partial \Psi_2}{\partial D_2} & \cdots & \dfrac{\partial \Psi_2}{\partial D_K} & \dfrac{\partial \Psi_2}{\partial \mu_{a1}} & \dfrac{\partial \Psi_2}{\partial \mu_{a2}} & \cdots & \dfrac{\partial \Psi_2}{\partial \mu_{aL}} \\[2ex] \cdots & \cdots & \ddots & \cdots & \cdots & \cdots & \ddots & \cdots \\[2ex] \dfrac{\partial \Psi_M}{\partial D_1} & \dfrac{\partial \Psi_M}{\partial D_2} & \cdots & \dfrac{\partial \Psi_M}{\partial D_K} & \dfrac{\partial \Psi_M}{\partial \mu_{a1}} & \dfrac{\partial \Psi_M}{\partial \mu_{a2}} & \cdots & \dfrac{\partial \Psi_M}{\partial \mu_{aL}} \end{bmatrix} \tag{10.20}$$

$$\Delta \chi = [\Delta D_1 \quad \Delta D_2 \quad \cdots \quad \Delta D_K \quad \Delta \mu_{a1} \quad \Delta \mu_{a2} \quad \cdots \quad \Delta \mu_{aL}]^T \tag{10.21}$$

$$\Psi^o = [\Psi_1^o \quad \Psi_2^o \quad \cdots \quad \Psi_M^o]^T \tag{10.22}$$

$$\Psi^c = [\Psi_1^c \quad \Psi_2^c \quad \cdots \quad \Psi_M^c]^T \tag{10.23}$$

and Ψ_i^o and Ψ_i^c are observed and calculated data for $i = 1, 2, \ldots M$ measurements, D_k for $k = 1, 2, \ldots, K$, and μ_{al} for $l = 1, 2, \ldots, L$ are the reconstruction optical parameters. In order for equation (10.19) to be invertible, a regularization method is used and expressed as

$$(J^T J + \lambda I) \Delta \chi = J^T (\Psi^o - \Psi^c) \tag{10.24}$$

where I is the identity matrix and λ is the regularization parameter.

EXPERIMENTAL TECHNIQUES

There are three types of experimental systems in DOT, including time domain (Benaron et al., 2000; Schmidt et al., 2000), frequency domain (Jiang et al., 1996) and CW domain (Colak et al., 1999; Grable, Rohler, and Sastry, 2004; Li and Jiang, 2004b). In a time domain system, photon fly times are recorded and the times are less than several nanoseconds. The time domain measurements require a short-duration pulse laser generator and very complicated and expensive photon detection instruments. In a frequency domain system, an amplitude-modulated laser with a modulation frequency of several hundred MHz provides the source, and both the amplitude and phase shift are measured. Because a frequency domain system measures only the signals at one frequency, the cost of the system is much lower than a time domain system. Compared with time and frequency domain systems, a CW domain system just needs a CW laser source and only the amplitude is measured. The implementation of a CW system is the simplest and the cost is lowest among the three types of systems.

Here we detail a photodiode-based 10-wavelength CW DOT system developed in our lab. The imaging system is schematically shown in Figure 10.1. Laser beams from ten laser modules are transmitted to the optical switch, which sequentially passes one of the beams to M1 (M1 = 64 for 3-D imaging or M1 = 16 for 2-D imaging) preselected points at the surface of the phantom or the breast via source fiber bundles. The ring structure or fiber optic/tissue interface holds the source and detection fiber bundles. Light from the detection fiber bundles is sensed by the detection

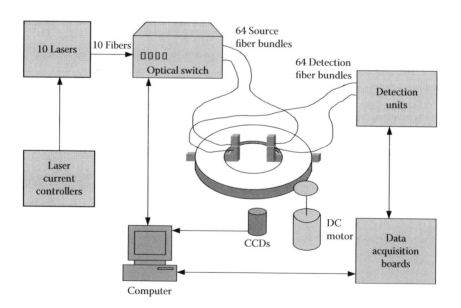

FIGURE 10.1 Schematic of the experimental system. (Reprinted from Li, C. and Jiang, H., *Journal of Optics A: Pure and Applied* Optics 6: 844–852, 2004a. With permission.)

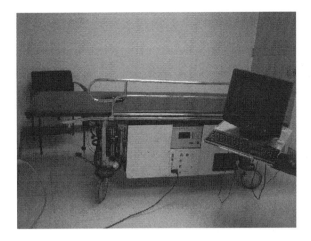

FIGURE 10.2 Photographs of the experimental system.

units, which convert the light intensity into voltage signals. The computer collects the signals through a data acquisition board. The DC motor near the ring is used to adjust the diameter of the ring. Two CCD cameras are mounted underneath the ring to monitor the contact between the tissue and fiber optics. Figure 10.2 shows the photograph of the entire imaging system where we can see that all the optical and electronic components are housed under the exam table.

A calibration procedure was developed in order to reduce systematic errors. For a 2-D imaging experiment, the calibration procedure is described by the following six steps:

1. Make a homogeneous phantom that has the same diameter as the heterogeneous phantom interested or the breast tissue examined.
2. Perform experiments with the homogeneous phantom. For 2-D imaging experiments, there are 16 illumination positions and 16 detection positions. Obtain a set of measured data D_{ij}, where i is the source position number from 1 to 16 and j is the detector number from 1 to 16. For each source i, light intensities from the 16 detectors are normalized.
3. Find the initial values of the absorption coefficient (μ_a), reduced scattering coefficient (μ_s'), and boundary conditions coefficient α using D_{ij}. These initial parameters are required by our nonlinear iterative algorithm.
4. Generate a 2-D finite element mesh with the same diameter as the phantom. Using a unit source intensity for the 16 illuminated positions, the 2-D photon propagation is simulated with the optical properties μ_a, μ_s' and the boundary conditions coefficient α identified in Step 3. This creates a new set of data D_{ij}^* from the simulation.
5. Obtain a factor matrix f_{ij} using the following equation:

$$f_{ij} = D_{ij}^* / D_{ij} \qquad i, j = 1 \ldots 16 \qquad (10.25)$$

6. Multiply f_{ij} by the data set (E_{ij}) from the heterogeneous phantom or the tissue to get the final data set for image reconstruction:

$$E_{ij}^* = f_{ij}E_{ij} \qquad\qquad i,j = 1...16 \qquad\qquad (10.26)$$

The above-described calibration method can be easily extended to 3-D imaging cases. We have found that the effectiveness of this calibration method depends on the size and optical properties of the reference phantom. A detailed study of this calibration method was reported elsewhere (Li and Jiang, 2004a).

INVERSE ALGORITHM FOR PARTICLE SIZING

Once $D(\lambda)$ is recovered using the DOT algorithm, the reduced scattering spectra can be obtained by the following relationship for turbid or highly scattering media:

$$(1-g)\mu_s(\lambda) \approx \frac{1}{3D(\lambda)} \qquad\qquad (10.27)$$

where $(1-g)\mu_s$ is the reduced scattering coefficient and g is the average cosine of scattering angles. Following Jiang et al. (Jiang, 1997; Jiang, 1998), the scattering spectra are correlated with particle size distribution and concentration through the following relationship under Mie Theory:

$$(1-g)\mu_s(\lambda) = \mu_s'(\lambda) = \int_0^\infty \frac{3Q_{scat}(x,n,\lambda)[1-g(x,n,\lambda)]}{2x}\phi f(x)dx \qquad (10.28)$$

where Q_{scat} is the scattering efficiency, x is the particle size, n is the refractive index of particles, ϕ is the particle concentration/volume fraction, and $f(x)$ is the particle size distribution. Both Q_{scat} and g can be computed with Mie Theory. In equation (10.28), we have assumed that particles act as independent scatterers without particle-particle interaction. In order to solve for $f(x)$ and ϕ from measured scattering spectra, an inversion of equation (10.28) must be obtained. Our numerical inversion is based on a Newton-type iterative scheme through least-squares minimization of the objective functional:

$$\chi^2 = \sum_{j=\lambda_1}^{\lambda_{10}} [(\mu_s')_j^o - (\mu_s')_j^c]^2 \qquad\qquad (10.29)$$

where $(\mu_s')_j^o$ and $(\mu_s')_j^c$ are the observed and computed reduced scattering coefficients at ten wavelengths, $j = \lambda_1, \lambda_2,...,\lambda_{10}$ (more wavelengths can be used, depending on the number of wavelength available from the experimental system). In the reconstruction, we have assumed a Gaussian particle size distribution in this study (*a priori* knowledge about the mode and distribution form of the particle size are usually available in a practical situation),

$$f(x) = \frac{1}{\sqrt{2\pi b^2}} e^{-\frac{(x-a)^2}{2b^2}}, \qquad\qquad (10.30)$$

where a is the average size of particles and b is the standard deviation. Substituting above $f(x)$ into equation (10.28), we obtain

$$(1-g)\mu_s(\lambda) = \mu_s'(\lambda) = \int_0^\infty \frac{3Q_{scat}(x,n,\lambda)[1-g(x,n,\lambda)]}{2x} \phi \frac{1}{\sqrt{2\pi b^2}} e^{-\frac{(x-a)^2}{2b^2}} dx \quad (10.31)$$

Now the particle sizing task becomes to recover three parameters a, b and ϕ. We have used a combined Marquardt-Tikhonov regularization scheme to stabilize the reconstruction procedure.

We first performed simulations to evaluate the sensitivity of particle sizing on the number of wavelengths used. "Measured" μ_s' spectra were generated using equation (10.31) with $a = 2.86$ μm, $b = 0.145$ μm and $\phi = 1.02\%$ for 10, 20, and 50 wavelengths between 600 and 1000 nm, respectively. When 5% noise was added to each set of measured μ_s' spectra, we found that the relative errors of recovering the particle parameters were within 14% using the 10-wavelength spectra, while such errors were as low as 4% when 50-wavelength spectra were used. The Mie Theory fittings using the extracted parameters at 10, 20, and 50 wavelengths are shown in Figure 10.3 where the exact spectra are also presented for comparison. We see that the 10-wavelength spectra are able to provide quantitatively accurate reconstruction. We also performed simulations to test the noise sensitivity when 1, 5, or 10% random noise was added to the 10-wavelength spectra. The relative errors of the recovered parameters (a and ϕ) were calculated to be 3, 14, and 17% for parameter a and 3, 11,

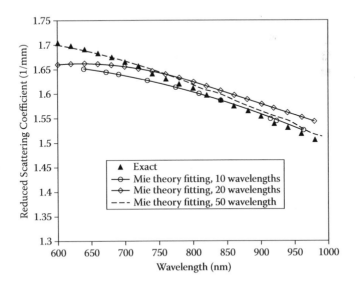

FIGURE 10.3 The spectra generated by equation (10.31) and the Mie fittings using recovered particle parameters from simulated data (5% noise) with 10, 20, and 50 wavelengths, respectively. (Reprinted from Li, C. and Jiang, H., *Optics Express* 12 (25): 6313–6318, 2004b. With permission.)

and 14% for the parameter ϕ. However, the recovery of the standard deviation was sensitive to noise, which had a relative error of 93% when 10% noise was added.

EXPERIMENTAL RESULTS

Phantom Experimental Results

Phantom experiments were conducted using our 10-wavelength DOT system (638, 673, 690, 733, 775, 808, 840, 915, 922, and 960 nm). Two sets of phantom experiments were conducted to demonstrate the overall approach for imaging particle size and concentration. The cylindrical background phantom had a radius of 25 mm, an absorption coefficient of 0.005/mm (India ink as absorber), and a reduced scattering coefficient of 1.0/mm (Intralipid as scatterer). A thin glass tube (9 mm in inner diameter, 0.4 mm in thickness) containing polystyrene suspensions (Polysciences, Warrington, PA) was embedded off-center in the background solid phantom. Two different types of polystyrene spheres were used in the experiments: one had a diameter of 2.06 μm and a concentration of 0.52%, and the other had a diameter 5.66 μm and a concentration of 2.62%. The refractive index of the spheres and their surrounding aqueous medium were 1.59 and 1.33, respectively.

A finite element mesh was used with the DOT reconstructions. To show the accuracy of the DOT reconstruction, Figure 10.4 depicts the recovered μ_s' spectra at a typical node location in the target area for the 2.06-μm polystyrene case, in comparison with the corresponding Mie Theory fitting using the extracted particle parameters.

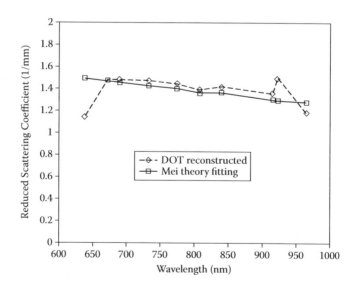

FIGURE 10.4 Experimental spectra DOT reconstructed at a typical node in the target area and the corresponding Mie fitting using recovered particle parameters for the 2.06-μm polystyrene case. (Reprinted from Li, C. and Jiang, H., *Optics Express* 12 (25): 6313–6318, 2004b. With permission.)

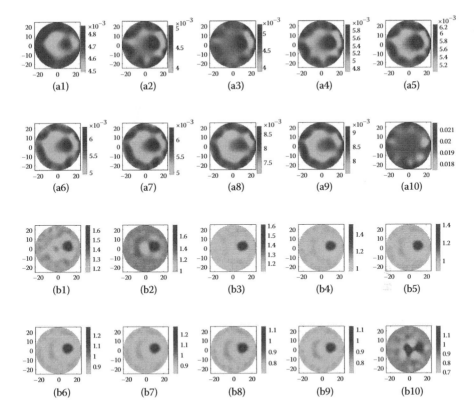

FIGURE 10.5 The DOT reconstructed absorption images (a1 to a10) and reduced scattering images (b1 to b10) at ten different wavelengths: a1/b1 at 638 nm; a2/b2 at 673 nm; a3/b3 at 690 nm; a4/b4 at 733 nm; a5/b5 at 775 nm; a6/b6 at 808 nm; a7/b7 at 840 nm; a8/b8 at 915 nm; a9/b9 at 922 nm and a10/b10 at 965 nm for the 2.06 m polystyrene case for all ten wavelengths. (Reproduced from Li, C. and Jiang, H., *Optics Express* 12 (25): 6313–6318, 2004b. With permission.)

And the DOT reconstructed absorption and reduced scattering images for the 2.06 μm polystyrene case for all ten wavelengths are shown in Figure 10.5. Figure 10.6 presents the reconstructed images of mean particle size and concentration for the 2.06 and 5.66 μm polystyrene cases. We immediately note that the particle size and concentration of both the target and background are quantitatively imaged. It should be noted that only the recovered μ_s' spectra were needed for particle sizing.

Ex vivo EXPERIMENTAL RESULTS

We have tested the potential of this technique for imaging cellular size and crowding in *ex vivo* breast tissue. Immediately after the mastectomy of the right breast of a 62-year-old female with a biopsy-confirmed infiltrating ductal carcinoma ($\approx 3.3 \times 1.7 \times 3.0$ cm); a portion of the breast with the tumors was sectioned transversely into an approximately 1.5-cm-thick section using a tissue slicer. From the section, a 3×2 cm

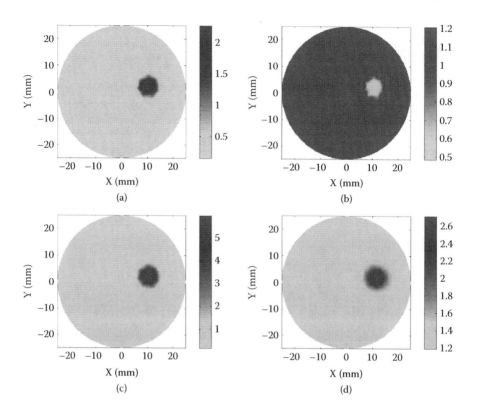

FIGURE 10.6 Reconstructed images of mean particle size (a, c) and concentration (b, d) for the 2.06-μm (a, b) and 5.66-μm (c, d) polystyrene cases, respectively. The axes (*left and bottom*) indicate the spatial scale, in millimeters, whereas the colorful scale (*right*) records the mean particle size or concentration, in micrometers or percentage. (Reprinted from Li, C. and Jiang, H., *Optics Express* 12 (25): 6313–6318, 2004b. With permission.)

fresh tissue slice with both tumor and normal tissues was obtained. We then placed this tissue slice in the hole (4 × 6 cm) of a 7 × 10 cm cylindrical solid background phantom composed of agar powder, Intralipid solution, and India ink for mimicking tissue scattering and absorption (Figure 10.7a). We further poured liquid phantom (at 39°C) with the same optical properties as the solid phantom into the hole/breast tissue and then cooled down the liquid phantom to 20°C to obtain a full tissue-containing solid phantom for imaging. After the optical measurement, a tissue sample from the same tissue slice was taken for histological examination. The microscopic sections of both tumor and normal tissues were obtained.

Figure. 10.7b and Figure 10.7c show the recovered cellular/subcellular size and crowding or volume fraction images, respectively, where we immediately note that the tumor, normal tissue, and background phantom are readily differentiated. The recovered average size of the scatterers in the tumor area was calculated to be 5.1 μm relative to the average nuclear size of 8.9 μm in the tumor region measured from

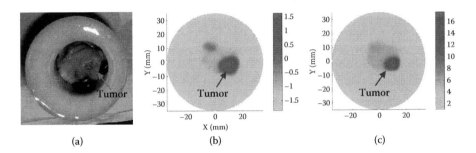

FIGURE 10.7 (a) Photograph of the sliced breast tissue inside the background phantom. (b) Recovered cellular size image, where the color scale refers to cellular/subcellular size in logarithm of micrometers. (c) Recovered cellular crowding image, where the color scale indicates the volume fraction in percentage.

the microscopic sections. This relatively large difference in size makes us to believe that what we reconstructed is most likely an effective size of combined nuclear and mitochondria (normally 1–2 μm) structure, which gives an expected effective size of approximately 5.0–5.5 μm from the microscopic sections. From Figure 10.7c, we found that the extracted average cellular volume fraction or crowding in the tumor is 17%, which is close to the nuclear crowding of 19% estimated from the microscopic sections.

DISCUSSION AND CONCLUSIONS

The simulations, phantom, and *ex vivo* studies show that it is possible to extract cellular/subcellular morphological information from breast tissue using multispectral DOT. In our particle sizing, the refractive indices of the polystyrene suspensions and the surrounding medium (water) are important parameters and have been assumed known as *a priori*. In a clinical situation, we can obtain the information empirically from the literature, or we can ultimately recover the refractive indices of scatterers as we reconstruct the PSD and concentration. Our phantom results have shown that the calibration method developed for correcting the initial scattering values is able to provide quality reconstruction of PSD and concentration.

While these assumptions do not pose problems in our phantom study, they may do so in real tissue where scatterers generally are not spherical. Interestingly, Mourant and colleagues (Mourant et al., 1998) presented Mie calculations of μ_s' in cultured cell suspensions, which showed excellent agreement between the Mie Theory and diffusion approximation. In a prior study, we have shown that accurate PSD can be recovered in concentrated TiO_2 suspensions where TiO_2 particles are not spherical (Jiang, 1998; Jiang, Marquez, and Wang, 1998). In another study, we have demonstrated that the PSD in KCl suspensions can still be well reconstructed when the concentration of KCl particles is as high as 40% (Jiang, 1998; Bartlett and Jiang, 2001). Thus, it is reasonable to believe that the approach described here for particle sizing would provide quality results from *in vivo* data, particularly if we focus on the reconstruction of only particle mean size and concentration. The *ex vivo* results presented have shown promise towards this goal.

In sum, this chapter has reviewed emerging optical techniques for imaging particle size and concentration with a focus on multi-DOT. Initial experimental results reported thus far have shown the potential of a multi-DOT technique for clinical applications. It is clear that research in this area is still in its infancy and more exciting results, especially *in vivo* results, are expected to appear in the literature in the near future.

ACKNOWLEDGMENT

The research reported in this article has been supported by the National Institutes of Health (NIH) (R01CA090533).

REFERENCES

Amelink A, Bard MPL, Burgers SA, et al., Single-scattering spectroscopy for the endoscopic analysis of particle size in superficial layers of turbid media, *Applied Optics* 42 (19): 4095–4101. 2003.

Anderson RR, Polarized-light examination and photography of the skin, *Archives of Dermatology* 127 (7): 1000–1005, 1991.

Arridge SR, Optical tomography in medical imaging, *Inverse Problems* 15 (2): R41–R93, 1999.

Arridge SR, Schweiger M, Hiraoka M, Delpy DT, A finite-element approach for modeling photon transport in tissue, *Medical Physics* 20 (2): 299–309, 1993.

Backman V, Gurjar R, Badizadegan K, Itzkan L, Dasari RR, Perelman LT, Feld MS, Polarized light scattering spectroscopy for quantitative measurement of epithelial cellular structures *in situ*, *IEEE Journal of Selected Topics in Quantum Electronics* 5 (4): 1019–1026, 1999.

Bartlett M, Huang G, Larcom L, Jiang HB, Measurement of particle size distribution in mammalian cells *in vitro* by use of polarized light spectroscopy, *Applied Optics* 43 (6): 1296–1307, 2004.

Bartlett M, Jiang H, Measurement of particle size distribution in concentrated, rapidly flowing potassium chloride (KCl) suspensions using continuous-wave photon migration techniques, *AIChE Journal* 47, 60–65, 2001.

Beauvoit B, Liu H, Kang K, Kaplan PD, Miwa M, Chance B, Characterization of absorption and scattering properties of various yeast strains by time-resolved spectroscopy, *Cell Biophysics* 23, 91–109, 1993.

Benaron DA, Hintz SR, Villringer A, Boas D, Kleinschmidt A, Frahm J, Hirth C, Obrig H, van Houten JC, Kermit EL, Cheong WF, Stevenson DK, Noninvasive functional imaging of human brain using light, *Journal of Cerebral Blood Flow and Metabolism* 20 (3): 469–477, 2000.

Boas DA, Culver JP, Stott JJ, Dunn AK, Three dimensional Monte Carlo code for photon migration through complex heterogeneous media including the adult human head, *Optics Express* 10 (3): 159–170, 2002.

Boas DA, Oleary MA, Chance B, Yodh AG, Sscattering of diffuse photon density waves by spherical inhomogeneities within turbid media—analytic solution and applications, *Proceedings of the National Academy of Sciences of the United States of America* 91 (11): 4887–4891, 1994.

Brunsting A, Mullaney P., Differential light scattering from spherical mammalian cells, *Biophysics Journal* 14: 439–453, 1974.

Cerussi AE, Jakubowski D, Shah N, Bevilacqua F, Lanning R, Berger AJ, Hsiang D, Butler J, Holcombe RF, Tromberg BJ, Spectroscopy enhances the information content of optical mammography, *Journal of Biomedical Optics* 7 (1): 60–71, 2002.

Colak SB, van der Mark MB, Hooft GW, Hoogenraad JH, van der Linden ES, Kuijpers FA, Clinical optical tomography and NIR spectroscopy for breast cancer detection, *IEEE Journal of Selected Topics in Quantum Electronics* 5 (4): 1143–1158, 1999.

Cotran RS, Robbins S, Kumar V, *Robbins Pathological Basis of Disease,* W.B. Saunders, Philadelphia, Pennsylvania, 1994.

Demos SG, Alfano RR, Temporal gating in highly scattering media by the degree of optical polarization, *Optics Letters* 21 (2): 161–163, 1996.

Demos SG, Alfano RR, and Dmons SJ, Optical polarization imaging, *Applied Optics* 36 (1): 150–155, 1997.

Grable RJ, Rohler DP and Sastry KLA, Optical tomography breast imaging, Proceedings SPIE 2979: 297–210, 2004.

Gurjar RS, Backman V, Perelman LT, Georgakoudi I, Badizadegan K, Itzkan I, Dasari RR, Feld MS, Imaging human epithelial properties with polarized light-scattering spectroscopy, *Nature Medicine* 7 (11): 1245–1248, 2001.

Hielscher AH, Klose AD, Scheel Ak, Moa-Anderson B, Backhaus M, Netz U, Beuthan J, Sagittal laser optical tomography for imaging in turbid media, *Physics in Medicine and Biology*, 49: 1147–1163, 2004.

Iftimia N, Jiang H, Quantitative optical image reconstruction of turbid media by use of direct-current measurements, *Applied Optics* 39: 5256–5261, 2000.

Jiang H, Enhanced photon migration methods for particle sizing in concentrated suspensions, *AIChE Journal* 44: 1740–1744, 1998.

Jiang H, Marquez G, Wang L, Particle sizing in concentrated suspensions using steady-state, continuous-wave photon migration techniques, *Optics Letters* 23: 394–396, 1998.

Jiang HB, Paulsen KD, Osterberg UL, Pogue BW, Patterson MS, Optical image reconstruction using frequency-domain data: simulations and experiments, *Journal of the Optical Society of America A—Optics Image Science and Vision* 13 (2): 253–266, 1996.

Jiang H, Pierce J, Kao J, Sevick-Muraca E, Measurement of particle-size distribution and volume fraction in concentrated suspensions with photon migration techniques, *Applied Optics* 36: 3310–3318, 1997.

Li C, Jiang H, A calibration method in diffuse optical tomography, *Journal of Optics A: Pure and Applied* Optics 6: 844–852, 2004a.

Li C, Jiang H, Imaging of particle size and concentration in heterogeneous turbid media with multispectral diffuse optical tomography, *Optics Express* 12 (25): 6313–6318, 2004b.

Mourant JR, Hielscher AH, Eick AA, Johnson TM, Freyer JP, Evidence of intrinsic differences in the light scattering properties of tumorigenic and nontumorigenic cells, *Cancer Cytopathology* 84: 366–374, 1998.

Mourant JR, Johnson TM, Freyer JP, Characterizing mammalian cells and cell phantoms by polarized backscattering fiber-optic measurements, *Applied Optics* 40 (28): 5114–5123, 2001.

Patterson MS, Chance B, Wilson BC, Time resolved reflectance and transmittance for the noninvasive measurement of tissue optical-properties, *Applied Optics* 28 (12): 2331–2336, 1989.

Paulsen KD, Jiang HB, Spatially varying optical property reconstruction using a finite-element diffusion equation approximation, *Medical Physics* 22 (6): 691–701,1995.

Perelman LT, Backman V, Wallace M, Zonios G, Manoharan R, Nusrat A, Shields S, Seiler M, Lima C, Hamano T, Itzkan I, J Van Dam, Crawford JM, Feld MS, Observation of periodic fine structure in reflectance from biological tissue: a new technique for measuring nuclear size distribution, *Physical Review Letters* 80: 627–630, 1998.

Ren K, Abdoulaev G, Bal G, Hielscher A, Algorithm for solving the equation of radiative transfer in the frequency domain, *Optics Letters* 29: 578–580, 2004.

Ripoll J, Ntziachristos V, Iterative boundary method for diffuse optical tomography, *Journal of the Optic Society of America A* 20: 1103–1110, 2003.

Schmidt FEW, Fry ME, Hillman EMC, Hebden JC, Delpy DT, A 32-channel time-resolved instrument for medical optical tomography, *Review of Scientific Instruments* 71 (1): 256–265, 2000.

Wang LH, Jacques SL, Zheng LQ, mcml—monte-carlo modeling of light transport in multi-layered tissues, *Computer Methods and Programs in Biomedicine* 47 (2): 131–146, 1995.

Yoo KM, Alfano RR, Time resolved depolarization of multiple backscattered light from random-media, *Physics Letters A* 142 (8–9): 531–536, 1989.

11 Surfactants/Hybrid Polymers and Their Nanoparticles for Personal Care Applications

P. Somasundaran and P. Deo

CONTENTS

INTRODUCTION

With new surface-active systems developed, there is a large potential for developing personal care products with special properties for controlled delivery and deposition of sensory attributes. Engineering of delivery and encapsulation systems has grown into an independent field, transcending the scope of traditional disciplines and capturing the interest of both academic and industrial researchers. The trend is to develop such systems capable of protecting, transporting, and selectively depositing attributes at desired sites and at desired rates. Inspiration for developing new classes of such products arises from an understanding of the mechanisms by which a "touch me not" plant folds up rapidly upon being assaulted or microbes depositing on surfaces. Many surfactants and polymers with appropriate modifications can be tuned to simulate such responses. In the personal care industry, surfactants are primarily used for emulsion stabilization as well as for imparting properties that include detergency, solubilization, conditioning, thickening, and emolliency (1,2). On the other hand, polymers are employed to enhance system viscosity and consistency of the formulation, which in turn affect the flow and smooth delivery of the formulations. Rheology of personal care products such as shampoos and shower gels is a key property to be controlled since consumers desire shear-thinning behavior of the liquid, characterized by slow flow from the containers, absence of long threads, and ease of distribution on the skin or hair. Such desirable properties are often obtained by including appropriate combinations of polymers and surfactants in formulations.

Development of new products depends upon understanding the interactions among surfactants, polymers, and proteins in relevant media. For example, new hybrid polymers and nanogels can be designed with nanodomains that can extract and deliver, at will, cosmetics or drugs, or extract sebaceous or toxic materials by making use of usual changes in pH, temperature, or ionic strength of personal care systems.

In this chapter, development of systems based on polymer/surfactant colloid chemistry is examined so as to acquire transport and release of cosmetic and pharmaceutical molecules at desired rates and at desired sites based on the above principles.

It is important to note that the effects of surfactants depend not only on how much is adsorbed, but also on the mode of adsorption. A water-wetted surface that is beneficial for displacement of sebaceous and oily materials can be obtained by manipulating the orientation of the adsorbed layers. The availability of modern equipment such as analytical ultracentrifuge, fluorescence, and electron spin resonance spectrophotometers offers an unprecedented opportunity for elucidating the mechanisms of adsorption and for designing optimum reagent schemes.

In many cases, mixed surfactants perform much better than single surfactants due to synergetic effects and ability to alleviate precipitation. Adsorption of simpler single systems will be discussed here first and then increasingly complex systems will be examined, steadily tending toward real systems.

SINGLE SURFACTANTS

Various mechanisms involving a combination of the following factors have been proposed for interactions among surfactants and substrates. They include combinations of electrostatic attraction/repulsion, ion-exchange, hydrophobic bonding, chemisorption, hydrogen bonding, and lateral chain-chain interactions. Much effort has been made to explore the adsorption effects of variables such as the chain length and branching; polar substitutions of the surfactants, oil, and polymers; and solution pH, salinity, and temperature.

The basic process of adsorption of surfactants can best be understood by examining the Somasundaran-Fuerstenau IV region adsorption isotherm, originally proposed for dodecyl sulfonate adsorption on a positively charged solid (3), which is illustrated in Figure 11.1.

Adsorption in various regions was accounted for by considering the electrostatic, hydrophobic, and micellar interactions in the system. In Region I, adsorption of the negatively charged dodecyl sulfonate on the positively charged surface sites takes

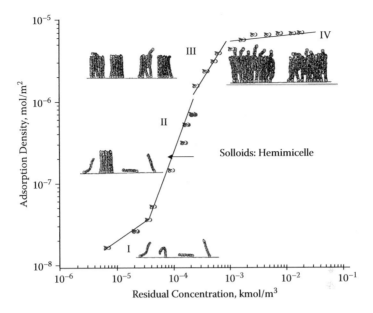

FIGURE 11.1   Schematic representation of the evolution in the adsorption layers at the solid-liquid interface.

place due to electrostatic attraction between the two. At the onset of Region II, the surfactants begin to self-assemble to form solloids (surface colloids, also termed hemimicelles in some cases), and in this region adsorption is due to electrostatic attraction between the surface sites and the oppositely charged surfactant species and lateral hydrophobic interactions between the hydrocarbon chains. Once the surface is electrically neutralized, further adsorption takes place due to chain-chain interactions alone countered by electrostatic repulsion that builds up as the surface begins to acquire now the same charge as the adsorbing surfactant species. Above the critical micelle concentration of the surfactant in Region IV, monomer activity is essentially constant and under these conditions adsorption also remains constant. It is important to note that the wettability of the surface and hence the displacement of oily material is determined by the orientation the surfactant assumes in various regions. Thus, in Region I, the surface is essentially water wetted and in Region II it is oil wetted, while in regions III and IV it would begin to become less oil wetted. It is clear that adsorption and wettability can be controlled by manipulating the structure of the surfactants.

SURFACTANT STRUCTURAL EFFECTS

Based on a full knowledge of the role of the structure of the surfactants in determining the adsorption, new surfactants can be designed for minimum adsorption as well as low interfacial tension and better salt tolerance for emulsification and other colloidal processes. The effects on adsorption of some structural variations of surfactants have been extensively studied (4–6):

- Increasing the chain length of the alkyl groups increases the adsorption by orders of magnitude.
- Incorporation of a phenyl group into the surfactant increases the effective chain length by three or four methylene groups and hence the adsorption.
- The position of the branching of the alkyl groups has a measurable effect on adsorption, due to the difficulties in the packing of the surfactant species in the two-dimentional colloidal aggregates.
- Even the apparently minor change in the position of the sulfonate and the methyl groups on the aromatic ring of the alkylaryl sulfonate showed a ten-fold effect on adsorption as well as the degree of wettability of the adsorbed layer (7).

Mixed Surfactants

Mixtures of surfactants normally yield improved solution and interfacial properties compared to their individual components. Adsorption of surfactant mixtures can be synergistic or competitive and can be manipulated for practical applications better than the adsorption of single surfactants. For example, our work has shown that the nonionic surfactant $C_{12}EO_8$ does not adsorb on solids, but in the presence of an anionic surfactant, synergetic adsorption of it can be obtained. In addition, the adsorption of the anionic surfactant is facilitated on similarly charged substrates when co-adsorbed from a mixture containing the nonionic ones, interaction between the hydrocarbon chains of the surfactants being the driving force. Adsorption of the anionic sodium dodecyl sulfate (SDS) and the nonionic octaethylene glycol mono n-dodecyl ether ($C_{12}EO_8$) from their mixtures revealed many interesting phenomena. For example, while at pH 5, the adsorption of the nonionic $C_{12}EO_3$ was markedly enhanced by the presence of the anionic dodecyl sulfate, suggesting cooperative adsorption through lateral interaction between the hydrocarbon chains; at pH 10 its adsorption was suppressed.

The phenomenon of mixed micellization of ionic and nonionic surfactants in homogeneous solutions has been addressed in the past where models based on electrostatic principles or regular solution theory showed deviations from ideal behavior. However, regular solution theory can be modified to model interactions between similar and dissimilar surfactants in the bulk solution, and obtain interaction parameters that enable the prediction of synergism or competition for mixtures.

The micellization behavior of cationic-nonionic hydrocarbon surfactant mixtures has been studied using the ultrafiltration technique, and the data suggest coexistence of mixed micelles (Figure 11.2) (7). The importance of such coexistence in guiding colloidal behavior of the micellar solutions should be noted. The results suggested that the micellization behavior can be divided into three regions. In Region I, the surfactant concentration is low and only monomers exist. In Region II, surfactants begin to form aggregates. In Region III, two types of coexisting mixed micelles are proposed to coexist in the solution. Recently, we have proven the coexistence

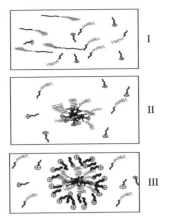

FIGURE 11.2 Schematic model for a binary surfactant system: Region I, no micelles in the mixed solution; Region II, one type of mixed micelles formed; Region III, two types of coexisting mixed micelles formed.

of two micelles in Region III in the same system using analytical ultracentrifugation (Figure 11.3). Here, the concentration of micellar species is plotted as a function of sedimentation coefficient, two peaks suggest the presence of two coexisting micelles. Coexistence of such multispecies has, in fact, practical implications in both system stability and efficacy.

Liposome/Surfactant Interactions

Many important natural biosystems are composed of mixed surface-active species of proteins, carbohydrates, cholesterols, lipids, and other components. Interactions among these components determine the stability of the systems and even simple combinations show interesting effects. Thus, when a surfactant such as dodecyl sulfonate

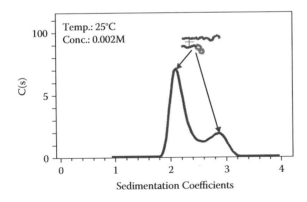

FIGURE 11.3 Analytical ultracentrifugal results of mixed surfactant systems: coexistence of two micelles.

is added to a liposome made up of phosphatidyl choline and phosphatidic acid, initially the size of the liposome increases and subsequently it is solubilized (8). Furthermore, while the addition of cholesterol stabilizes liposomes, proteins destabilize them. The reasons for this effect are still far from evident (9). Electron spin resonance studies have shown the polarity and viscosity of liposomes to change to that of micelles of dodecyl sulfonate at sufficiently high concentrations of the surfactant (Figure 11.4A).

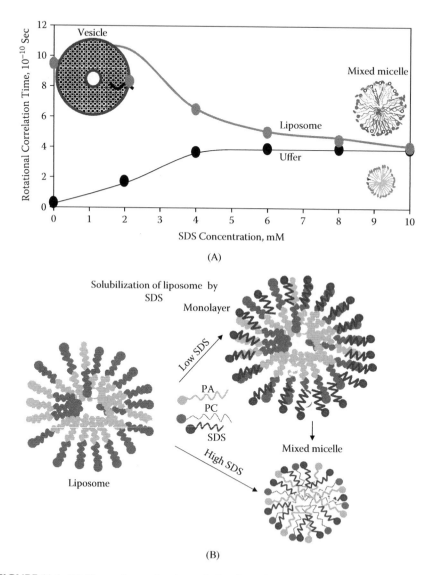

FIGURE 11.4 (A) Change in rotational correlation time of 5-doxyl stearic acid due to interactions with SDS. Buffer: increased correlation time (T)—transfer into SDS micelle. Liposome: decreased T—transfer into less rigid mixed micelle. (B) Depiction of solubilization of liposomes under low and high SDS concentrations.

These results suggest that liposome stabilization by dodecyl sulfonate involves the adsorption of sulfonate on them, leading to an increase in size, and their subsequent disintegration into mixed micelles composed of liposome components and dodecyl sulfonate, resulting in a drastic decrease in size (Figure 11.4B).

The actual processes by which such disintegration takes place are not known. Our investigation suggests that phosphatidic acid would exit first, leading to the weakening of the liposome structure and its dissolution (8,9). Similarly, we have seen

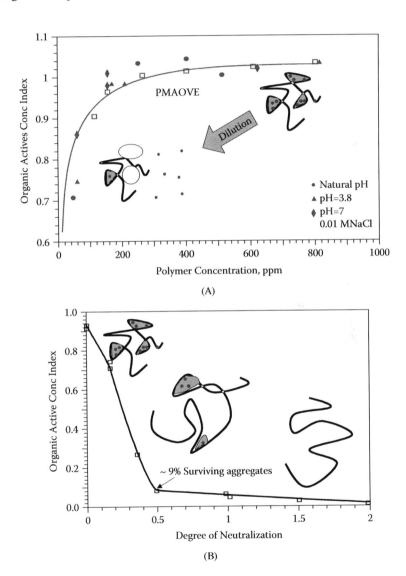

(A)

(B)

FIGURE 11.5 (A) Effect of pH of the medium on the uncoiling of poly(maleic acid/octyl vinyl ether) (PMAOVE) for the release of organic actives. (B) Effect of pH of the medium on the uncoiling of PMAOVE for the release of organic actives.

cholesterol to stabilize liposomes and proteins to destabilize them, but the mechanisms by which this occurs are also not known.

Hydrophobic Polymers

Unusual rheological properties (10) that result from the association of hydrophobic groups are the main attractive features of these polymers. Hydrophobically modified (hybrid) polymers, also known as polysoaps (11–15), have the advantage that they have features of both polymers and surfactants; and due to the associative nature of the hydrophobic groups, hybrid polymers can form intramolecular nanodomains at all concentrations and enhance viscosity above certain concentrations.

The ability of the hydrophobically modified polymers to form nanodomains can be utilized for the release of organic sensory attributes. Thus, the poly(maleic acid/octyl vinyl ether) (PMAOVE) system forms hydrophobic nanodomains that can solubilize and release organic molecules by changes in pH or salinity (16–18). The effect of dilution and pH on such release is shown in Figure 11.5A and Figure 11.5B. These figures show that hydrophobically modified polymers uncoil on dilution or on increasing the pH of the medium and releasing the trapped hydrophobes.

Such polymers can also release entrapped material upon deposition on solids. Figure 11.6 shows results obtained when PMAOVE is deposited on alumina particles. It can be seen that organics entrapped is less when the PMAOVE is in contact with the solid than when it is in micelles above 200 ppm. Evidently, the hydrophobic polymeric entanglements loosen up on coming in contact with a surface and facilitates the release of entrapped moieties.

The potential use of hybrid polymer for formulation of personal care products is illustrated in Figure 11.7. This figure shows that in the presence of a polar solvent, the hybrid polymers will form hydrophobic domains inside which hydrophobic

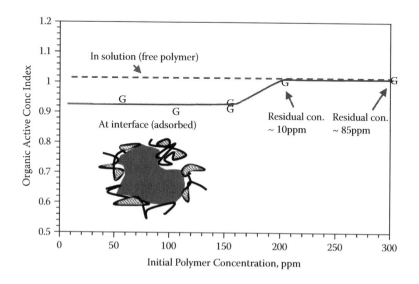

FIGURE 11.6 Uncoiling of hydrophobic domain in contact with a surface.

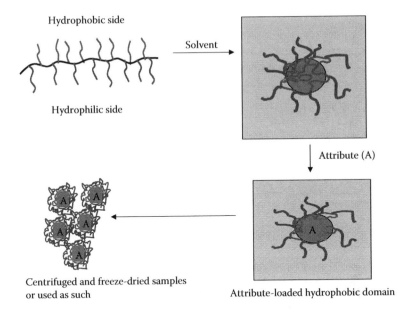

Hydrophobic side

Solvent

Hydrophilic side

Attribute (A)

Centrifuged and freeze-dried samples
or used as such

Attribute-loaded hydrophobic domain

FIGURE 11.7 Use of hybrid polymer for encapsulation of cosmetic attributes.

attributes can be incorporated and used as such or can be centrifuged and freeze-dried for the end use.

NANOGELS AS ENCAPSULATING AGENT

Water-soluble, lightly cross-linked polymers can act as a thickening agent. Molecular weight distribution, cross-linking type and density largely determine rheological performance in this case. Cross-linked poly(acrylic acid) and its copolymers have been considered the preeminent thickening resins. If these crosslinked polymers are reduced to nano- or micron-size particles, apart from being used as thickeners they can also be used as vehicles for the release of cosmetic attributes such as fragrances and antimicrobial agents at a rate and duration designed to accomplish the intended effect. Other than fragrances, sunscreen, sebum oil absorber, and antiaging compounds can also be formulated in these controlled release systems. The release from these formulations is usually triggered by shear force (rubbing), temperature, pH, or dilution.

Nanogels developed recently by reverse microemulsion polymerization include polyacrylamide, poly(acrylic acid), and starch nanogels modified for extraction and subsequent slow release of fragrances and overdosed toxic drugs. Our AFM (atomic force microscopy) study (Figure 11.8) revealed that the dry particle size of the polyacrylamide nanogels is 40 nm and on swelling the size increases to 140 nm, giving the nanogels the capability to contain and release materials.

The potential of polyacrylamide and starch nanogels to extract vanillin was recently evaluated. We have also examined the extraction and release properties

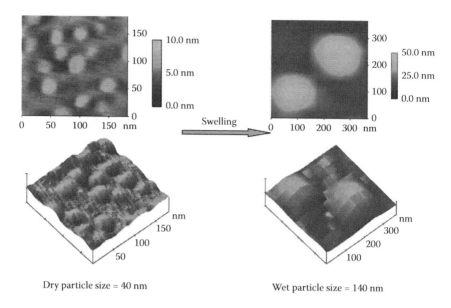

FIGURE 11.8 AFM image of dry polyacrylamide nanogels (40 nm) and wet polyacrylamide nanogels (140 nm).

of the linalyl acetate fragrance from poly(acrylic acid) and hydrophobically modified poly(acrylic acid) nanogels. As shown in Figure 11.9, hydrophobically modified poly(acrylic acid) nanogels extracted 50% linalyl acetate (LA) in 4 hours whereas the unmodified one could extract 45% LA in 4 hours.

At pH 7, unmodified poly(acrylic acid) nanogels release more fragrance as compared to the modified nanogels (Figure 11.10). This suggests that appropriate structural modification is required for optimum carrier and delivery properties.

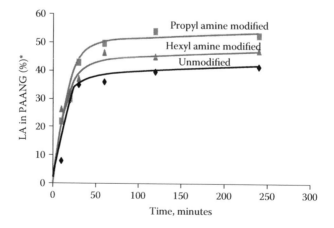

FIGURE 11.9 Extraction of linalyl acetate (LA) by poly(acrylic acid) nanogels (PAANG) in methanol. Conditions for extraction: PAANG: 10 mg; LA: 10 μL; methanol: 10 mL.

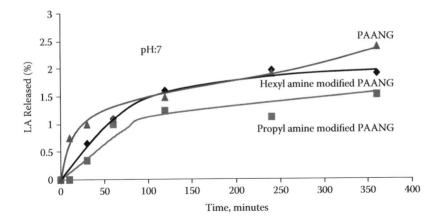

FIGURE 11.10 Release of linalyl acetate (LA) from unmodified and hydrophobically modified poly(acrylic acid) nanogels (PAANG) in water. Conditions for extraction: linalyl acetate (LA) incorporated PAANG : 10 mg; water: 10 mL.

Clearly, it is important to have a knowledge of the mechanisms controlling the rate of extraction and release. Toward this purpose, a technique based on surface plasmon resonance technique was successfully developed recently for monitoring both short-term and long-term dynamics (19,20) Results indicate that release of 50% of organic attribute from polyacrylamide nanogels could be detected in less than a minute using this technique (Figure 11.11).

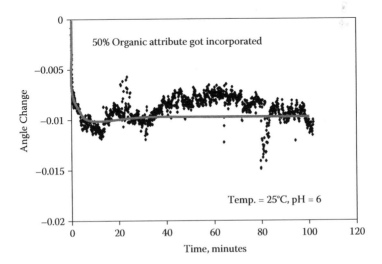

FIGURE 11.11 Interaction of organic attribute with polyacrylamide nanogels as studied by surface plasmon resonance (SPR) technique.

ZWITTERIONIC LATEX PARTICLES

A deposition of appropriate actives can provide good conditioning properties to the hair. Properties such as combability, flyaway, and curl retention are affected by deposition of polymers on hair strands. Based on the consideration that nano- or microparticles with different types of tethers can deposit on surfaces if there are at least some tethers that are complementary to some sites on the substrate, we examined negative zwitterionic latex particles for deposition on negative surfaces. Deposition of anionic, cationic, and zwitterionic latex particles on a negatively charged surface was studied as a function of deposition time, pH, and particle concentration. It was discovered that zwitterionic particles deposit better than all the others, even under hostile conditions (21).

CONCLUDING REMARKS

It is clear that research involving synthesis and characterization of hydrophobically modified polymers and polymeric nanogels and study of their interactions with surfactants and relevant biosurfaces hold considerable promise for developing better personal care products. This will depend primarily on a full understanding of the relationship between the chemical structure and performance of the surface-active molecules and the particles. We have demonstrated that hybrid polymers and polyacrylamide and poly(acrylic acid) nanogels can be used to encapsulate attribute molecules and release upon dilution. The nanogels can be included as an ingredient in shampoo or conditioner formulations. In this type of formulation, the nanogels, owing to their small size, can achieve good dispersion in the formulation, imparting a thickening property to the product and at the same time delivering a pleasant smell of the fragrance over a long period of time or against to prevent sensing odors.

ACKNOWLEDGMENTS

The authors acknowledge the financial support by the National Science Foundation, Industrial/University Cooperative Research Center (IUCR) at Columbia University and at University of Florida Engineering Research Center and the industrial sponsors of the IUCR Center.

REFERENCES

1. Balzer, D., Varwig, S., and Weihrauch, M. *Colloids and Surfaces A: Physicochemicals and Engineering Aspects* 99, 233–246, 1995.
2. Kostarelos, K. *Adv. Colloids Interf. Sci.*, 106, 147–168, 2003.
3. Somasundaran, P., Healy, T.W., and Fuerstenau, D.W. *J. Phys. Chem.*, 68, 3562, 1964.
4. Somasundaran, P., Middleton, R., and Viswanathan, K.V. *Relation between Structure and Performance*, American Chemical Society, Washington, D.C., p. 269, 1984.
5. Sivakumar, A. and Somasundaran, P. *Langmuir*, 10, 131–134, 1994.
6. Somasundaran, P., Fu, E., and Xu, Q. *Langmuir*, 8, 1065–1069, 1992.
7. Huang, L. and Somasundaran, P. *Langmuir*, 12, 5790–5795, 1996.

8. Deo, N. and Somasundaran, P. *Colloids and Surfaces: A,* 186, 33–41, 2001.
9. Deo, N. and Somasundaran, P. *Colloids and Surfaces: B*, 25, 225–232, 2002.
10. Hsu, J.-L., and Strauss, U.P. *J. Phys. Chem.*, 91, 6238–6241, 1987.
11. Barbieri, B.W., and Strauss, U.P. *Macromolecules,* 18, 411, 1985.
12. Zdanowicz, V.S., and Strauss, U.P. *Macromolecules,* 26, 4770, 1993.
13. Bokias, G., Hourdet, D., and Iliopoulos, I. *Macromolecules,* 33, 2929, 2000.
14. Moore, W.R., *Prog. Polym. Sci.*, 1, 3, 1967.
15. Gittings, M.R., Lal, J., Dérian, P.J., and Marques, C. *ACS Symposium Series 737*, El-Nokaly, M.A. and Soini, H.A. eds., pp. 287–299.
16. Deo, P., Jockusch, S., Ottaviani, M.F., Moscatelli, A., Turro, N.J., and Somasundaran, P., *Langmuir*, 19, 10747–10752, 2003.
17. Deo, P., Deo, N., and Somasundaran, P., *Langmuir*, 21, 9998–10003, 2005.
18. Qiu, Q., Lou, A., Somasundaran, P., and Pethica, B.A., *Langmuir,* 18, 5921, 2002.
19. Sarkar, D. and Somasundaran, P., *Langmuir*, 18 (22), 8271–8277, 2002.
20. Sarkar, D. and Somasundaran, P., *J. Colloid Interf. Sci.*, 197–205, 2003.
21. Deo, N. and Somasundaran, P., *J. Colloid Interf. Sci.*, 23, 1269–1279, 2002.

12 FloDots for Bioimaging and Bioanalysis

G. Yao, Y. Wu, D.L. Schiavone,
L. Wang, and W. Tan

CONTENTS

INTRODUCTION

In the past years, luminescent (including fluorescent) probes have been widely used for bioimaging and bioanalysis in the fields of chemistry, biology, medical science, disease diagnosis, and biotechnology. They are designed to localize within a specific region of a biological specimen or to respond to a specific stimulus (1). In other words, these probes can monitor the chemical and physiological state of biomolecules by presenting their optical properties in response to changes in the environment.

ORGANIC FLUOROPHORES

Among the most commonly used luminescent probes are organic fluorophores, such as the fluorescein series (FITC, FAM), Rhodamine series (R6G, TMR), Alexa Fluor series, Cy3, Cy5, and Texas Red. These dye molecules span the visible spectrum and can fluoresce when properly excited. The *Handbook of Fluorescent Probes and Research Chemicals*, published by Molecular Probes, Inc. in Eugene, Oregon, is a rich source of information about these dye molecules.

Although organic fluorophores have been extensively applied in bioanalysis and greatly helped the understanding of many chemical and biological problems, they still suffer from some inherent limitations:

1. *Photobleaching*. Most fluorophores photobleach rapidly, causing great difficulties in bioimaging and poor reproducibility in bioanalysis. It is also generally difficult to predict the necessity for and effectiveness of the remedy because photobleaching rates are dependent to some extent on the environment (2,3).

2. *Low sensitivity*. Only single or a few fluorophores are used to express each biomolecule recognition event, resulting in limited signal amplification and poor detection limits.

3. *Hard bioconjugation*. Each fluorophore has to use a specific mechanism for conjugation with biomolecules and there is no universal functional group(s) for biolabeling. All of these limitations have precluded the use of organic fluorophores in many applications, especially for sensitive biochemical sample detections, where sample amount/volume is small, or the spatial areas are limited, or real-time monitoring with continuous light excitation is desired.

LUMINESCENT NANOPROBES

Introduction of various luminescent probes have highly improved the detection and identification capability of analytical methods with an enhancement in sensitivity, selectivity, and stability. Nanotechnology, the process to generate, manipulate, and deploy nanomaterials, has also opened a promising field in the development of a new generation of luminescent probes. The size of the nanomaterials distinguishes them from bulk materials in that the nanomaterials have unique optical properties, high surface-to-volume ratio, and other size-dependent qualities. All these properties, together with surface modification, offer researchers diverse opportunities to develop nanoprobes for highly selective and ultrasensitive bioimaging and bioanalysis (4–18). Numerous works have been published on the applications of different luminescent nanomaterials, such as quantum dots (4–6,19–21), fluorescent latex prticles (7–9), and dye-doped nanoparticles (10–18).

QUANTUM DOTS

Quantum dots are made of atoms of group II-VI or III-V of the periodic table of elements. They are defined as particles with physical dimensions smaller than the exciton Bohr radius (4–6). Because of quantum confinement, both the absorption and emission of quantum dots shift to short wavelength as the size of the particles decreases. By changing the size and composition of quantum dots, the emission wavelength can be precisely tuned from blue to near-infrared (19). When compared to organic fluorophores, quantum dots are reported to be 20 times as bright, 100 times as photostable, and one-third as wide in spectral width (4). Quantum dots also have the advantage of long fluorescence lifetime and negligible photobleaching, so that quantum dots and their bioconjugates have made a significant impact on bioanalytical and biomedical research for multiple-color, multiple-analyte analysis (20) and biological imaging (21). Nonetheless, quantum dots have the disadvantages of poor solubility, easy agglutination, strong blinking (intermittent), and low quantum yield. Although efforts have recently been made to solubilize quantum dots in water by coating them with silica layer (22), mercaptoacetic acid (4), or protein molecules (5), other problems are still hard to solve.

FLUORESCENT LATEX PARTICLES

Fluorescent latex particles, such as fluorescent polystyrene particles or fluorescent polymethacrylic particles, have been employed in some biological applications (7–9). However, because of their large size, low solubility, aggregation, swelling, and dye leakage, these latex particles are not very suitable for ultrasensitive bioanalysis (10). Nevertheless, fluorescent latex particles provide the information that (a) highly luminescent particles can be obtained by doping luminescent dye into a polymer matrix; (b) if the dye molecules can chemically react with the polymer or strongly hydrophobic interact with the polymer, dye leakage will be minimized or eliminated; and (c) low porosity due to crosslinking in the polymerization will result in less diffusion of oxygen and hence less photobleaching.

FloDots

The Tan research group at the University of Florida has developed and extensively studied the preparation and application of the luminescent dye-doped silica nanoparticles, which consist of luminescent organic or inorganic dye molecules trapped inside the silica matrix. The commercialization of these particles is now being conducted in a Florida-based company, Life Sciences, Inc., in St. Petersburg. Since the major research work has been performed in Florida, we name these novel nanoparticles FloDots.

Silica has been found to be an appropriate matrix to prepare luminescent nanoparticles due to the following reasons. (a) Silica is chemically and physically inert so that it will not affect the reaction at the surface or target samples. (b) Silica shell is transparent so that the dye molecules doped inside can be effectively excited and the luminescence light can be effectively emitted out. (c) Silica particle is not subject to swelling or porosity change with the altering of environment conditions. (d) Well-developed silica chemistry makes it easy to modify the silica surface with different functional groups.

FloDots have significant advantages over organic fluorophores and quantum dots. With tens of thousands of luminescent dye molecules being encapsulated inside a silica particle ranging from 2 nm to 100 nm in diameter (10–12), the FloDots are extremely bright, which leads to high signal amplification (16). The silica matrix shielding effect protects the doped dye molecules from outside environmental oxygen, solvent molecules, and free radicals, enabling the luminescence to be stable and thus providing accurate measurements (12,17). The silica matrix is also an excellent

---

TABLE 12.1
Comparison of Fluorophore, Quantum Dot and FloDot

	Fluorophore	Quantum Dot	FloDot
Signal amplification	Low	~20 times of 1 fluorophore	Extremely high, up to 100,000 times
Photostability	Poor	Good	Excellent
Ability to bioconjugate	Easy, non-universal	Under investigation, progress being made	Easy, universal
Aqueous solubility	Mostly excellent, but some low	Low	Excellent
Multiplex analysis capability	Difficult	Excellent	Possible & under investigation
Toxicity	Some to severe	Minimal	None
Environmental influence	Some to severe	Minor	None
Manufacturing & reproducibility	N/A	Difficult	Easy and batch production
Compatibility with existing detection	Excellent	Excellent	Excellent
Overall feasibility for biolabeling	Good	Great potential, under investigation	Excellent

substrate for surface modification and biomolecule immobilization. In addition, the nanoparticles are nontoxic, environmentally friendly, and size-tunable with a narrow size distribution (±2%) (11). The comparisons among organic fluorophores, quantum dots, and FloDots are listed in Table 12.1.

In this chapter, we provide an overview of the synthesis, characterization, modification, and bioimaging and bioanalysis application of FloDots in our research group.

SYNTHESIS OF FloDots

A variety of techniques have been developed to prepare nanoparticles, and two major routes are employed to make luminescent dye-doped silica nanoparticles. One is the reverse microemulsion method (10,11,13–18) and the other is the Stöber method (12). The reverse microemulsion method is mainly used for the preparation of hydrophilic dye-doped FloDots and the Stöber method is mostly used for the synthesis of hydrophobic dye-doped FloDots.

REVERSE MICROEMULSION METHOD

Reverse microemulsion, or water-in-oil (W/O) microemulsion, has been widely used as a powerful tool to synthesize various kinds of monodisperse nanoprticles (23–27). Reverse microemulsion is an isotropic and thermodunamically stable single-phase system made of water, oil, and surfactant. The surfactant molecules lower the interfacial tension between water and oil, which results in the formation of a transparent uniform solution. Water is then surrounded by surfactant and forms nanodroplet dispersed in continuous bulk oil solvent. The shape of the water droplet is spherical and the water droplet can serve as confined reaction media for the formation of discrete nanoparticles.

FloDots have been synthesized using this method, showing its flexibility for the preparation of different types and sizes of silica nanoparticles. As shown in Figure 12.1, the reaction solution was prepared by mixing adequate amounts of surfactant, cosurfactant, organic solvent, water, aqueous solution of dye, silica precursor, and aqueous ammonia, where aqueous ammonia acts as both a reactant (H_2O) and a catalyst (NH_3) for the hydrolysis of silica precursor. Four main steps are involved in the particle formation: association of silica precursor with microemulsion, hydrolysis of silica precursor and formation of monomers, nucleation, and particle growth. In a typical reverse microemulsion method for the fabrication of FloDots (17,18), a water-in-oil microemulsion was prepared by mixing 1.77 mL of TX-100, 7.5 mL of

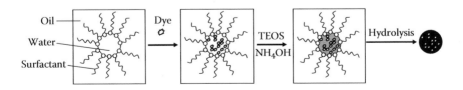

FIGURE 12.1 Schematic diagram of reverse microemulsion method.

cyclohexane, 1.8 mL of n-hexanol, and 480 μL of the dye solutions. Then, 100 μL of tetraethylorthosilicate (TEOS) was added as a precursor for silica formation, followed by the addition of 60 μL $NH_3 \cdot H_2O$ to initiate the polymerization process. The reaction was allowed to continue for 24 h at room temperature. Acetone was then added to break the microemulsion and isolate the nanoparticles. The FloDots were centrifuged and washed carefully with ethanol and water to remove any surfactant and free dye molecules. The final products will be stored in suitable buffer at room temperature or 4°C and avoid freezing.

The reverse microemulsion system contains seven different chemicals in the reaction solution, each of which will influence the final FloDots. The size of the water droplet greatly influences the nanoparticle size, so that the size of the nanoparticles can be controlled and tuned by changing the size of the water droplet, which is predominantly determined by the water-to-surfactant molar ratio (W_0) (10). In general, the higher the W_0 value, the larger the water droplet size, hence the larger the nanoparticle size. By changing the water-to-surfactant molar ratio (W_0), inorganic dye tris(2,2′-bipyridyl)dichlororuthenium(II) hexahydrate (Rubpy) doped Flodots were prepared with different sizes (11). The effects of the nature of surfactant molecules, the concentrations of TEOS and ammonium hydroxide, the water-to-surfactant molar ratio, the cosurfactant-to-surfactant molar ratio on the fluorescence spectra, nanoparticle size, and size distribution of FloDots were also systemically examined (13) and are listed in Table 12.2.

When the reverse microemulsion method was applied for FloDot preparation, the luminescent dye needs to be well dissolved in the water droplet and then doped inside the hydrophilic silica shell. Most organic molecules, however, are hydrophobic and hence are difficult to be easily doped inside the nanoparticle with the reverse microemulsion method. Although organic dye molecules usually have higher quantum yields (more than 90%) than the inorganic dye molecules, organic dye-doped FloDots made by reverse microemulsion method always have a major problem of dye leakage from the silica particles after dispersing in aqueous solutions for bioanalysis. This problem was solved by using a dye-dextran complex as the dye source and using acidic solution, instead of pure water, to dissolve the dye-dextran (16). Dextran molecules are highly hydrophilic with low toxicity and therefore make organic dye molecules water soluble in the format of dye-dextran. The relatively large size of the dextran molecule makes dye-dextran easily entrapped inside the silica matrix with reduced leakage. The acidic solution in the water droplet will create an electrostatic attraction between the dye molecules and the silica matrix, helping the organic dye molecules remain inside the silica shell. Results have revealed that commonly used fluorescent dyes tetramethylrhodamine (TMR), fluorescein, and Alexa Fluor 647 can be successfully doped into the silica nanoparticles without leakage when those FloDots were immersed in water (16).

Stöber Method

The Stöber method is a relatively simple method to make silica spheres ranging from nanometer to micrometer in size with a high degree of uniformity (28). The particles are synthesized by adding silica precursor into a solution of ethanol, water, and ammonia.

TABLE 12.2
Effect of Reverse Microemulsion Components on FloDot Synthesis

Working Condition		Emission Wavelength (nm)	FloDot Size (nm)
Surfactant	AOT (a)	595	29 ± 8
	NP-5 (b)	588	14 ± 2
	AOT+NP-5 (c)	597	130 ± 8
TEOS (mM)	0.025	590	82 ± 15
	0.1	590	82 ± 15
$NH_3 \cdot H_2O$ (wt%)	0.5	590	82 ± 13
	1.0	591	54 ± 5
	1.5	594	52 ± 6
	2.0	596	50 ± 4
Water-to-surfactant molar ratio	5	594	178 ± 29
	10	591	82 ± 14
	15	590	69 ± 4
	20	587	—
Cosurfactant-to-surfactant molar ratio	2.7	590	97 ± 14
	5.5	590	82 ± 13
	7.7	591	72 ± 6
	11	592	—

Note: Experiments were based on preparation of Rubpy FloDots in Triton X-100/cyclohexane/hexanol/
water system, except (a) in AOT/heptane/water system, (b) in NP-5/cyclohexane/water system, and (c) in
AOT+NP-5/heptane/water system.

Three main steps are involved in the particle formation: hydrolysis of silica precursor,
polymerization, and particle formation. The ammonia catalyzes the hydrolysis of the
silica precursor, such as TEOS, to silicic acid. When the concentration of silicic acid
is above its solubility in ethanol, the silicic acid will homogeneously nucleate as nanometer-sized particles. The size of the particles will be determined by the amount of
TEOS, water, and ammonia. Generally, a lower concentration of water and ammonia
produces smaller-sized particles. It is critical that the reaction be performed without
any other sites for silica nucleation, such as dust or a rough reaction vessel. Otherwise,
the particle size will not be uniform and large particles will form. Therefore, before
particle preparation, the reaction vessel should be washed with DI water, an acid such
as 1 M HCl, and with a base such as 1 M NaOH or a concentrated ammonia solution.
The solvents, especially the ethanol, should be filtered to remove particulates.

Hydrophilic organic dyes such as Nile Blue (29) or water-soluble porphyrin (30) have
been doped into a silica matrix to make dye-doped microspheres with the Stöber method,
but making hydrophobic dye-doped silica nanoparticles with the Stöber method was
still difficult. Two modifications have been applied to the Stöber method to synthesize

FloDots. The first one was to use a combination of both hydrophobic and hydrophilic precursors to make the FloDots. The hydrophobic nature keeps the organic dye in the silica matrix and the hydrophilic nature allows the resulting FloDots to be dispersed in aqueous solutions. Rhodamine 6G (R6G)-doped silica nanoparticles were successfully prepared by this approach with hydrophobic phenyltriethoxysilane (PTES) and hydrophilic tetraethoxysilane (TEOS) (12). These R6G FloDots exhibited high fluorescence intensity, excellent photostability, minimal dye leakage when stored in aqueous solution, and easy bioconjugation. Increased fluorescence intensity of FloDots was observed with increased R6G trapped inside, and the organic R6G dye molecules doped inside the FloDot were found to be proportional to the amount of the hydrophobic PTES. But it should also be noted that a high amount of PTES was found to make the particle hydrophobic and caused difficulties in surface modification for bioanalysis.

The second modification to the Stöber method to synthesize FloDots is to chemically conjugate the organic dye to a silane molecule. The succinimidyl esters or isothiocyanate group of the organic dye molecules are reactive with amine silanes such as aminopropyl triethoxysilane (APTS), and maleimide groups of the dye are reactive with thiol silanes such as (3-mercaptopropyl)triethoxysilane (MPTS). The ethoxy groups on the silanes and the dye functional groups are water reactive, so the dye and silanes must be coupled in an anhydrous solvent. Bubbling the solvent with nitrogen or argon is best to remove water. After adding a light excess of silane into anhydrous DMF or DMSO containing the dissolved organic dye, blow out the air with nitrogen or argon and tightly seal the container. Shake contents for at least 1 h or up to 24 h at room temperature to obtain dye-coupled silane. The dye-coupled silane is then added directly into the mixture of ethanol and ammonia solution, followed by immediate addition of TEOS. Close the reaction vessel to prevent dust entering the system and react for 5 h. The size of the FloDots is mainly determined by the concentration of ammonia in the reaction solution. After the synthesis of FloDots, centrifuge them from the ethanol solution and add water or buffer with immediate sonication to redisperse the particles. If the particles have been pelleted in ethanol and water is added, sonication must be performed quickly; otherwise permanent aggregation will take place due to solvent exchange and osmotic pressure.

CHARACTERIZATION OF FLODOTS

Freshly prepared FloDots are usually characterized by particle size, luminescence intensity, and photostability, which are all very important in evaluating whether the FloDots can be used in bioimaging and bioanalysis.

Particle Size

Size is an important parameter for nanoparticle analysis. It will help us to predict particle behavior, calculate surface area, and determine the handling method. FloDot sizes are generally measured by transmission electron microscopy (TEM), scanning electron microscopy (SEM), or dynamic light scattering. As shown in Figure 12.2, FloDots are usually found to be spherical. Due to the heavy metal ion in Rubpy, the dye molecules can be clearly found from the high resolution of the TEM image as black dots embedded inside the silica matrix (10,11).

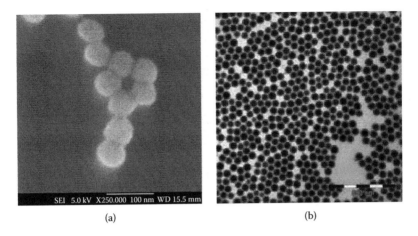

SEI 5.0 kV X250.000 100 nm WD 15.5 mm

(a)   (b)

FIGURE 12.2 FloDot image obtained from (a) scanning electron microscopy (SEM) of 50 nm FloDots (scale bar 100 nm) and (b) transmission electron microscopy (TEM) of 50 nm FloDots (scale bar 100 nm).

Luminescence Properties

The luminescence properties of the Flodots are generally tested by a spectrofluorometer. The FloDots have very similar absorption and emission spectra as those of the embedded dye molecules, and the maximal excitation and emission wavelengths of the FloDots may be slightly shifted from those of the pure dye molecules. As shown in Figure 12.3, the TMR FloDots prepared by the Stöber method under certain conditions have maximum excitation/emission wavelengths at 556 nm/576 nm, while the pure TMR molecules have maximum peaks at 551 nm/572 nm. The Rubpy FloDot prepared by the reverse microemulsion method under certain conditions and the pure Rubpy have the same excitation maximum, but the emission maximum of the Rubpy FloDot shifts by 7 nm toward longer wavelength due to aggregation of dye molecules inside the nanoparticle (10). Compared to the maximal ex/em wavelengths of Rhodamine B dye at 555 nm/575 nm, the maximal ex/em of Rhodamine B FloDot made by the Stöber method is located at 562 nm/580 nm. Experiments also found that FITC FloDots have an emission maximum between 509 nm and 514 nm, a blue shift instead of a red shift from the maximal emission wavelength of pure FITC at 516–518 nm.

Characterization of the luminescence property also confirmed the high emission intensity of the FloDots, which is the foundation that FloDots can be used in bioimaging and ultrasensitive bioanalysis. Each FloDot contains a large number of luminescent dye molecules, so that the FloDot can produce a strong emission signal when it is properly excited. Experiments have been performed to compare the luminescence intensity of Rubpy FloDots with commercially obtained quantum dots QD605 (from Auantum Dot Corp., Hayward, CA) and commonly used organic dye Texas Red. At corresponding optimal excitation and emission wavelengths, the luminescence intensities of serially diluted samples were recorded and plotted versus FloDot

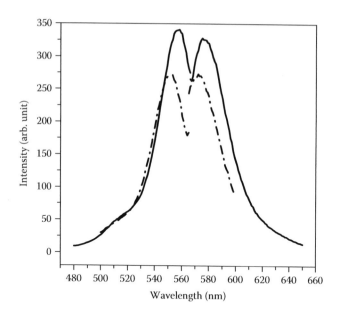

FIGURE 12.3 Luminescence excitation and emission spectra recorded in the aqueous phase for the pure TMR (dash curve) and TMR FloDot (solid curve).

or dye concentration. Each 70-nm Rubpy FloDot was calculated to have the same luminescence intensity as that of 39 particles of QD605, 1290 molecules of Texas Red, or 72,413 molecules of Rubpy (17). Experiments also found that the fluorescence intensity ratio of one TMR-dextran-FloDot to that of one TMR dye molecule was 10,000 (Zhao04b). Clearly, FloDots are highly luminescent.

Photostability

Photostability is very important if a luminescent probe is to be used in bioimaging and bioanalysis, especially for surface study and real-time study. Easy photobleaching of traditional organic fluorophores confines their application. Photostability of the FloDot was evaluated by measuring the emission intensity of FloDots with respect to time under continuous excitation.

Aqueous solutions of inorganic dye Rubpy-doped FloDot, quantum dot Q605, pure Rubpy, and pure Texas Red were taken for the photostability study in the solution phase by exciting continuously at each optimal excitation wavelength, respectively, for 50 min with a Spectrafluor microplate reader (17). The luminescence intensity was measured every 2 min, and the results suggest clearly that the FloDot is as photostable as QD 605, and much more photostable than Texas Red and Rubpy dyes.

Organic dye-doped FloDots in aqueous solution also have good photostability (16). After continuous excitation at 545 nm for 1200 s with a 150-W Xe lamp of a spectrofluorometer, practically no photobleaching can be observed for the 100 μL TMR-dextran–doped FloDot solution, while TMR solution alone showed an 85% decrease in fluorescence intensity under the same experimental conditions. Figure 12.4

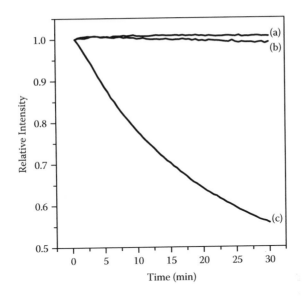

FIGURE 12.4 Photostability comparison of (a) RITC FloDot, (b) antibody-modified RITC FloDot, and (c) pure RITC in aqueous solution.

shows the photostability comparison between Rhodamine isothiocyanate (RITC), RITC FloDots, and antibody-labeled RITC FloDots. It can be clearly found that both FloDots and biomolecule-modified FloDots are very photostable in aqueous solution.

The luminescence samples have Brownian motion in the solution phase, so that they are maybe not really exposed to the excitation light all the time. A rigid experiment was designed to monitor the photostability of these luminescent probes at solid state and was performed in a thin-film format under a 0.503-W argon ion laser excited at 488 nm (10). A 0.1% poly(methyl methacrylate) (PMMA) solution in toluene was used as the film matrix by spin coating on a microscope coverslip. The thin film embedded with testing sample was mounted on the microscope for the laser excitation and the emission was collected by an ICCD camera. Four different luminescent samples, pure R6G, pure Rubpy, Rubpy FloDot, and post-silica–coated Rubpy FloDot, were compared with the same experimental method. Results show that the Rubpy FloDot is much more photostable than pure Rubpy or pure R6G. However, the Rubpy FloDot still has little photobleaching under the intense excitation by a laser beam. Two possible reasons may contribute to the photobleaching of the Rubpy FloDot in solid state. One is that a minute amount of oxygen probably penetrated into the silica matrix, and the other is that few Rubpy molecules are located close to the FloDot surface. These two possible reasons can be minimized by adding another silica layer to the FloDot. No photobleaching of post coated Rubpy FloDot can be observed over more than 400 s of intensive laser excitation.

Organic dye-doped FloDots were sandwiched between two coverslips for the photostability study (12). The samples were continuously illuminated for 1000 s and the fluorescence intensities were measured with a solid-state spectrofluorometer. The intensity of the pure R6G decreased rapidly, whereas the fluorescence intensity of the same R6G inside the FloDots did not change significantly under the same conditions.

The aforementioned results confirmed that both inorganic and organic dye-doped FloDots are very photostable in solution or solid state. The much improved photostability of the dye molecules in the FloDots makes it possible to minimize phobleaching of bioimaging and improve the accuracy of ultrasensitive bioanalysis.

SURFACE MODIFICATION AND BIOCONJUGATION OF FLODOTS

To be employed in biological applications, FloDots need to be conjugated with biomolecules for analyte recognition. The availability of surface functional groups on the FloDots can be confirmed by various means, such as the positive fluorescamine assay (10,31). A variety of surface modification and biomodification methods have been utilized to conjugate the FloDots to various biomolecules, including nucleic acids, antibodies, enzymes, and other proteins.

Surface Functionalization of FloDots

In order to immobilize the biomolecules, the FloDot surface must have functional groups that are available for bioconjugation. The FloDot surface has the same properties as that of a silica glass and thus provides a versatile substrate for surface immobilization with existing silica chemistry to attach desired functional groups (10). Those functional groups are often obtained by coating the FloDot with another layer of functionalized alkoxysilane by means of the gel-sol technique (22,28,32–34).

AMINO GROUP-MODIFIED FLODOTS

Amino groups can be modified on FloDots with aminopropyltriethoxysilane (APTS, available from Acros or Aldrich) at pH 4. The FloDot is only slightly negative at this pH and the amino groups should be neutral. Therefore, the very low surface charge will cause particle aggregation. This problem was solved by mixing 3-trihydroxysilylpropyl methyl-phosphonate, sodium salt, 42% in water (THPMP, from Gelest Inc.) and aminosilane with FloDots in phosphate buffer to modify the FloDot surface with both phosphate and amino groups (18). The excess phosphate groups should provide a highly negative surface charge for well-dispersed particles, although the number of amino groups is reduced. The amino/phosphate group-modified FloDots are stored in 10 mM, pH 5.6, MES buffer.

Amino-functionalized FloDots can also be prepared by continuously stirring FloDots with 1% N′-(3-(trimethoxysilyl)-propyl)diethylenetriamine (DETA) in 1 mM acetic acid at room temperature for 30 min and thoroughly washed three times with deionized water (10,15). Successful modification of amino group on the FloDot surface can be confirmed by the fluorescamine assay (10).

CARBOXYL GROUP-MODIFIED FLODOTS

Carboxyl modification is accomplished by using carboxyethylsilanetriol, sodium salt, 25% in water (CTES) (18) or with N-(trimethoxysilylpropyl) ethylenediamine, triacetic acid, sodium salt, 45% in water (both from Gelest Inc., Morrisville, PA). The first attaches one carboxylic acid group per molecule and the other three. Both of them are insoluble in ethanol, so the freshly prepared FloDots must be washed after synthesis and dispersed in aqueous solution. Reaction starts with adding carboxyl-silane into the FloDots solution in 10 mM pH 7.4 phosphate buffer, and finishes in several hours. Although a high temperature will enhance silane attachment, it will also increase the solubility of the silica matrix and hence cause the leaking of doped dye; so, the carboxyl modification of the FloDot is also generally performed at room temperature.

In an alternative method, amino-functionalized FloDots were first washed with N,N-dimethylformamide (DMF) and then mixed with 10% succinic anhydride in DMF solution under nitrogen for 6 h with continuous stirring. This reaction will also form carboxyl groups on the silica surface of FloDots (15,35,36).

OTHER FUNCTIONAL GROUP-MODIFIED FLODOTS

FloDots can also be surface modified with other functional groups. After the FloDot was activated with 2 M sodium carbonate, dropwise adding a solution of CNBr in acetonitrile into the FloDot suspension while stirring at room temperature will result in cyanate ester (–OCN) group modified FloDots (10). Further treatment like washing the cyanate ester modified FloDots should be performed at 4°C. Silanizing FloDots with 3-mercaptopropyl trimethoxysilane (MPTS) (1% in 95% ethanol) and acetic acid (16 mM, pH 4.5) for 30 min at room temperature followed by washing the particles once with 95% ethanol and 16 mM acetic acid will produce thiol-functionalized silica particles (37).

Bioconjugation of FloDots

Surface-functionalized FloDots can be further conjugated with biomolecules through physical absorption, chemical binding, or hydrophobic interaction. It is important to note that optimization will be required for each different bioconjugation step, such as the density of biomolecules on FloDot, the composition of the reaction buffer, and the structure/activity change of biomolecules. Here we just provide the basic information that has been used for the bioconjugation of FloDots.

PHYSICAL ABSORPTION

Physical absorption is a simple and flexible method to attach biomolecules onto a solid surface. The mechanism for absorption is based on ionic interaction (electrostatic interaction). The ionic strength and pH of reaction or storage buffer can affect the absorption.

Physical absorption is primarily used to modify FloDots with avidin molecule (12,14). Avidin, synthesized in the hen oviduct, is a glycoprotein of MW 68,000 daltons with an isoelectric point (pI) of 10.5 (38). It is a tetrameric protein composed of four identical subunits, each of which has one binding site for biotin. Biotin-avidin interaction has been widely applied for bioconjugation because (a) this interaction is the strongest known noncovalent biological coupling with a constant of 10^{15} L/mol (39), and (b) biotin is a small molecule and can be easily conjugated with other biomolecules without significantly changing their biological properties.

In most buffer conditions, the silica surface of FloDot is negatively charged and therefore it can be easily attached to the positively charged avidin. No surface functionalization is needed for the FloDots before the absorption step, and avidin molecules can be coated just by incubating with FloDots. The avidin-absorbed FloDots are usually treated with glutaraldehyde, which forms a covalent bond between avidin molecules to stabilize the avidin layer.

CHEMICAL BINDING

Chemical binding, or covalent coupling, can permanently conjugate the biomolecules on FloDots that have been surface functionalized with thiol (-SH) (37), cyanate ester (-OCN) (10), amino (-NH₂) (12,18,35,), or carboxyl (-COOH) (15,17,18) groups. A schematic diagram is shown in Figure 12.5.

Thiol or disulfide coupling is a simple and efficient procedure to directly immobilize disulfide-containing biomolecules on the solid surface through a thiol/disulfide exchange reaction (40). This procedure is a simple approach for bioconjugation and does not require any pretreatment of the thiol group-modified FloDots. Due to the high specificity of disulfide bonds, other functional groups on the surface will not affect the coupling. This method has been used to covalently immobilize presynthesized

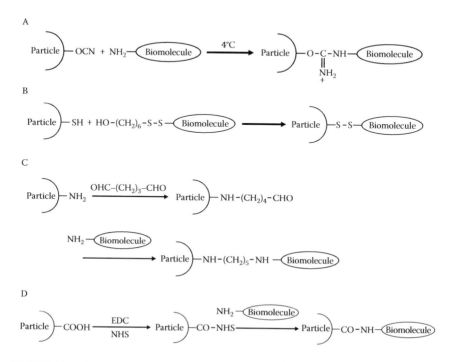

FIGURE 12.5 Schematic diagram of chemical binding for FloDots bioconjugation.

DNA probes on thiol-modified silica surface in pH 9 sodium carbonate buffer (37); however, it cannot be used under strong reducing conditions where the disulfide bond is very unstable. Whereas most proteins having amino groups can also be easily linked to the 5' or 3' terminal of DNA sequences, cyanate ester, amino, and carboxyl coupling are commonly used bioconjugation methods for FloDots.

Cyanate ester coupling is also a simple step (10). Soon after the FloDots are treated with CNBr and washed with ice-cold water and buffer, biomolecules will be added to the surface-functionalized FloDot solution, and stirring will be continued for 24 h at 4°C. The biomecule-conjugated FloDots will incubate with 0.03 M glycine solution for 30 min. The final product will be washed and resuspended in suitable buffer, and stored at 4°C for further experiments.

Amino-functionalized FloDots can be modified with biomolecules through bifunctional crosslinkers such as glutaraldehyde (12). Glutaraldehyde should be added in large excess so that the amino groups on FloDots will be saturated, thus avoiding the crosslinking of FloDots prior to biomolecule conjugation. But too much glutaraldehyde may change the native structure of biomolecules and reduce their biological activity, so that the amount of glutaraldehyde needs to be optimized. After surface activation, FloDots should be washed thoroughly to remove the excess reagents before adding biomelocules. Amino-functionalized FloDots can also easily react with the NHS group for surface conjugation, such as making biotin-labeled FloDots by the reaction between amino-modified FloDots and sulfo-NHS-LC-biotin (18).

Covalent coupling of biomolecules to carboxylated FloDots occurs via water-soluble carbodiimide (WSC) chemistry. Typically (18), COOH-modified FloDots react with 10 mM sulfo-NHS and 4 mM EDC in pH 5.5 MES buffer first to form NHS-modified FloDots, followed by careful washing. The activated FloDots will then incubate with biomolecules in 10 mM pH 7.4 phosphate buffer at room temperature for 2 h with gentle agitation. The remaining free NHS esters on the FloDot surface will be quenched by adding 40 mM Tris-HCl with 0.05% BSA. Bioconjugated FloDots will be purified by alternately washing and centrifugation, and resuspended in 10 mM pH 7.4 phosphate buffer with 1% BSA and stored at 4°C.

HYDROPHOBIC INTERACTION

Many biological samples have hydrophobic regions in their structure, and hydrophobic interaction can hence be used to conjugate FloDots with those biomolecules. Amino-modified FloDots were treated with lauroyl chloride in dry tetrahydrofuran in an argon atmosphere (11,35), and the hydrophobic lauroyl group will thus be introduced on the FloDot surface. The lauroyl-functionalized FloDots are useful as membrane probes to bind the hydrophobic part of the membrane (such as cell surface), which consists mainly of phospholipids bilayers.

BIOIMAGING APPLICATIONS OF FloDots

Most bioimaging is obtained through fluorescence microscopy and is very important in biological and biomedical research. Unlike the conventional immunoassay where each antibody is labeled with, at most, several fluorophores for signaling, every biomodified FloDot can bring significant signal amplification due to thousands of fluorescent dyes doped inside each particle. Because of the excellent photostability, FloDots can also withstand continuous excitation. These two merits make FloDots a superior probe for bioimaging.

BIOIMAGING FOR IMMUNOCYTOCHEMISTRY

Immunolabeling of the cell surface with fluorescent probes is commonly used in cell biology, immunology, and clinical laboratories. Mouse antihantihuman CD10 antibody labeled FloDots have been used for cellular imaging of human leukemia cells (10). The antibody was immobilized on FloDots through cyanate ester coupling. The antibody-labeled FloDots were incubated with a suspension of leukemia cells in the cell culture media for 2 h, followed by washing away the unbound FloDots with pH 6.8 PBS buffer. The cell suspension was then imaged with both optical microscopy and fluorescence microscopy. The optical images and the fluorescence images were found to be very well matched. The bright spots on the fluorescence image indicated that FloDots were labeled on leukemia cells. In one control experiment where bare FloDots (no antibody-labeled) were incubated with leukemia cells, no bright spots could be found in the fluorescence, which meant that bare FloDots could not be attached on leukemia cells. In another control experiment where mouse antihantihuman CD10 antibody-labeled FloDots were incubated with PTK2 cells, also no FloDots could be found to bind on the cell surface due to the lack of the specific antigens on the cell membrane (10,35). All these results clearly demonstrated that antibody-labeled FloDots could be used as specific probes for cell imaging. It is also noteworthy that in the traditional method using FITC-labeled secondary antibody for leukemia cell detection, a signal-to-background ratio of 100 is regarded as a significantly high signal. Here the FloDots method provided a ratio of over 500, which means sensitive antigen detection could be achieved with FloDots, which is very useful for trace amount detection when antigen expression is at an early stage of disease developmemt.

FloDots have also been reported for the immunocytochemistry study of lymphocytes (B cell). IgM-positive lymphocytes were tested by using biotinylated anti-IgM and streptavidin-labeled FloDots (17). Streptavidin was conjugated on the FloDot surface through carboxyl coupling. Purified human peripheral blood mononuclear cells were first incubated with biotinylated mouse antihuman IgM and then incubated with streptavidin-conjugated FloDots or streptavidin-conjugated phycoerythrin (PE). After washing away the unbound reagents, the cells were applied to slides by cytospin centrifugation, and then stained with DAPI to visualize the nuclei. Combined with the cell morphology, FloDot-labeled lymphocytes could be clearly observed under a fluorescence microscope as shown in Figure 12.6, and similar results with weaker intensity were obtained with PE labeling.

BIOIMAGING FOR IMMUNOHISTOCHEMISTRY

In immunohistochemistry research and clinical laboratories, tissue sections are routinely labeled with fluorescent dyes. FloDots have been reported to image cholinergic neurons (17). Specific labeling of cholinergic neurons is important in both

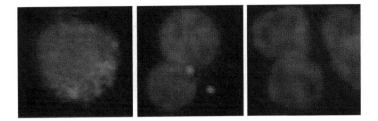

FIGURE 12.6 Labeling of B cell surface IgM molecules with FloDot *(left)*, PE *(middle)*, and control result *(right)*. Human peripheral blood mononuclear cells (PBMC) were first incubated with biotinylated mouse antihIgM and then with streptavidin-NP or avidin-PE (red). The PBMC were also stained with DAPI to visualize the nuclei (blue). (*Anal. Biochem.* 2004, 334, 135–144.)

clinics and research of Alzheimer's disease. Since choline acetyltransferase (ChAT) only exists in the target cholinergic neurons in the brain, it is therefore the best cholinergic marker (17,41,42). In the experiment, mouse brain section was first incubated with biotinylated goat anti-ChAT antibody and detected with streptavidin-labeled FloDots. From Figure 12.7, the population of whole cells can be found by the nuclei stained with DAPI, and the cell body of target cholinergic neurons was stained by FloDots, as judged from both the morphology and location of the FloDots stained cells (17,41,42).

ULTRASENSITIVE BIOANALYSIS

Compared to luminescent dye molecules, FloDots have similar luminescent properties, so that bioconjugated FloDots have the potential to be used as luminescent probes for bioanalysis. Meanwhile, the enhanced luminescent intensity emitted by

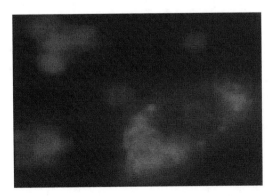

FIGURE 12.7 Labeling of choline acetyltransferase (ChAT) using FloDot. Mouse brain section was first incubated with biotinylated goat anit-ChAT Ab and then detected with streptavidin-NP (red). Cells were also stained with DAPI (blue). (*Anal. Biochem.* 2004, 334, 135–144.)

FloDots will highly increase the sensitivity of current analytical methods, which will result in the detection of minute changes in a biological system and early diagnosis of diseases. The excellent photostability will further ensure that accurate and reproducible results can be obtained from each analysis. All these observations indicate that FloDots, upon carefully modification and optimization, can be applied as powerful probes for ultrasensitive bioanalysis.

Fluorescence-Linked Immunosorbent Assay

Fluorescence-linked immunosorbent assay (FLISA) is an ELISA-like assay except that a fluorescence probe takes the place of the enzyme. Because FloDots have extremely high intensity, it is expected that FloDot-based FLISA will have high sensitivity.

R6G FloDot was first applied for immunosorbent assay (12). R6G FloDots were prepared by the Stöber method using cohydrolysis of TEOS and PTES. Avidin was physically absorbed on FloDots and crosslinked by glutaraldehyde. Glass slides were coated with avidin using the same method and treated with different concentrations of biotinylated bovine serum albumin (BSA), were where each BSA molecule has an average of nine biotin molecules. Avitin-conjugated FloDots were then added on the glass surface. After washing away the unbound FloDots, the glass slide was measured under a fluorescence microscope with 520-nm excitation and 550-nm emission. The amount of surface-bound FloDots will be dependent on the amount of available biotin, and hence will be dependent on the concentration of biotinylated BSA. Images in Figure 12.8 shows that the number of FloDots bound on the glass slide (hence the fluorescence intensity) will increase with the concentration of biotinylated BSA, which clearly demonstrates that FloDots can be used for immunosorbent assay.

Application of FloDots for immunosorbent assay has been investigated by Lian and others (17). Rubpy FloDots prepared with reverse microemulsion method were used for the study. Carboxyl groups were first modifies on FloDots through DETA and succinic anhydride, and avidin was then conjugated with FloDots through EDC and NHS. Mixtures of biotinylated hIgG and regular hIgG were immobilized on 96-well microplates in serial dilutions. The coated plates were blocked with BSA and probed with avidin-conjugated FloDots. Results show that the relationship between the luminescence intensity and biotin-hIgG concentration was linear in the range of 20 nm/mL to 2.5 mg/mL (17). Experiments also revealed that under the same assay conditions, FloDot-based FLISA resulted in the lowest detection limit, 1.9 ng, compared to 62.5 ng for quantum dot and 250 ng for Texas Red (17). In the indirect FLISA assay where the biotinylated target molecule is not available, FloDot-based assay also provided the linear relationship between the luminescence intensity and the target concentration with a higher sensitivity.

FloDot-Based DNA Microarrays

Sensitive detection of nucleic acids is extremely important in clinical diagnostics and gene therapy, and new biotechnologies have been developed to improve the sensitivity and selectivity for gene analysis. DNA microarray has been widely used for simultaneous detection of multiple different DNA sequences in a single experiment

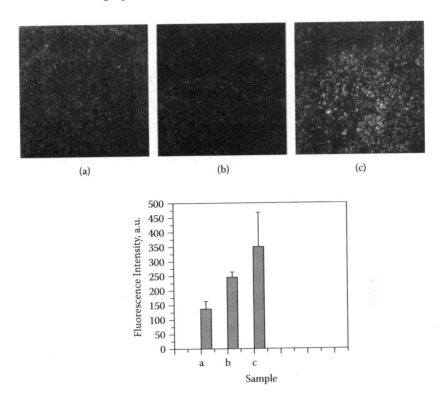

FIGURE 12.8 Bioassay for biotinylated BSA. Biotin interaction of avidin-modified FloDots: fluorescence images of FloDots on the samples treated with (a) BSA, control; (b) 1 mg/mL biotinylated BSA; (c) 2 mg/mL biotinylated BSA. All three images are in the same display range. The graph is a quantitative representation of the images. (*J. Nanosci. Nanotech.*, 2002, 2, 405–409)

(43–51). Most microarray techniques are based on nucleic acid hybridization where the fluorescent dye is utilized to signal the DNA pairing (52–55). However, each DNA target can only hybridize with one DNA probe labeled with only one or a few fluorescent dye molecules; low concentration of target DNA will be hardly accurately detected due to the weak fluorescent signal and serious photobleaching.

Application of FloDots to DNA microscopy has highly increased the detection sensitivity. Although each target DNA still hybridized with one DNA probe, each probe DNA was attached with one FloDot that was loaded with large amounts of dye molecules. Therefore, each hybridization will be reported by the integrity of dye molecules embedded in each FloDot.

A FloDots-based DNA array was first realized in a sandwich structure (14). A biotinylated capture DNA, complement of a portion of the target sequence, is first immobilized through avidin-biotin linkage to the glass substrate. Part of the target sequence will be bound with the capture DNA, while the remaining part of the target sequence can hybridize with a probe DNA conjugated with the FloDot. Typically, a 12-mer biotinylated DNA1 (5′ TAA CAA TAA TCC T-biotin 3′) was used as the capture DNA and was immobilized on an avidin-coated glass substrate to hybridize to one end of a 27-mer target DNA2 (5′ GGA TTA TTG TTA AAT TTA GAT AAG GAT 3′). The remaining 15 bases of the target were hybridized with the detection probe DNA3 (5′ biotin- TAT CCT TAT CTA AAT T 3′) attached to the TMR FloDot. The TMR FloDot was made via the reverse microemulsion method using TMR-dextran and acidic droplet. After physical absorption and crosslinking of avidin, TMR FloDots were attached to the biotinylated probe DNA3 with biotin-avidin coupling. Hybridization between probe DNA with target DNA will leave the FLoDot, thus thousands of dye molecules, on the substrate for signaling. By monitoring the luminescent intensity from the surface-bound FloDots, DNA target molecules can be detected with high sensitivity (0.8 fM detection limit) and good selectivity (little nonspecific binding of one-base mismatch DNA or random DNA).

In another work, biotinylated *P. aeruginosa* DNA was hybridized with the target genomic DNA from *P. aeruginosa* that had been printed on the CMT-GAPS slides (Corning). Avidin-conjugated Rubpy FloDots or streptavidin-conjugated Cy3 was then used as the signaling probe. Although the slides were finally scanned with the optimal excitation and emission wavelengths for Cy3, at which the Rubpy FloDot only emitted 5% of its maximum, DNA array labeled with FloDot was much brighter than that with Cy3, as shown in Figure 12.9. Results also showed that the detected luminescence intensity is proportional to the amount of target DNA spotted on the slide. All these data suggested that sensitive and quantitative results can be obtained with FloDot-based DNA microassay.

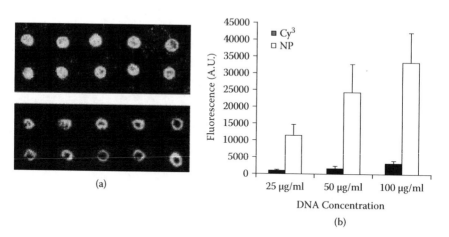

(a)   DNA Concentration

(b)

FIGURE 12.9 Labeling of DNA microarray with FloDot or Cy3. (a) DNA chips spotted with the same amount of *Pseudomonas aeruginosa* genomic DNA were hybridized with biotinylated probes generated from *P. aeruginosa* genomic DNA and detected with either avidin-labeled FloDot (upper slide) or streptavidin-labeled Cy3 (lower slide). The images were scanned with optimal excitation and emission wavelengths for Cy3. (b) Quantitative detection of target DNA on the microarray with avidin-FloDot (white bar) or with streptavidin-Cy3 (black bar) after scanning under optimal excitation and emission wavelengths for Cy3. (*Anal. Biochem.* 2004, 334, 135–144).

FLODOT-BASED PROTEIN MICROARRAYS

Protein microarray has become an increasingly important tool in genomic and proteomic study such as protein expression, profiling, and interaction (56–58). This technique is anticipated to replace many of the traditional techniques like filter binding, column chromatography, and gel-shift assay (58). Although all the basic principles established for DNA microarray can be extended to protein microarray, the latter has to overcome its own challenges due to the much higher complexity of proteins. In the DNA microarray study, the target DNA can always be amplified first by polymerase chain reaction (PCR) before the hybridization on array chips, while no method is now available for protein amplification. Therefore, highly improved sensitivity must be achieved for protein microarray.

Highly luminescent FloDot was proved to increase the detection capability of protein microarray (17). Serial dilutions of a mixture of hIgG-biotin and regular HIgG from 0.5 mg/ml to 100 ng/ml were spotted on SuperEposy microarray slides (Eric Scientific), followed by blocking with 1% BSA. The slide was then probed with avidin-conjugated Rubpy FloDots and imaged with a GenePix 4000B scanner. A high labeling specificity was observed by the highly bright target spots and negligible signal of the negative control. As shown in Figure 12.10, the luminescence intensity of each spot was also found to be proportional to the amount of hIgG-biotin applied. The sensitivity of FloDot-based protein microarray was determined by the comparison of luminescence signal intensities of protein array labeled with either Rubpy FloDots or FITC, and the analysis results showed that FloDot-based protein array is about 80 times more sensitive than FITC-based protein array. Experiments with protein microarray have also been performed to detect mouse IgG (mIgG) with sandwich structure. Slides spotted with goat anti-mIgG were hybridized with

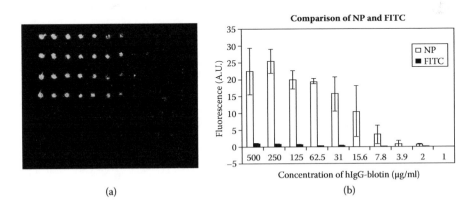

(a)   (b)

FIGURE 12.10 Labeling of protein chips with the FloDots or FITC. (a) Top four rows were spotted with serial dilution (from left to right) of biotinylated hIgG while lower four rows were spotted with that of hIgG. Avidin-FloDot was used to detect the fluorescence. (b) Comparison of fluorescent signal intensities of protein chips labeled with either FloDots or FITC. (*Anal. Biochem.* 2004, 334, 135–144.)

different concentrations of mIgG (antigen) and detected with excess of biotinylated goat anti-mIgG followed by avidin-conjugated FloDots. A linear relationship was found between the luminescence intensity and antigen concentration, and a detection limit of below 1 pg/mL could be achieved. All these results confirm that FLoDots are applicable to the ultrasensitive analysis of protein microarray.

FLoDot-Based Single Bacterium Detection

Simple, sensitive, effective, and rapid detection of pathogenic bacteria is of critical importance to human safety, environment protection, diagnosis of disease, elimination of infection. and antibioterrorism. Due to their complicated procedures of amplification or enrichment of target bacteria in the testing sample, traditional methods are laborious and time-consuming. Although many attempts had been made to improve the sensitivity, rapid bacteria detection at single-cell level was still a challenge.

With the development of bionanotechnology, single bacterial cells have been detected by FloDots (15). In conventional immunoassay, the bacterium was detected through antibody-antigen recognition where only several dye molecules were labeled on the antibody. Each FloDot is hundreds of times smaller than the bacterium, so that as high as 10,000 antibody-conjugated FloDots can bind to the surface of each bacterium. Since the luminescence intensity of each FloDot is more than thousands of times higher than the single dye, a FloDot-based method will provide the highly amplified signal for ultrasensitive sample detection.

In the newly reported method, Rubpy FloDot was first used for the detection of single *Escherichia coli* (*E. coli*) O157:H7 bacterium. *E. coli* O157:H7 is one of the most dangerous agents of food-borne disease. The presence of even a single *E. coli* O157:H7 bacterium may pose a serious outbreak that can lead to the death of children and early populations. In the experiment, FloDots were synthesized through the reverse microemulsion method, and further surface-functionalized with carboxyl group. Antibody against *E. coli* O157:H7 was modified on the carboxylated FloDot using carbodiimide chemistry. Bacterial samples were mixed with antibody-conjugated FloDots in a buffer for 10 min, and unbound antibody-conjugated FloDot were removed by centrifugation and washing. The bacteria, surface labeled with antibody-conjugated FloDots, were resuspended in the buffer and measured with a spectrofluorometer or flow cytometer. Results confirmed that a single bacterium could be detected accurately in less than 20 min. Furthermore, the FloDot based assay was successfully employed for high-throughput detection of multiple samples using a 384-well microplate format, which made possible bioanalysis for multiple pathogens. In addition, the practical use of this assay has been demonstrated by the accurate and reliable detection of 1–400 *E. coli* O157:H7 bacteria in spiked ground beef samples, as shown in Figure 12.11.

FLoDot-Based Multiplexed Bioanalysis

Detection of multiple targets such as different proteins and genes is important for patent recognition in disease diagnosis and biomedical research. Development of the multiplexed bioassay has accelerated in recent years (59–62). Although microarray

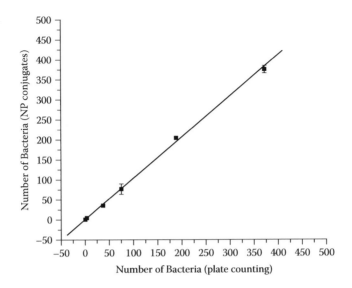

FIGURE 12.11 Single-bacterium detection with beef sample. Detection of *E. coli* O157:H7 in spiked ground beef was done with the plate-counting method and the antibody-conjugated, nanoparticle (NP)-based method. Bacteria in the range of 1–400 cells per sample were detected. The two methods had linear correlation, with an *R* value of 0.99. Total detection time for the beef sample was ≈20 min for the antibody-conjugated, nanoparticle-based method, and that for the plate-counting method was >1 day (*PNAS*, 2004 101, 15027–15032).

can handle a high degree of multiplexed detection (63–65), it cannot be easily applied on a routine basis or in real-time imaging of biological samples. Polymer microbeads, doped with either fluorophores or quantum dots, have offered several advantages (62,66,67). However, many biological systems are naturally nanostructured, so that it is important to develop nanometer scale probes with multiplexed capability.

Although different dye molecules can be used to make FloDots, they cannot be easily applied to multiplexed bioanalysis with a single-excitation-multiple-emission setup. Development of dual-luminophore–doped FloDots has solved the problem (18). Two luminophores, Rubpy and Osbpy, were doped together inside the silica matrix at precisely controlled ratios. Rubpy and Osbpy have a broad range of overlapped excitation wavelength, and two distinct maximum emission wavelengths with Rubpy at 610 nm and Osbpy at 710 nm. Upon excition at a single wavelength, the dual-luminophore–doped FloDot will have two well-resolved emission peaks, and the luminescence intensity ratios of these two peaks were found to be correlated to the molar ratio of the two dyes doped inside the FloDot. Changing the molar ratio of the doped-dye molecules will produce FloDots with a different intensity ratio, and hence made the multiplexed bioanalysis possible and practical.

In order to prove that those dual-luminophore–doped FloDots can be used for multiplexed bioanalysis, FloDots with intensity ratios (610 nm / 710 nm) of 9:1 and 2:1 were synthesized via the reverse microemulsion method and functionalized with carboxyl group with CTES, followed by modification with human IgG or mouse

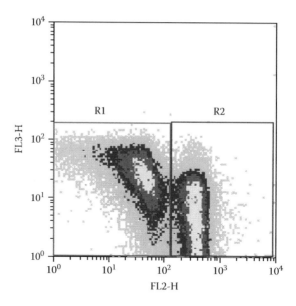

FIGURE 12.12 Two-dimensional dot plot showing classification of the two microsphere sets based on simultaneous analysis of logarithmic orange fluorescence (FL2) and logarithmic red fluorescence (FL3). (*Nano. Lett.* 2005, 5, 37–43)

IgG, respectively, through the EDC/NHS method. The same number of antimouse IgG and antihuman IgG coated microspheres (1.62×10^{-17} mol, respectively) were mixed with these two kinds of IgG coated NPs (8.1×10^{-15} mol, respectively) to form a cocktail. This cocktail was then diluted with buffer and flown through a flow cytometer. Figure 12.12 is a two-dimensional dot plot showing classification of the two microsphere sets based in simultaneous analysis of logarithmic orange luminescence (FL2) and red luminescence (FL3) in a flow cytometer. The dots in the R1 region represent antimouse IgG microspheres coated with mouse IgG NPs (2:1 ratio), and those in the R2 region represent antihuman IgG microspheres coated with human IgG NPs (9:1 ratio). Experimental results show that the population distribution is 46.56% in R1 region and 53.42% in R2 region, which correlates well with the expected value (equally distributed due to the same number of those two microspheres). This result demonstrated a model for the multiplexed detection of microspheres by bioconjugated FloDots, and provided the possibility to apply these smart FloDots for multiplexed target recognition. With the same concept, the dual-dye–doped FloDots can also be used for multiplexed bacterial pathogens detection and related experiments are now ongoing.

CONCLUSION

FloDots, namely luminescent dye-doped silica nanoparticle, have been successfully synthesized and applied for bioimaging and ultrasensitive bioanalysis. Those novel luminescent probes have many advantages over current, commonly used fluorophores due to the extremely high intensity, excellent photostability, and easy surface modification. Currently, we are working on the development and optimization of FloDots, which are smaller and uniform, brighter, and photostable, as well as developing protocols for easy batch production and biomolecule conjugation.

ACKNOWLEDGMENTS

We thank our colleagues at the University of Florida for their help with the work reported in this manuscript. This work was supported by NIH grants and NSF NIRT.

REFERENCES

1. Haugland, P., *Handbook of Fluorescent Probes and Research Chemicals*, 9th ed., Molecular Probes, Inc., Eugene, Oregon.
2. Song, L., Hennink, E.J., Young, I.T., Tanke, H.J., Photobleaching kinetics of fluorescein in quantitative fluorescence microscopy, *Biophys. J.* 68 (1995) 2588–2600.
3. Benson, D.M., Bryan, J., Plant, A.L., Gotto, Jr., A.M., Smith, L.C., Digital imaging fluorescence microscopy: Spatial heterogeneity of photobleaching rate constants in individual cells, *J. Cell Biol.* 100 (1985) 1309–1323.
4. Chan, W.C.W., Nie, S., Quantum dot bioconjugates for ultrasensitive nanisotopic detection, *Science* 285 (1998) 2016–2018.
5. Chan, W.C.W., Maxerll, D.J., Gao, X., Bailey, R.E., Han, M., Nie, S., Luminescent quantum dots for multiplex biological detection and imaging, *Curr. Opin. Biotechnol.* 13 (2002) 40–46.
6. Emory, S.R., Nie, S., Screening and enrichment of metal nanoparticle with novel optical properties, *J. Phys. Chem. B* 102 (1998) 493–497.
7. Bourel, D., Rolland, A., Leverge, R., Genetet, B., A new immunoreagent for cell labeling-CD3 monoclonal-antibody covalently coupled to fluorescent polymethacrylic nanoparticles, *J. Immunol. Methods* 106 (1988) 161–167.
8. Adler, J., Jayan, A., Melia, C.D., A method for quantifying differential expansion within hydrating hydrophilic matrixes by tracking embedded fluorescent microspheres, *J. Pharm. Sci.* 88 (1999) 371–377.
9. Taylor, J.R., Fang, M.M., Nie, S., Probing specific sequences on single DNA molecules with bioconjugated fluorescent nanoparticles, *Anal. Chem.* 72 (2000) 1979–1986.
10. Santra, S., Zhang, P., Wang, K., Tapec, R., Tan, W., Conjugation of biomolecules with luminophore-doped silica nanoparticles for photostable biomarkers, *Anal. Chem.* 73 (2001) 4988–4993.
11. Santra, S., Wang, K., Tapec, R., Tan, W., Development of novel dye doped silica nanoparticles for biomarker application, *J. Biomed. Optics*, 6:2 (2001)160–166.
12. Tapec, R., Zhao, X.J., Tan, W., Development of organic dye-doped silica nanoparticles for bioanalysis and biosensors, *J. Nanosci. NanoTechnol.* 2 (2002) 405–409.
13. Bagwe, R.P., Yang, C., Hilliard, L.R., Tan, W., Optimization of dye-doped silica nanoparticles prepared using a reverse microemulsion method, *Langmuir* 20 (2004) 8336–8342.
14. Zhao, X., Tapec-Dytioco, R., Tan, W., Ultrasensitive DNA detection using highly fluorescent bioconjugated nanoparticles, *J. Am. Chem. Soc.* 125 (2003) 11474–11475.
15. Zhao, X., Hilliard, L.R., Mechery, S.J., Wang, Y., Bagwe, R.P., Jin, S., Tan, W., A rapid bioassay for single bacterial cell quantitation using bioconjugated nanoparticles, *Proc. Natl. Acad. Sci. USA* 101 (2004) 15027–15032.

16. Zhao, X., Bagwe, R.P., Tan, W., Development of organic-dye-doped silica nanoparticles in reverse microemulsion, *Adv. Mater.* 16:2 (2004) 173–176.

17. Lian, W., Litherland, S.A., Badrane, H., Tan, W., Wu, D., Baker, H.V., Gulig, P.A., Lim, D.V., Jin, S., Ultrasensitive detection of biomolecules with fluorescent dye-doped nanoparticles, *Anal. Biochem.* 334 (2004) 135–144.

18. Wang, L., Yang, C., Tan, W., Dual-luminophore-doped silica nanoparticles for multiplexed signaling, *Nano Lett.* 5:1 (2005) 37–43.

19. Hines, M.A., Guyot-Sionnest, P., Bright UV-blue luminescent colloidal ZnSe nanocrystals, *J. Phys. Chem. B.* 102 (1998) 3655–3657.

20. Bruchez, M.J., Moronne, M., Alivisatos, A.P., Weiss, S., Semiconductor nanocrystals as fluorescent biological labels. *Science* 281(1998) 2013–2016.

21. Dahan, M., Laurence, T., Pinaud, F., Chemla, D.S., Alivisatos, A.P., Sauer, M., Weiss, S., Time-gated biological imaging by use of colloidal quantum dots. *Optics Letters* 26 (2001) 825–827.

22. Gerion, D., Pinaud, F., Williams, S.C., Parak, W.J., Zanchet, D., Weiss, S., Alivisatos, A.P., Synthesis and properties of biocompatible water-soluble silica coated CdSe/ZnS semiconductor quantum dots. *J. Phys. Chem. B* 105, (2001) 8861–8871.

23. Li, T., Moon, J., Morrone, A.A., Mecholsky, J.J., Talham, D.R., Adair, J.H., Preparation of Ag/SiO$_2$ nanosize composites by a reverse micelle and Sol-Gel technique, *Langmuir* 15 (1999) 4328–4334.

24. Shiojiri, S., Hirai, T., Komasawa, I., Immobilization of semiconductor nanoparticles formed in reverse micelles into polyurea via *in situ* polymerization of diisocyanates, *Chem. Commun.* 14 (1998) 1439–1440.

25. Chang, S.Y., Liu, L., Asher, S.A., Preparation and properties of tailored morphology, monodisperse colloidal silica-cadmium sulfide nanocomposites, *J. Am. Chem. Soc. 116* (1994) 6739–26. Chang, S.Y., Liu, L., Asher, S.A., Creation of templated complex topological morphologies in colloidal silica, *J. Am. Chem. Soc. 116* (1994) 6745–6747.

27. Stathatos, E., Lianos, P., Del Monte, F., Levy, D., Tsiourvas, D., Formation of TiO$_2$ nanoparticles in reverse micelles and their deposition as thin films on glass substrates, *Langmuir* 13, (1997) 4295–4300.

28. Stöber, W., Fink, A., Bohn, E., Control growth of monodisperse silica spheres in micron size range, *J. Colloid Interface Sci.* 26 (1968) 62–69.

29. Shibata, S., Yano, T., Yamane, M., Formation of dye-doped silica particles, *Jpn. J. App. Phys.* 37 (1998) 41–44.

30. Shibata, S., Taniguchi, T., Yano, T., Yamane, M., Formation of water-soluble dye-doped silica particles, *J. Sol-Gel Sci. Technol.* 10 (1997) 263–268.

31. Chung, L.A., A fluorescamine assay for membrane protein and peptide samples with non-amino-containing lipids, *Anal. Biochem.* 248 (1997) 195–201.

32. Liz-Marzan, L.M., Giersig, M., Mulvaney, P., Synthesis of nanosize gold–silica coreshell particles, *Langmuir* 12 (1996) 4329–4335.

33. Ung, T., Liz-Marzan, L.M., Mulvaney, P., Controlled method for silica coating of silver colloids. Influence of coating on the rate of chemical reactions, *Langmuir* 14 (1998) 3740–3748.

34. Buining, P.A., Humbel, B.M., Philipse, A.P., Verkleij, A.J., Preparation of functional silane stabilized gold colloids in the (sub)nanometer size range, *Langmuir* 13 (1997) 3921–3926.

35. Qhobosheane, M., Santra, S., Zhang, P., Tan, W., Biochemically functionalized silica nanoparticles, *Analyst* 126 (2001) 1274–1278.

36. Tan, W., Wang, K., He, X., Zhao, X.J., Drake, T., Wang, L., Bagwe, R., Bionanotechnology based on silica nanoparticles, *Medicinal Res. Rev.* 24 (2004) 621–638.

37. Hilliard, L.R., Zhao, X., Tan, W., Immobilization of oligonucleotides onto silica nano-particles for DNA hybridization studies, *Anal. Chim. Acta* 470 (2002) 51–56.
38. http://www.affiland.com/avidin.html
39. Blakenburg, R., Meller, P., Ringsdorf, H., Salesse, C., Interaction between biotin lipids and streptavidin in monolayers—Formation of oriented two-dimensional protein domains induced by surface recognition, *Biochemistry* 28 (1989) 8214–8221.
40. Rogers, Y.H., Jiang-Baucom, P., Huang, Z.J., Bogdanov, V., Anderson, S., Boyce-Jacino, M.T., Immobilization of oligonucleotides onto a glass support via disulfide bonds: a method for preparation of DNA microarrays, *Anal. Biochem.* 266 (1999) 23–30.
41. Houser, C.R., Cholinergic synapses in the central nervous system: Studies of the immunocytochemical localization of choline acetyltransferase, *J. Electron Microsc. Tech.* 15 (1990) 2–19.
42. Quirion, R., Cholinergic markers in Alzheimer disease and the autoregulation of acetylcholine release, *J. Psychiatry Neurosci.* 18 (1993) 226–234.
43. Tamayo, P., Slonim, D., Mesirov, J., Zhu, Q., Kitareewan, S., Dmitrovsky, E., Lander, E.S., Golub, T.R., Interpreting patterns of gene expression with self-organizing maps: Methods and application to hematopoietic differentiation, *Proc. Natl. Acad. Sci. U.S.A.* 96 (1999) 2907–2912.
44. Salama, N., Guillemin, K., McDaniel, T.K., Sherlock, G., Tompkins, L., Falkow, S., A whole-genome microarray reveals genetic diversity among *Helicobacter pylori* strains, *Proc. Natl. Acad. Sci. U.S.A.* 97 (2000) 14668–14673.
45. Gasch, A.P., Spellman, P.T., Kao, C.M., Carmel-Harel, O., Eisen, M.B., Storz, G., Botstein, D., Brown, P.O., Genomic expression programs in the response of yeast cells to environmental changes, *Mol. Biol. Cell* 11 (2000) 4241–4257.
46. Schena, M., Shalon, D., Davis, R.W., Brown, P.O., Quantitative monitoring of gene-expression patterns with a complementary-dna microarray, *Science* 270 (1995) 467–470.
47. Alizadeh, A.A., Eisen, M.B., Davis, R.E., Ma, C., Lossos, L.S., Rosenwald, A., Boldrick, J.C., Sabet, H., Tran, T., Yu, X., Powell, J.I., Yang, L., Marti, G.E., Moore, T., Hudson, Jr., J., Lu, L., Lewis, D.B., Tibshirani, R., Sherlock, G., Chan, W.C., Greiner, T.C., Weisenbuerger, D.D., Armitage, J.O., Warnke, R., Staudt, L.M., Distinct types of diffuse large B-cell lymphoma identified by gene expression profiling, *Nature* 3 (2000) 503–511.
48. Golub, T.R., Slonim, D.K., Tamayo, P., Huard, C., Gaasenbeek, M., Mesirov, J.P., Coller, H., Loh, M.L., Downing, J.R., Caligiuri, M.A., Bloomfield, C.D., Lander, E.S., Molecular classification of cancer: class discovery and class prediction by gene expression monitoring, *Science* 286 (1999) 531–537.
49. Debouck, C., Goodfellow, P.N., DNA microarrays in drug discovery and development, *Nat. Genet.* 21 (1999) 48–50.
50. Grifantini, R., Bartolini, E., Muzzi, A., Draghi, M., Frigimelica, E., Berger, J., Ratti, G., Petracca, R., Galli, G., Agnusdei, M., Giuliani, M.M., Santini, L., Brunelli, B., Tettelin, H., Rappuoli, R., Randazzo, F., Grandi, G., Previously unrecognized vaccine candidates against group B meningococcus identified by DNA microarrays, *Nat. Biotechnol.* 20 (2002) 914–921.
51. Fan, J.B., Chen, X. X., Halushka, M.K., Berno, A., Huang, X., Ryder, T., Lipshutz, R.J., Lockhart, D.J., Chakravaiti, A., Parallel genotyping of human SNPs using generic high-density oligonucleotide tag arrays, *Genome Res.* 10 (2000) 853–860.
52. Weiss, S., Fluorescence spectroscopy of single biomolecules, *Science* 283 (1999) 1676–1683.
53. Fang, X., Liu, X., Schuster, S., Tan, W., Designing a novel molecular beacon for surface-immobilized DNA hybridization studies, *J. Am. Chem. Soc.* 121 (1999) 2921–2922.

54. Yao, G., Fang, X., Yokota, H., Yanagida, T., Tan, W., Study of dynamics of molecular beacon DNA probe hybridization at single molecule level, *Chem. A Euro. J.* 9 (2003) 5686–5692.

55. Yao, G., Tan, W., A molecular beacon based array for sensitive DNA analysis, *Anal. Biochem.* 331 (2004) 216–223.

56. Belov, L., de la Vega, O., dos Remedios, C.G., Mulligan, S.P., Christopherson, R.I., Immunophenotyping of leukemias using a cluster of differentiation antibody microarray, *Cancer Res.* 61 (2001) 4483–4489.

57. Wegner, G.J., Lee, H.J., Marriott, G., Corn, R.M., Fabrication of histidine-tagged fusion protein arrays for surface plasmon resonance imaging studies of protein-protein and protein-DNA interactions, *Anal. Chem.* 75 (2003) 4740–4746.

58. Stears, R.L., Martinsky, T., Schena, M., Trends in microarray analysis, *Nat. Med.*, 9 (2003) 140–145.

59. Eriksson, S., Vehniainen, M., Jansen, T., Meretoja, V., Saviranta, P., Pettersson, K., Lovgren, T., Dual-label time-resolved immunofluorometric assay of free and total prostate-specific antigen based on recombinant Fab fragments. *Clin. Chem.* 46 (2000) 658–666.

60. Plowman, T.E., Durstchi, J.D., Wang, H.K., Christensen, D.A., Herron, J.N., Reichert, W.M., Multiple-analyte fluoroimmunoassay using an integrated optical waveguide sensor, *Anal. Chem.* 71 (1999) 4344–4352.

61. McBride, M.T., Gammon, S., Pitesky, M., Brien, T.W.O., Smith, T., Aldrich, J., Langlois, R.G., Colston, B., Venkateswaran, K.S., Multiplexed liquid arrays for simultaneous detection of simulants of biological warfare agents, *Anal. Chem.* 75 (2003) 1924–1930.

62. Xu, H., Sha, M.Y., Wong, E.Y., Uphoff, J., Xu, Y., Treadway, J.A., Truong, A., Brien, E.O., Asquith, S., Stubbins, M., Spurr, N.K., Lai, E.H., Mahoney, W., Multiplexed SNP genotyping using the Qbead™ system: A quantum dot-encoded microsphere-based assay, *Nucl. Acids Res.* 31 (2003) e43.

63. S. L. Beaucage, Strategies in the preparation of DNA oligonucleotide arrays for diagnostic applications. *Curr. Med. Chem.* 8 (2001) 1213–1244.

64. Stimpson, D.I., Knepper, S.M., Shida, M., Obata, K., Tajima, H., Three-dimensional microarray platform applied to single nucleotide polymorphism analysis, *Biotechnol. Bioeng.* 87 (2004) 99–103.

65. Qiu, J., Gurpide, J.M., Misek, D.E., Kuick, R., Brenner, D.E., Michailidis, G., Haab, B.B., Omenn, G.S., Hanash, S., Development of natural protein microarrays for diagnosing cancer based on an antibody response to tumor antigens, *J. Proteome Res.* 3 (2004) 261–267.

66. Spiro, A., Lowe, M., Brown, D., A bead-based method for multiplexed identification and quantitation of DNA sequences using flow cytometry, *Appl. Environ. Microbiol.* 66 (2000) 4258–4265.

67. Brodsky, A.S., Silver, P.A., A microbead-based system for identifying and characterizing RNA-protein interactions by flow cytometry, *Mol. Cell. Proteom.* 1 (2002) 922–929.

13 Photocatalytic Particles for Biocidal Applications

G. Pyrgiotakis and W. Sigmund

CONTENTS

INTRODUCTION

In the last few decades, the demand for safer environmental conditions has dramatically increased. One of the major reasons is the constantly growing biological threat from emerging pathogens that can be seen in every aspect of daily life, ranging from cases as simple as food bacterial contamination (*E. coli* and *salmonella*) to epidemic outbreaks (Ebola and SARS) and biological warfare (anthrax and smallpox). The need for effective and efficient disinfection is driving the development of a wide range of methods. These can be divided into three major categories:

Chemical disinfectants: Chemical-based disinfectants are the oldest and most widely used. Most of them are chlorine-, alcohol-, or ammonium-based products. These are in liquid form and therefore are limited to surface treatments. The majority is used to disinfect contaminated surfaces but they are not suitable to be used as agents to avoid new contamination of surfaces. Although their use is relatively simple and straightforward, they are inherently toxic to humans, animals, and plants. Chemical disinfectants in gaseous form are difficult in application and highly corrosive, attacking or even destroying the objects that need to be cleaned.

Radiation-based disinfection: Radiation is a very effective technique since it can immediately inactivate the majority of the contaminants without damaging the surroundings. Still, however, the use is limited since it usually requires expensive equipment and exposure of humans to the applied radiation can be hazardous under certain conditions. Furthermore, the radiation might be absorbed in the top layers and not penetrate deep enough into ducts, crevasses, or porous structures.

Passive disinfectants: Passive disinfectants are characterized as those that do not require a certain application (chemicals) or operation (irradiation), but constantly purify and clean surfaces, air, and water. Activated carbon filters are probably the best known and most widely used for water and air treatment. However, they do not deactivate the contaminants so constant replacement is required. If they are not replaced regularly they can become a source of contamination rather than a treatment. Furthermore, material disposal along with transport and handling provide more hazards and incineration can only be done in approved facilities.

Therefore, novel approaches for treatment and especially prevention are needed. One of the most promising and rapidly emerging fields is photocatalysis. Photocatalysis is the type of reaction that takes place on the surface of a semiconductor in the presence of a very specific range of radiation. There are many materials that can display this type of reaction, but the most widely used is titanium dioxide, TiO_2, or titania. Titania in addition to its high efficiency is widely abundant and used in many other applications. Therefore, there are large-scale production facilities available that provide a variety of titanias at low cost. Furthermore, numerous studies have demonstrated that titania is environmentally safe (1,2). However, there are significant limitations to the application of titania for disinfection at this point since the efficiency is not yet high enough or at least not on a competitive level with chemical disinfectants (1,3).

The one area where all disinfection techniques have severe problems are spores. Here, an improvement in all types of treatments is needed since the multilayer structure of the spores protects them from the most widely used decontamination techniques. As shown in the illustration in Figure 13.1, the different coatings protect spores from extreme temperatures, UV exposure, and harsh chemical environments. Spores, in addition, are known to be able to survive for long periods of time and only when the right conditions are detected can they germinate into vegetative cells (4). The reversible transition to the dormant form makes the spores one of the top biological threats. Well-known spores such as anthrax have been used in the past as compounds in biological weapons and recently for bioterrorism.

To further the understanding of photocatalysis and enhancing of photocatalysts' application for the destruction of spores, this chapter introduces the major principles and theories of photocatalysis and summarizes the latest advances in the field of

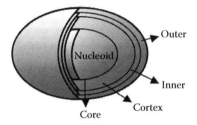

FIGURE 13.1 Illustration of the multilayer spore structure; the cortex is the hardest layer and protects the core from extreme conditions such as freezing, heat, and radiation.

photocatalytic disinfection. The authors like to point out that there have been excellent reviews published on photocatalysis covering, in detail, all the basic principles and steps in the photocatalytic process (1,5,6). Therefore, this chapter will mainly focus on the most recent advances in the field that have not been addressed by the general review articles.

SEMICONDUCTOR PHOTOCATALYSIS

Many authors use the term "photocatalysis" for titania. However, strictly speaking, it is a special case of photocatalysis, which is best described by the term semiconductor photocatalysis (1). Photocatalysis itself is defined as the initiation of chemical reactions in the presence of light only (7). This does not accurately describe the case for semiconductor photocatalysis, since here the presence of the semiconductor is equally important. We therefore use the term "semiconductor photocatalysis" in this chapter to define the process of chemical reactions that take place on the surface of a semiconductor in the presence of a certain range of radiation.

The first report on photocatalytic activity dates back to 1839 when Becquerel observed voltage and electric current on a silver chloride electrode that was exposed to sunlight while immersed in an electrolyte solution (8). Technically, all semiconductors can display photocatalytic properties but typically the oxides and compound semiconductors demonstrate significantly better results (5,6,9). The ability of a semiconductor to undergo photocatalytic oxidation is governed by the band energy positions of the semiconductor and redox potentials of the acceptor species (9). The latter is thermodynamically required to be below (more positive than) the conduction band potential of the semiconductor.

The potential level of the donor needs to be above (more negative than) the valence band position of the semiconductor in order to donate an electron to the vacant hole. Figure 13.2 shows some of the most popular semiconductor photocatalysts represented with their band energy positions. The internal energy scale is given on the left for comparison to the Normal Hydrogen Electrode (NHE). The positions are derived from the flat band potential in contact with a solution of aqueous electrolyte at pH 0. Among them, TiO_2 is the most popular. It is, efficient, effective, requires shallow UV radiation, and is easily incorporated with other materials (1). In 1972, Honda and Fujishima discovered that UV irradiation of titania yields the photocatalytic splitting of water (10).

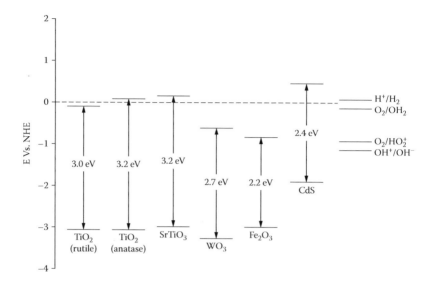

FIGURE 13.2 Schematic diagram representing the main photocatalysts with their bandgap energy. In order to photoreduce a chemical species, the conductance band of the semiconductor must be more negative than the reduction potential of the chemical species; to photooxidize a chemical species, the potential of the valence band must be more positive than the oxidation potential of the chemical species. The energies are shown for pH 0.

This and many other exciting findings triggered a large amount of research in the understanding of the overall mechanism and the specific reaction chain.

CRYSTAL STRUCTURE OF TITANIA

Titanium dioxide (titania) exists, in principle, in eight phases: rutile, anatase, brookite, columbite, baddeleyite, fluourite, pyrite, and cotunnite (11). The columbite, baddeleyite, fluourite, pyrite, and cotunnite phases can be generated only under very high temperatures and/or pressures, which is the reason those phases do not occur naturally, but they still possess some very interesting properties (12–14). Cotunnite, for example, is the hardest polycrystalline material known to exist (11,13). Of those eight phases the thermodynamically more stable ones are rutile, anatase, and brookite, with rutile being the most stable (15,16). Since photocatalytic activity is demonstrated only from rutile and anatase, the following analysis will focus on those two structures only. The major reason for the lack of photocatalytic activity in the other compounds is the electronic structure, which for the brookite, columbite, baddeleyite, fluourite, pyrite, and cotunnite has an indirect gap (17–19). In addition, all of the those structures except brookite are generated only under very high pressures and temperatures (11).

Figure 13.3a shows the crystal structure of anatase. It is tetragonal with $a = b = 3.782$ Å and $c = 9.502$ Å and has a D_{4h}^{19} -I4$_1$/amd symmetry. The building block of anatase is TiO$_6$, which forms a slightly deformed octahedron (Figure 13.3c). The Ti atom that is in line with the two oxygen atoms (apical oxygen atoms) has a bond length of 1.966 Å

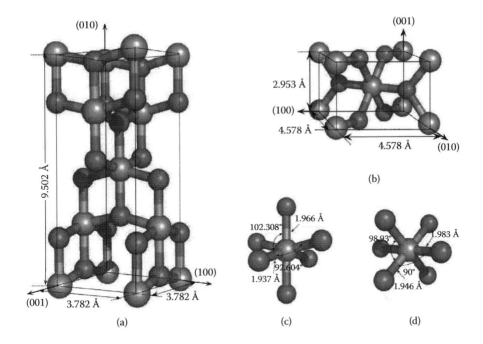

FIGURE 13.3 The two basic titania structures (a) anatase and (b) rutile. The distorted octahedra shown are used to construct the (c) anatase and (d) the rutile.

and the other four oxygen atoms (equatorial oxygen atoms) have Ti–O bond lengths of 1.937 Å. The widest angle of those two bonds Ti–O equatorial and Ti–O apical is 102.308°. The angle between two consecutive equatorial bonds is 92.604° or 87.394° (11,19,20). All the bond lengths and angles given above represent the structure at room temperature. Anatase is an unstable structure and it transforms to rutile at approximately 800°C (11,16). While the temperature increases, the bond lengths change and gradually anatase turns into rutile. Rutile has a more compact structure and therefore energy wise is more favorable. The transformation to rutile is an irreversible process.

Rutile has also a tetragonal structure (Figure 13.3b), but it is more compact compared to anatase. The tetragonal structure has a = b = 4.578 Å and c = 2.953 Å. It has D^{15}_{4h} -P42/mmm symmetry. Again the building block of the crystal structure is an octahedron that is slightly distorted (Figure 13.3d). The apical oxygen atoms have Ti–O bond length of 1.983 Å and the equatorial Ti–O bond is 1.946 Å. The equatorial and apical Ti–O bonds form a right angle while the largest angle between the two equatorial bonds is 98.93° (11,19,20).

The bond difference between the two phases is the reason for several differences in the electronic structure. The anatase, which has a smaller bond length, has a larger band gap compared to rutile. The structure of the other phases results in an indirect band gap and that is the reason that only rutile and anatase display photocatalytic activity.

Beyond the necessity of the crystal structure, specific defects in the crystal structure impact the electronic properties of titania. Titania is an oxygen-deficient material and usually it is considered an n-type semiconductor (5,6). The Fermi-level, therefore, is not at a fixed value since the production method will determine the level of oxygen deficiency and, therefore, the Fermi-level shift. This is true for both anatase and rutile. In addition, one of the most common defects in titania is the Ti^{+4} substitution by Ti^{+3} (and often Ti^{+2} and Ti^{+1}), which also creates a charge imbalance that, beyond the electrical properties, can affect spectroscopic techniques that rely on the electronic charge, such as XPS. Those Ti cations can be generated by annealing, sputtering, or chemical reduction (21,22).

BASIC PRINCIPLES OF SEMICONDUCTOR PHOTOCATALYSIS

Figure 13.4 schematically represents the steps of semiconductor photocatalysis. Initially when a photon of proper energy ($hv \geq E_g$) strikes the surface of the semiconductor, it generates an electron-hole pair ($h^+ - e^-$). Both electrons and holes either recombine or migrate to the surface, where they proceed with chemical reactions. The holes generate (OH•) and the electrons H_2O_2 (1). A very important factor for those processes is the required time. Here are summarized the main reactions and the time required for each one (5). The required time has been measured by laser flash photolysis (5,23,24):

Charge-Carrier Generation

$$TiO_2 + hv \rightarrow h_{vb}^+ + e_{cb}^-, \quad 10^{-15} \text{ s} \qquad \text{(Ref. 5)}$$

Charge-Carrier Trapping

$$h_{vb}^+ + > Ti^{IV}OH \rightarrow \{> Ti^{IV}OH^\bullet\}^+, \quad 10 \times 10^{-9} \text{ s} \qquad \text{(Ref. 24)}$$

$$e_{cb}^- + > Ti^{IV}OH \rightarrow \{> Ti^{III}OH^\bullet\}, \quad 100 \times 10^{-9} \text{ s} \qquad \text{(Ref. 24)}$$

$$e_{cb}^- + > Ti^{IV} \rightarrow Ti^{III}, \quad 10 \times 10^{-9} \text{ s Irreversible} \qquad \text{(Ref. 24)}$$

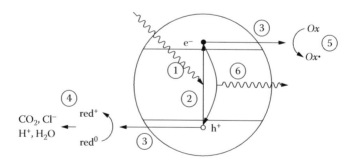

FIGURE 13.4 Schematic representation of the reactions taking place in titania. (1) Light strikes the semiconductor. (2) An electron-hole pair is formed. (3) Electrons and holes are migrating to the surface. (4) The holes initiate oxidation, leading to CO_2, Cl^-, H^+, H_2O. (5) The conduction band electrons initiate reduction reactions. (6) Electron and hole recombination to heat or light.

Charge-Carrier Recombination

$$e_{cb}^- + \{> \text{Ti}^{IV}\text{OH}^\bullet\}^+ \rightarrow\, > \text{Ti}^{IV}\text{OH}, \quad 100 \times 10^{-9} \text{ s} \qquad \text{(Ref. 23)}$$

$$h_{vb}^+ + \{> \text{Ti}^{III}\text{OH}\} \rightarrow\, > \text{Ti}^{IV}\text{OH}, \quad 10 \times 10^{-9} \text{ s} \qquad \text{(Ref. 23)}$$

Oxidation or Reduction

$$\{> \text{Ti}^{IV}\text{OH}^\bullet\}^+ + \text{Red}^0 \rightarrow\, > \text{Ti}^{IV}\text{OH} + \text{Red}^{\bullet+}, \quad 100 \times 10^{-9} \text{ s} \qquad \text{(Ref. 23)}$$

$$e_{tr}^- + \text{Ox}^0 \rightarrow\, > \text{Ti}^{IV}\text{OH} + \text{Ox}^{\bullet+}, \quad 100 \times 10^{-9} \text{ s} \qquad \text{(Ref. 23)}$$

According to the above-proposed mechanism, the overall quantum efficiency depends on two major types of reactions: the carrier recombination and $(\text{OH}^\bullet)/\text{H}_2\text{O}_2$ generation. The dominant reaction is the recombination of e^- and h^+ (1 ns), followed by the reduction reaction (10 ns) and oxidation (1 ms). Since the recombination is also assisted by localized crystal defects, the remaining carriers are not enough for an efficient photocatalytic reaction.

A very crucial parameter is the diameter of the photocatalytic particle. For particles less than 10 nm, the traditional energy band diagrams change. The conduction and valence band are no longer continuous but consist of energy levels that are distinctively separated. For titania, this limit is approximately 5 nm (25,26). This effect moves the edge of the valence band to lower energies and the edge of the conduction band to higher energies. Therefore, the required energy to excite an electron is larger compared to the energy of bulk titania and yields higher redox potentials. Based on a theoretical approach, the increase in the band gap can be calculated

$$\Delta E_g = \frac{h^2 \pi^2}{2R^2} \left(\frac{1}{m_e} + \frac{1}{m_h} \right) - 1.786 \frac{e^2}{\varepsilon R} - 0.248 E_{RY}^*$$

where h is Planck's constant, R is the particle radius, E_{RY}^* is the effective Rydberg energy calculated to be 4.3×10^{-39} J, ε is the dielectric constant of anatase TiO_2, which is 86, and m_e and m_h are the electron and hole masses, respectively. Reddy, Manorama, and Reddy calculated the ΔE_g, for 5 and 10 nm particles as 0.2 and 0.1 eV, respectively (25). So for 5 nm particles, a minimum of 346 nm is required. According to So and colleagues, however, in order to effectively assist photocatalysis, UV lamps with a peak wavelength of 305 nm is required (27). This deviation is attributed to crystal defects that were not accounted for when the previous equation was derived.

Another effect that occurs when particle size is reduced is modification of the surface levels. The surface of the semiconductor is slightly different compared to the bulk since the termination of the periodicity forces the energy bands to bend upwards (p-type semiconductors) or downwards (n-type semiconductors) and generates extra energy levels on the surface (surface levels). These levels are located just below the conduction band and their number depends on the surface of the semiconductor and the plane they are terminated on. Those levels can act as trapping levels where the excited electrons can be trapped for a significant amount of time. For large particles (>10 nm), the number of the surface energy states is negligible compared to the number of states in the conduction band and therefore for these particles they do not play an important role. However, for very small particles, this number becomes significant and therefore acquires a very important role. In this case, the number of electrons that can be trapped at those levels is significant and this can significantly retard the recombination. The phenomenon can take effect in all titania particles but for particles larger than 10 nm, the number of trapped electrons is not significant compared to the number of the excited electrons.

ENHANCEMENT OF SEMICONDUCTOR PHOTOCATALYSIS

It is necessary to enhance the photocatalytic efficiency of titania to obtain a more effective material. Time-wise, the oxidation coming from the holes is the fastest degrading reaction (23,24,27). It is reasonable, therefore, to favor this reaction over the reduction reaction initiated by the electrons. Since the cause of reduced efficiency is the recombination between the h^+ and e^-, all previous research focused on either scavenging the electrons from the system to prevent recombination, or just retarding the recombination so holes will generate (OH$\cdot$) (5,6,9,28). The major methods are doping with metallic and nonmetallic elements coupling with a metal and coupling with a semiconductor. There has been extensive work toward all three types of photocatalytic enhancement for semiconductors, with the titania/semiconductor and titania/metal coupling more dominant since they are easier to achieve (1,6,29).

Doping of Titania

A great deal of work has been done the last few decades to dope titania with transition metals, and nonmetallic elements such as N (30) and C (31,32). In general, transition metals are incorporated into the structure of titania and occupy substitutional or interstitial positions. It is a very common defect in the case of semiconductors since it generates trap levels in the bandgap. Figure 13.5a shows the electronic

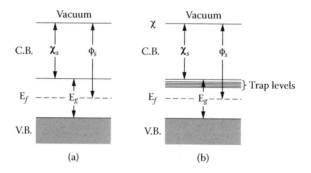

FIGURE 13.5 Titania band structure (a) before and (b) after doping. The transition metals are interstitial or substitutional defects in the structure of titania and generate trapping levels in the bandgap.

structure of titania before doping. After doping (Figure 13.5b), the bandgap has been modified with the addition of the trap levels. The trap levels are usually located slightly below the lower edge of the conduction band and usually are in the form of a narrow band.

There are several advantages to this modification. Before the modification, the required photon energy had to satisfy the condition $h\nu \geq E_g$. After the modification, the required energy is going to be $h\nu \geq (E_g - E_t)$, where E_t is the lower edge of the trapping level band. In addition, the electrons that are excited at those levels are trapped, and the holes have sufficient time for OH· generation.

Even in the case that $h\nu \geq E_g$ and the electron is excited to the conduction band, then during the de-excitation process the electron is going to be transitioned from the conduction band to the trap levels and then to the valence band, which again retards the recombination and therefore increases the overall efficiency. The most common transition metals used are Fe^{+3}, Cr^{+3}, and Cu^{+2}. Fe^{+3} doping of titania has been shown to increase the quantum efficiency for the reduction of N_2 (31,33,34) and methylviologen (31), and inhibit electron-hole recombination (23,24,35). In the case of phenol degradation, Sclafani, Palmisano, and Schiavello (33) and Palmisano and others (36) reported that Fe^{+3} had little effect on the efficiency.

Enhanced photoreactivity for water splitting and N_2 reduction has been reported with Cr^{+3} (36–39) doping while other reports mention the opposite result. Cr^{+3} can, in addition, modify the surface by creating recombination sites. Negative effects have also been reported with Mo and V doping, while Grätzel and Howe reported inhibition of electron-hole recombination. Finally, Karakitsou and Verykios noted a positive effect on the efficiency by doping titania with cations of higher valency than Ti^{+4} (40). Butler and Davis (40,41) and Fujihira, Satoh, and Osa (42) reported that Cu^+ can also inhibit recombination.

COUPLING WITH A METAL

In photocatalysis, the addition of metals can affect the overall efficiency of the semiconductor by changing the semiconductor surface properties. The addition of metal, which is not chemically bonded to TiO_2, can selectively enhance the generation of holes by scavenging electrons. The enhancement of the photocatalysis by metal was first observed using the Pt/TiO_2 system, yielding an increase in the splitting of H_2O to H_2 and O_2 (43,44). In some particular cases, the addition of metal can affect the reaction products. The addition of silver especially, which acts as a disinfection agent that can further increase the efficiency. In addition, it was found that metals could react with some of the oxidation species and create compounds that are harder to decompose.

Figure 13.6 shows a band structure model for titania. Here, the titania band structure is coupled with a metal. In general, when a semiconductor that has a work function ϕs is compared to a metal with work function of $\phi_m > \phi_s$, the Fermi level of the semiconductor, E_{fs}, is higher than the Fermi level of the metal E_{fm} (Figure 13.6a). So when the two materials are brought in contact (Figure 13.6b), there will be electrons flowing from the semiconductor to the metal until the two

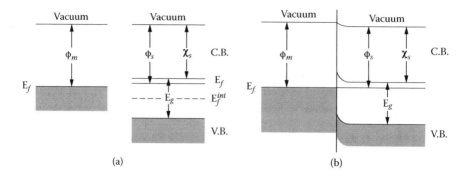

FIGURE 13.6 The principles of rectifying contact between titania ($E_g = 3.2$ eV) and a metal with work function (ϕ_m), in this example, 5 eV, greater than the affinity (χ_s) of titania. (a) Before the contact, and (b) after the contact, where a barrier is formed to prevent the electrons from crossing back to the semiconductor. The E_f^{int} is the Fermi level if titania is an intrinsic semiconductor and E_f is the Fermi level of an oxygen-deficient material.

Fermi energy levels come to equilibrium. The electrons that transition will generate an excess of positive charge in the titania, which causes an upward band bending. This bending creates a small barrier (on the order of 0.1 eV) that only excited electrons can cross and be transported to the metal. Electrons that migrated to the metal cannot cross back since the barrier for this action is larger and therefore the electrons will remain in the metal. Currently the most effective metal/TiO$_2$ interface is achieved by colloidal suspension (45). It was found that in the case of the Pt/TiO$_2$ system, the Pt particles are gathered in the form of clusters on the surface of TiO$_2$ (46). Other metals have also been investigated. Ag has been found to increase the efficiency (47). Although in principle all metals can be used, noble metals are preferred since they have a higher work function and better electrical conductivity. In all cases, high solids loading of the particles in suspension will affect the kinetics of the system, the distribution of irradiating light, and eventually decrease the overall efficiency (48).

COUPLING A SEMICONDUCTOR PHOTOCATALYST WITH A SEMICONDUCTOR

Coupling a semiconductor with a semiconductor photocatalyst is a very interesting way of assisting semiconductor photocatalysis since both species can act as photocatalysts. Figure 13.7 demonstrates the principles of anatase being coupled with rutile. When two semiconductors are brought together, as in the previous case for the metal-semiconductor combination, the Fermi levels tend to balance; that is, electrons flow from the semiconductor with the highest Fermi level to the semiconductor with the lowest.

This charge transfer will create an excess of positive charge in the semiconductor with the highest Fermi level and an excess of negative charge in the semiconductor

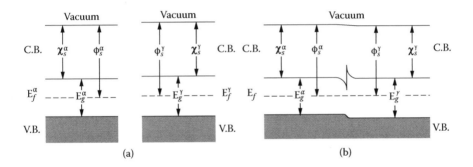

FIGURE 13.7 The principles of rectifying contact between anatase (α) titania ($E\alpha_g = 3.2$ eV) and and rutile (r) titania ($Er_g = 3.0$ eV). (a) Before the contact, and (b) after the contact, where a barrier is forming to prevent the electrons created in anatase crossing to the rutile. On the other hand, holes created in anatase can migrate to rutile. So the couple of anatase-rutile is creating an effective electron-hole separation.

with the lowest Fermi energy (Figure 13.7b). When irradiated by light, e^--h^+ pairs are generated in both semiconductors. The barrier that forms separates the electrons in the conduction band, but at the valence band the holes are free to move; and based on the energy diagram, they move from the semiconductor with the larger gap to the one with the smaller. In this case, the composite material acts as a charge separator. The holes are gathered in the rutile where they form an excess of holes, and despite the fact that the recombination is still the main process, the excess of holes will be enough to photooxidize the organic molecules.

In addition to this specific mechanism for rutile/anatase, semiconductors can be used as a hole or electron injectors. In order to achieve optimum results, a candidate semiconductor has to satisfy the following criteria in comparison with the semiconductor photocatalyst:

- Have a proper band-gap.
- Have a proper position of the Fermi energy level.
- Have proper relative position of the conduction and valence band to the vacuum level.

The combination of the bandgap and Fermi level will determine if there are holes or electrons that will be injected and toward which direction. Thus, in order to couple a semiconductor with titania for enhancement of photocatalysis, the semiconductor must have very specific properties. This is the reason that this technique, despite its simplicity, ease of manufacturing, and very promising results, is not very widely applied. Systems that have been developed are the TiO_2/CdS (49), TiO_2/RuO_2 (50), and anatase-TiO_2/rutile-TiO_2 (51,52). The last one is a system commercially available from DEGUSSA, known as Aeroxide P25™, and is the most powerful commercial, particle photocatalytic system. The excellent and uniform properties have established it as a benchmark material to compare photocatalytic efficiencies.

ENVIRONMENTAL APPLICATIONS OF
SEMICONDUCTOR PHOTOCATALYSIS

During the last few decades, the environmental applications of TiO_2 have attracted a great deal of attention since titania can be the base of low maintenance systems. So far their main focus is on water and air treatment and the objectives are primarily the removal of organic contaminants (5,51–55) and secondarily the removal of bio-cidal hazards (1,4,56–59). Although the systems can equally target biological con-taminants, the effectiveness is less than or equal to other competitive technologies (chemical disinfection). So the biological problems, although they are unique and interesting, are not widely addressed yet by photocatalysts.

Besides the engineering of the bandgap, several other engineering aspects remain, such as solids loading and light absorption with minimum scattering effects. For this purpose, several reactor configurations have been developed for the most effective removal of contaminants (9,60,61). One of the most popular configurations, mainly for experimental application, is the slurry reactor, where contaminated water is mixed and agitated with titania particles under the presence of UV radiation. The main advan-tage of this configuration is the large specific surface area of the particle systems that allows faster processing. The main disadvantage is the separation of the particles after the reaction, which is not economically favorable. They can be separated by filtration, centrifugation, coagulation, and flocculation. An alternative to the slurry reaction is the flat-bed reactor where the particles are immobilized on a ceramic membrane. The efficiency is lower compared to the slurry reaction due to the lower specific surface area, but the system does not need any kind of separation, which adds to the overall efficiency. Recently, in order to increase the surface area of titania particles, they have been coated on tubes, glass beads, and fiber or woven into glass (62).

BIOLOGICAL APPLICATIONS OF
SEMICONDUCTOR PHOTOCATALYSTS

Potential applications of photocatalytic semiconductors for disinfection first focused on bacteria. Since bacteria are much larger than the molecular scale and complex structures, the destruction of that biohazard was found less efficient compared to the destruction of pollutants that are on the molecular level. However, in 1985, Matsunaga and others were the first to combine titania particles with platinum par-ticles and achieved 90% bacterial removal in a slurry system, compared to 50% by pure titania. They used the platinum particles as electron scavengers as described earlier (56,57).

Since then a large volume of publications and patents have dealt with the use of titania as a sterilizing agent. The overall observed disinfection of water or surfaces from bacteria with titania, however, is a very complicated process. For molecular contaminants, the destruction rate depends on the UV intensity, the amount of titania, the specific surface area, the particle size and size distribution, the phase, the pH, the temperature, and the chemistry of the pollutant. In biological applications, however, there are even more parameters. Switching between constant and intermittent irradia-tion has an impact on the photocatalytic efficiency. Furthermore, the results in this case do not allow drawing conclusions in either direction. Increase, decrease, or no change in microbial rate was observed under intermittent irradiation when the results are compared to continuous radiation. In addition, a very critical factor is the response of the bacteria or spores to photocatalysis. Bacteria, in general, are more susceptible to destruction compared to spores. As explained in the introduction, spores have a more tolerant structure. However, depending on the conditions, they can germinate into bacteria, which can lead to easier destruction. The presence of oxygen and hydrogen and their concentration can also affect the system. In addition, the types of interactions between the titania and the bacteria can significantly affect the results.

It is therefore a very tedious process to evaluate the activity of photocatalyitc particles based on those biological tests. Most researchers use chemical methods instead for the evaluation, such as monitoring the concentration of an organic contaminant, that is, an organic dye.

LANGMUIR-HINSHELWOOD REACTION KINETICS

The study of the reaction kinetics is very important in order to understand and quantify the efficiency of the particles. Dry state photocatalysis is complicated and therefore there is not a general theory available to describe it. In liquid media, however, most experimental results can be interpreted in agreement when the rate of photocatalytic oxidation of organic molecules is approximated by the Langmuir-Hinshelwood (L-H) model (63–72). With some assumptions this is also valid for bacteria and spores removal. The model assumption states that the rate will depend on the adsorption of the dye molecule on the TiO_2 particle and its oxidation reaction. So if it is assumed that k is the reaction constant and K is the adsorption constant, then according to the L-H kinetics model the oxidation rate is:

$$r = -\frac{dC}{dt} = \frac{kKC}{1+KC}$$

where k and r are in mg/L min, K is in L/mg, and C is the dye concentration in mg/L. This model is nonlinear but it can be further simplified:

$$\frac{1}{KC_0}\ln\left(\frac{C}{C_0}\right)+\left(\frac{C}{C_0}-1\right)=\frac{kt}{C_0}$$

where C_0 is the initial dye concentration. With the assumption that $C_0 \rightarrow 0$, then the previous equation will become:

$$\frac{1}{KC_0}\ln\left(\frac{C}{C_0}\right)=-kKt$$

which yields a simple exponential decay:

$$C(t)=C_0 e^{-k_{app}t}=C_0 e^{-t/\tau}$$

The parameter t is the inverse reaction constant and physically describes the time required for 63% reduction of the pollutant. It is widely used to compare activities between several photocatalytic particles. It has to be emphasized that the Langmuir-Hinshelwood model is built for a single reaction (AB $\leftrightarrow$ A + B), which is not true for the case of the dye degradation. As described before for this certain dye, there are a lot more reactions involved during the degradation. In this case it is just assumed that k refers to the slowest reaction step.

RECENT APPROACHES

As seen, photocatalysis is mainly a surface reaction. Therefore the available surface area is equally important as the electronic properties of the materials. In general, increasing the available surface area will increase the overall efficiency of the system. This concept was, until recently, interpreted by the solids loading, where increasing the solids loading will increase the surface area of the system. However, nanotechnology introduced the ability to achieve the same surface area with significantly less solids loading simply by reducing the diameter of the particles. In the last decade the general trend has been to develop systems of dimensions usually less than 100 nm. At the same time, several self-assembled nanostructures allowed the incorporation of other materials in order to further increase the functionality of the final composites. Briefly, those additions can enhance photocatalysis by increasing the corresponding light wavelength, increasing the specific surface area, or by adding other functionalities such us magnetic core for better collection and dispersion.

NOVEL MATERIALS

Carbon can be used in combination with titania in several forms (such as elemental carbon, activated carbon, and in the form of carbon nanotubes). Elemental carbon is usually used as dopant. The incorporation of carbon has been proven to reduce the energy of the photon required to initiate the reaction. In addition, it changes the white color of titania to shades of grey, which results in a reduction in the reflectivity of titania and therefore it adsorbs the visible light more efficiently. The most usual incorporation of carbon in the structure of titania is done by the addition of carbon-containing substances during sol-gel (hydrolysis) synthesis (73). They are chemically bonded to the structure of the amorphous titania and during the heat treatment, the excess material is burned and typically final concentrations range from 0.2 to 3 mass% carbon. The band-gap reduction has been repeatedly measured and it has been found that the reduction arises from the upward shifting of the valence band, while the conduction band is slightly changing. The same results have been theoretically confirmed by Arashi and Neumann and colleagues (12,73) who showed the existence of a continuous band as an extension of the valence band. The results are very similar to the results obtained by addition of other elements such as metals; however, the fact that the color is also changing and enhancing the visible light absorption gives elemental carbon an additional advantage. Carbon incorporation is also an easy task and it is achieved using inexpensive materials (30,48).

However, the best results of carbon and titania combination come from the use of carbon as a carrier of the titania (53). The most widely form used for this application is activated carbon. Activated carbon is a material that has an exceptionally large specific surface area, typically determined by gas adsorption, and it contains significant microporosity. Sufficient activation for useful applications may come solely from the high surface area, although further chemical treatment is often used to enhance the absorbing properties of the material. It has been used with great success in water treatment equipment.

The major disadvantage of activated carbon is that the adsorbed hazards are not destroyed. In order to eliminate this problem, activated carbon is usually coated with titania. A simple way to couple those two materials was suggested by Tao and others (74–78). Activated carbon (AC) was coated with commercially available photocatalyst by a spray desiccation method. Laboratory-scale experiments were conducted in a fixed-bed reactor equipped with an 8-W black light UV lamp (peak wavelength at 365 nm) at the center. The photocatalyst loaded onto the AC had no significant impact on the adsorption capacity of the carbon. High humidity was found to greatly reduce the material's capacity in the adsorption and simultaneous adsorption and photocatalytic oxidation of methanol. The photocatalytic regeneration process is limited by the desorption of the adsorbate. Increasing the desorption rate using purge air greatly increased the regeneration capacity. When the desorption rate was greater than the photocatalytic oxidation rate, part of the methanol was directly desorbed without degradation. With this technology, the porosity of the activated carbon increases the specific surface area of titania, which impacts the overall efficiency.

The surface of the activated carbon may attract contaminants and bring them in close proximity to the titania particles, thus enhancing photocatalytic activity. Gao and Liu immobilized TiO_2 and activated carbon (AC) particles on silicone rubber (SR) film (79). Its photocatalytic activity was tested by photodegradation of acridine dye and phenol in aqueous solution. A silicone rubber film that was used as support showed many advantages: It is transparent to light, recyclable, and can be suspended in aqueous media to increase the collision possibility of pollutants with TiO_2 photocatalysts. The activated carbon used as adsorbent helps to accelerate the degradation of organic substrates. This immobilized hybrid catalyst 90TAS (90 wt% TiO_2–10 wt% AC/SR) is efficient in photoinduced mineralization of organic substrates at dilute concentrations and can be recycled.

Recently the field of the nitrogen-modified titania has become very vibrant since the obtained titania can work in the presence of visible light. Janczarek and others investigated the differences between the bulk-modified titania and the surface-modified titania (80–82). The bulk-modified titania was prepared from thiourea and $TiCl_4$ in ethanol. The surface-modified titania was prepared by impregnation of titanium hydroxide with urea in water. The particles had specific surface areas of 148 and 173 m^2/g, respectively, as measured by BET. In photocatalytic evaluation tests with 4-chlorophenol, bulk-modified titania showed better results by degrading 69% of the pollutant in 1 hour while the surface-modified titania in the same time period degraded 55%. The authors attribute this difference to the fact that for the surface-modified titania, TiO_2 particles were used directly and not $Ti(OH)_4$. The fact that the N dopants were organic compounds caused the formation of C, which according to the authors is not important for visible light catalysis. The most important result is that the Ti-N bond and molecular nitrogen were not present, indicating that N doping is not substitutional. The two particles were also compared based on their band gap. The band gap of the bulk-modified titania was measured as 2.91 eV (400 nm) and for the surface-modified titania was found to be 2.48 eV. They also concluded that for the deposition on glass and immobilization of the catalyst, the bulk-modified titania behaved better since it had superior thermal stability compared to the surface-modified titania.

Recently developed and promising materials are the titania-coated magnetic core particles. The use of the magnetic core is not directly related to the electronic properties of the material, although in principle it can affect the motion of the carriers. The magnetic core, however, enhances the dispersion and recollection of the particles. Lee and colleagues have synthesized magnetic core titania nanoparticles (83). The process started with barium ferrite particles that were coated with a silica layer since it was found that the leaching of ions from the magnetic core into the photocatalyst degrades the photocatalytic activity. Afterward, the particles were coated with the anatase phase of titania via sol-gel chemistry. The XRD and TEM confirmed the structure of the anatase titania. In addition, the particles were tested under a magnetically agitated stirrer. The authors used three kinds of particles: titania particles, magnetic core particles, and magnetic core particles that were heat-treated at 500°C. The heat-treated particles displayed less activity compared to the untreated particles. This was attributed to the effect that the heat treatment has on the magnetic properties of the core. With prolonged heat treatment, the domains are losing their orientation and the magnetic agitation is not as effective anymore.

NOVEL MECHANISMS

Alternatives to the use of new materials several new mechanisms have been proposed as photocatalysis enhancers. One of the recently proposed mechanisms includes that the UV necessary for the photocatalytic degradation could further influence the adsorption mechanism. Khan and Mazyck (84), similar to Schwarzenbach, Gschwend, and Imboden (85), assumed that an aromatic molecule, with bonding or nonbonding electrons, could be altered to a molecule possessing antibonding electrons via irradiation. Furthermore they proposed that this new species, i*, can exhibit different adsorption behavior than its parent molecule, i, due to enhanced/impeded π–π interactions between the molecule (i*) and the activated carbon, whereby the activated carbon surface chemistry dictates whether the adsorption is heightened or reduced. Based on this hypothesis, the calculations showed that the adsorption of Methylene Blue by activated carbon is influenced by the presence of UV light. It was also found that this trend occurs with Procion Red MX-5B, and it is expected that the same will occur with other aromatic compounds. They also hypothesized that activated carbons may have different capacities for irradiated molecules.

A relatively new approach is to combine titania with carbon nanotubes. A carbon nanotube can be thought of as a graphene sheet rolled seamlessly into a cylinder. It usually has 10–40 carbon atoms in circumference and is capped. Carbon nanotubes have been discovered by Iijima in 1991 and since their discovery they have attracted a great deal of attention due to their exceptional electronic, thermal, and mechanical properties. Iijima reported the creation of multiwall carbon nanotubes (MWNT) with outer diameters up to 5.5 nm and inner diameters down to 2.3 nm. The first attempt at combining titania with carbon nanotubes was done in the form of a composite that had the nanotubes dispersed in a titania matrix. Recently, however, a novel approach was taken where the titania was applied as a coating. Since Lee and Sigmund (4) published the first report about coated nanotubes, there has been a wide variety of approaches all based on sol-gel chemistry (86,87). However, there are only few reports about the activity of those particles.

Pyrgiotakis and others (88,89) has reported that the anatase-coated nanotubes gave results better than the Degussa P25 particles. The inverse reaction constant was 25.1 min for Degussa while for the anatase-coated carbon nanotubes the result was 21.1 min. This behavior is attributed so far to the specific surface area of the carbon nanotubes that, similar to the case of the activated carbon, increases the available sites for OH˙ generation. The two composites are different, with the anatase carbon nanotubes being photocatalytically more effective (in direct comparison to Degussa P25). That result suggests that the carbon nanotubes have another property that further enhances the photocatalysis.

Pyrgiotakis (88) investigated carbon nanotubes as photocatalytic templates in depth with Raman, FTIR, and XPS. Raman spectroscopy was chosen since it has been widely applied before to characterize nanotubes and it is a fast nondestructive technique. Pyrgiotakis examined two types of carbon nanotubes. One type of nanotube was synthesized by arc discharge and the other one was grown by chemical vapor deposition (CVD). Initially, the anatase-coated tubes were evaluated regarding their photocatalytic ability and the arc discharge tubes showed strong results (19.5 min for destruction of a dye) while the CVD grown nanotubes had a weaker performance (120 min).

Figure 13.8 shows the Raman spectra prior and after the coating. The major bands that were examined by Raman were the D and G bands, both characteristic bands, for single and multiwall nanotubes. In addition to the carbon nanotubes' peak analysis, the Raman peaks for titania were also compared to anatase nanoparticles (diameter 5 nm). The reason is that titania's major peak at 144 cm^{-1} has been found to change shape and position when the size of the particle is smaller than 10 nm. The peaks were found to be in good agreement with literature values. The most interesting result came from the spectral comparison before and after the coating where all the peaks, both for titania and for CNTs, were significantly shifted. The second important result was that all the peaks have shifted differently not only in magnitude, but in direction too. This observation basically eliminated the fact that the shift may have been caused by a miscalibration of the instrument. The authors contend that this is a very good indication for the existence of a bond between the carbon nanotubes and the titania coating. This can be a direct bond C-Ti or indirect with the two elements bonded to the oxygen (C-O-Ti). In addition it was observed that the G band for the case of arc-discharge nanotubes shows a split (G$^-$ and G$^+$) while the CVD grown tubes did not display similar behavior. The split was not only very obvious with the two peaks having almost similar intensity, but the G$^-$ fit better with the Breit-Wigner-Fano lineshape compared to Lorentz. The Breit-Wigner-Fano

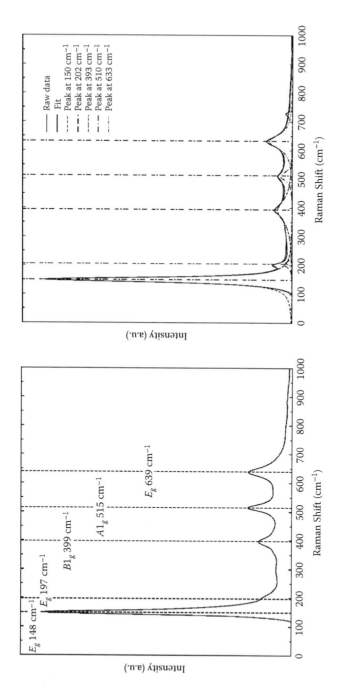

FIGURE 13.8 Collective representation of the Raman spectra regarding the titania-coated carbon nanotubes. The picture on the *left* is obtained from pure anatase particles while the picture on the *right* is the result of the titania as it is coated on the nanotubes. While the shape of the peaks has not changed, all the titania peaks seem to be shifted. This is an indication of a bond that has formed between the titania coating and the underlying carbon nanotubes.

line shape is reported in the literature as the lineshape of the G^- that represents metallic nanotubes. So while both tubes showed indications of the TiO_2-CNT bond, only the arc discharge showed metallic behavior. Based on the previous discussion, the titania can act as a catalyst when it is bonded to carbon. In addition, when there is an electron sink present in the system, the electrons are scavenged and therefore the system becomes more efficient. Since only the metallic properties are present in the arc discharge system, it was concluded that this is the main reason for the exceptional performance of the system. However, both the metallic nature as well as the bond are important. Therefore, the authors focused the investigation on the nature of the bond. This was done by XPS. Again, the anatase nanoparticles were used as reference and the tubes were tested prior to and after coating with titania. XPS initially showed all titania peaks (Ti2p and O1s) to be shifted to higher energies. This means that they are bonded to a metallic species, which in this case can only be the CNT. Furthermore, the change in the C1s peak is very interesting. Prior to coating, the elemental carbon peak and the C-OH and C=O peaks are seen in the XPS spectrum as a result of the strong acid treatment. However, after coating the tubes, those peaks shifted, which again indicates, bonding to a nonnative element. This shift was accompanied by the appearance of a new peak at 289.5 eV. In the literature, this value has not yet been reported as being relevant to the system. It was therefore attributed to the CNT/TiO_2 bond. Furthermore, since the C-Ti bond appears at 281 eV, it is concluded that this bond is of the form of Ti-O-C. In addition, the peak intensity of that peak is proportional to the C=O/C-OH peak, which supports the argument of the Ti-O-C peak.

Lee and others used the CNT/titania nanotubes on spores of *Bacillus cerius* (4). The survival ratio is presented as a function of irradiation time. The constants that are used to characterize the sample performance are the values LD_{90} and the D-value. LD_{90} represents the required exposure time for the 0.1 survival ratio and the D-value represents the time for one log reduction in bacteria. All the values were obtained by the linear portion of the logarithmic graphs in Figure 13.6b. The calculated values are summarized in Table 13.1.

The UV alone was found to be able to destroy the spores, but it requires longer exposure times (LD_{90} of 151 min and D-value of 169 min). It is hypothesized that the inactivation mode is based on the genetic damage that can be caused by prolonged exposure to UV radiation, which eventually prevents the spores from replicating. As

TABLE 13.1
Summary of the Different Biocidal Tests
Carried Out under 350-nm UV

System	LD_{90} (min)	D-Value (min)
UV	151 ± 41	169 ± 40
UV + P25	198 ± 41	144 ± 5
UV + ANT	84 ± 29	72 ± 20

Explanation of Major Symbols and Notations

Symbol	Explanation
a, b, c	Lattice parameters
D^{19}_{4h} -I4$_1$	Tetrahedral symmetry
D^{15}_{4h} -P42	Tetrahedral symmetry
h^+	Hole
e^-	Electron
$(OH^{\cdot})$	Hydroxyl radicals
ΔE_g	Change in band gap energy
h	Planck's constant
m_e	Electron mass
m_h	Hole mass
e	Electron charge
E^*_{RY}	Rydberg energy
R	Particle radius
v	Light frequency
E_g	Band gap energy
K	Reaction rate
K	Adsorption rate
k_{app}	Apparent reaction rate
C	Dye concentration
T	Time
i, i^*	Molecule
Π	Orbital

can be seen from Table 13.1, Degussa did not have a significant impact on the deactivation time (LD_{90}) compared to UV alone. Experiments conducted under the same conditions and following the same protocols showed that P25 was effective for *E. coli* bacteria, by causing their destruction within 1 hour. The failure to demonstrate similar performance with the spores is attributed to the very strong resistance of the spores in oxidizing environments, as explained previously. On the other hand, the nanocomposite performed excellent and had an LD_{90} value of 84 minutes. Since the experiments were carried out under the same conditions and on the same surface area base, the results suggest that the reason for the exceptional performance is the conductive nature of the MWNTs, which inhibits the recombination of the e-h pair. After the first 2 hours, a sudden increase is observed in the spore population that decayed again later. The reason for this unique behavior is still unknown.

CONCLUSION

Overall, photocatalysis is the most promising technique to effectively decontaminate water air and surfaces without leaving toxic waste. Among all the photocatalytic materials, TiO_2 is the most promising material. It is inexpensive, widely available, and environmentally friendly. The efficiency (measure of the time required to reduce a certain percentage of the contaminant), however, of the pure material is not satisfying, especially when titania is used for biological decontamination. The reason is that the recombination rate of the electrons and holes is one order of magnitude faster than the redox reaction time. Several methods have been studied to improve the efficiency. One of the first methods used was the coupling of titania with a metal. The metal will act as an electron sink and scavenge the electron and therefore inhibit the recombination. A similar method involves the combination of titania with a semiconductor and in this case the semiconductor will act as carrier separator, which will again inhibit the recombination. Finally, a more recently proposed method is the doping of titania with several transition metals, N, and C. In this case, the modification will directly affect the electronic structure of the titania, which will not directly impact the recombination rates but will make titania function in a different radiation range closer to visible light. The most promising element for doping titania seems to be C. In the last few years, novel titania-based composite materials have been introduced. Most of these composites use carbon, such as in carbon nanotubes or activated carbon. Both forms of carbon, offer the advantage of high specific surface areas while carbon nanotubes under certain conditions can also influence the electronic structure of the composite material. Photocatalysis has been successfully used to destroy several pollutants and biological contaminants. Due to the complexity of the biocidal applications, the use of titania has not yet spread. However, with recent advances in the field, photocatalysis is starting to appear as a very promising technique, even in the case of destruction of spores, which poses one of the biggest biological hazards and is one of the most difficult contaminants to deactivate.

REFERENCES

1. Fujishima, A., Rao, T.N., and Tryk, D.A., Titanium dioxide photocatalysis. *Journal of Photochemistry and Photobiology C: Photochemistry Reviews*, 2000. 1(1): 1–21.
2. Fujishima, A. and Zhang, X., Titanium dioxide photocatalysis: present situation and future approaches. *Comptes Rendus Chimie*, 2006. 9(5–6): 750–760.
3. Tryk, D.A., Fujishima, A., and Honda, K., Recent topics in photoelectrochemistry: achievements and future prospects. *Electrochimica Acta*, 2000. 45(15–16): 2363–2376.
4. Lee, S.H. and Sigmund, W.M., Inactivation of bacterial endospores by photocatalytic nanocomposites. *Abstracts of Papers of the American Chemical Society*, 2004. 227: U1275–U1276.
5. Hoffmann, M.R., et al., Environmental applications of semiconductor photocatalysis. *Chemical Reviews*, 1995. 95(1): 69–96.
6. Linsebigler, A.L., Lu, G.Q., and Yates, J.T., Photocatalysis on TiO_2 surfaces—Principles, mechanisms, and selected results. *Chemical Reviews*, 1995. 95(3): 735–758.
7. Suppan, P., *Chemistry and Light*, 1994. Royal Society of Chemistry, Cambridge.
8. Becquerel, E., Memoire sur les effets électriques produits sous l'influence des rayons solaires. *Comptes Rendus de l'Académie des Sciences,* 1839. 9: 561–567.
9. Mills, A. and Lehunte, S., An overview of semiconductor photocatalysis. *Journal of Photochemistry and Photobiology A—Chemistry*, 1997. 108(1): 1–35.
10. Fujishima, A. and Honda, K., Electrochemical photolysis of water at a semiconductor electrode. *Nature*, 1972. 238(53–58): 37–40.

11. Muscat, J., Swamy, V., and Harrison, N.M., First-principles calculations of the phase stability of TiO$_2$. *Physical Review B*, 2002. 65(22): 224112–224127.

12. Arashi, H., Raman-spectroscopic study of the pressure-induced phase-transition in TiO$_2$. *Journal of Physics and Chemistry of Solids*, 1992. 53(3): 355–359.

13. Mammone, J.F., Nicol, M., and Sharma, S.K., Raman-spectra of II TiO$_2$-, TiO$_2$-III, SnO$_2$, and GeO$_2$ at high-pressure. *Journal of Physics and Chemistry of Solids, 1981.* 42(5): 379–384.

14. Mammone, J.F., Sharma, S.K., and Nicol, M., Raman-study of rutile (TiO$_2$) at high-pressures. *Solid State Communications*, 1980. 34(10): 799–802.

15. Navrotsk, A., Jamieson, J.C., and Kleppa, O.J., Enthalpy of transformation of a high-pressure polymorph of titanium dioxide to rutile modification. *Science*, 1967. 158(3799): 388–&.

16. Mitsuhashi, T. and Kleppa, O.J., Transformation enthalpies of the TiO$_2$ polymorphs. *Journal of the American Ceramic Society*, 1979. 62(7–8): 356–357.

17. Glassford, K.M. and Chelikowsky, J.R., Optical-properties of titanium-dioxide in the rutile structure. *Physical Review B*, 1992. 45(7): 3874–3877.

18. Glassford, K.M., et al., Electronic and structural-properties of TiO$_2$ in the rutile structure. *Solid State Communications, 1990.* 76(5): 635–638.

19. Mo, S.D. and Ching, W.Y., Electronic and optical-properties of 3 phases of titanium-dioxide - rutile, anatase, and brookite. *Physical Review B*, 1995. 51(19): 13023–13032.

20. Calatayud, M., et al., Quantum-mechanical analysis of the equation of state of anatase TiO$_2$. *Physical Review B*, 2001. 64(18): 184113:1–184113:4.

21. Kurtz, R.L., et al., Synchrotron radiation studies of H$_2$O adsorption on TiO$_2$ (110). *Surface Science*, 1989. 218(1): 178–200.

22. Pan, J.M., et al., Interaction of water, oxygen, and hydrogen with TiO$_2$ (110) surfaces having different defect densities. *Journal of Vacuum Science & Technology A-Vacuum Surfaces and Films*, 1992. 10(4): 2470–2476.

23. Martin, S.T., et al., Time-resolved microwave conductivity. 1. TiO$_2$ photoreactivity and size quantization. *Journal of the Chemical Society-Faraday Transactions*, 1994. 90(21):3315–3322.

24. Martin, S.T., Herrmann, H., and Hoffmann, M.R., Time-resolved microwave conductivity. 2. Quantum-sized TiO$_2$ and the effect of adsorbates and light-intensity on charge-carrier dynamics. *Journal of the Chemical Society-Faraday Transactions*, 1994. 90(21): 3323–3330.

25. Reddy, K.M., Manorama, S.V., and Reddy, A.R., Bandgap studies on anatase titanium dioxide nanoparticles. *Materials Chemistry and Physics*, 2003. 78(1): 239–245.

26. Wang, Y. and Herron, N., Nanometer-sized semiconductor clusters—Materials synthesis, quantum size effects, and photophysical properties. *Journal of Physical Chemistry,* 1991. 95(2): p. 525–532.

27. So, C.M., et al., Degradation of azo dye Procion Red MX-5B by photocatalytic oxidation. *Chemosphere*, 2002. 46(6): 905–912.

28. Fox, M.A. and Dulay, M.T., Acceleration of secondary dark reactions of intermediates derived from adsorbed dyes on irradiated TiO$_2$ powders. *Journal of Photochemistry and Photobiology A—Chemistry,* 1996. 98(1–2): 91–101.

29. Liqiang, J., et al., Review of surface photovoltage spectra of nano-sized semiconductor and its applications in heterogeneous photocatalysis. *Solar Energy Materials and Solar Cells*, 2003. 79(2): 133–151.

30. Khan, S.U.M., Al-Shahry, M., and Ingler, W.B., Efficient photochemical water splitting by a chemically modified n-TiO$_2$. *Science*, 2002. 297(5590): 2243–2245.

31. Moser, J., Gratzel, M., and Gallay, R., Inhibition of electron-hole recombination in substitutionally doped colloidal semiconductor crystallites. *Helvetica Chimica Acta*, 1987. 70(6): 1596–1604.

32. Sakthivel, S. and Kisch, H., Daylight photocatalysis by carbon-modified titanium dioxide. *Angewandte Chemie-International Edition,* 2003. 42(40): 4908–4911.

33. Sclafani, A., Palmisano, L., and Schiavello, M., Phenol and nitrophenols photodegradation carried out using aqueous TiO₂ anatase dispersions. *Abstracts of Papers of the American Chemical Society,* 1992. 203: 132-ENVR.

34. Soria, J., et al., Dinitrogen photoreduction to ammonia over titanium-dioxide powders doped with ferric ions. *Journal of Physical Chemistry,* 1991. 95(1): 274–282.

35. Choi, W.Y., Termin, A., and Hoffmann, M.R., Effects of metal-ion dopants on the photocatalytic reactivity of quantum-sized TiO₂ particles. *Angewandte Chemie-International Edition in English,* 1994. 33(10): 1091–1092.

36. Palmisano, L., et al., Activity of chromium-ion-doped titania for the dinitrogen photoreduction to ammonia and for the phenol photodegradation. *Journal of Physical Chemistry,* 1988. 92(23): 6710–6713.

37. Herrmann, J.M., Disdier, J., and Pichat, P., Effect of chromium doping on the electrical and catalytic properties of powder titania under UV and visible illumination. *Chemical Physics Letters,* 1984. 108(6): 618–622.

38. MU, W., Herrmann, J.M., and Pichat, P., room-temperature photocatalytic oxidation of liquid cyclohexane into cyclohexanone over neat and modified TiO₂. *Catalysis Letters,* 1989. 3(1): 73–84.

39. Sun, B., Reddy, E.P., and Smirniotis, P.G., Effect of the Cr6+ concentration in Cr-incorporated TiO₂-loaded MCM-41 catalysts for visible light photocatalysis. *Applied Catalysis B—Environmental,* 2005. 57(2): 139–149.

40. Karakitsou, K.E. and Verykios, X.E., Effects of altervalent cation doping of TiO₂ on its performance as a photocatalyst for water cleavage. *Journal of Physical Chemistry,* 1993. 97(6): 1184–1189.

41. Butler, E.C. and Davis, A.P., Photocatalytic oxidation in aqueous titanium-dioxide suspensions—The influence of dissolved transition-metals. *Journal of Photochemistry and Photobiology A—Chemistry,* 1993. 70(3): 273–283.

42. Fujihira, M., Satoh, Y., and Osa, T., Heterogeneous photocatalytic reactions on semiconductor-materials. 3. Effect of pH and Cu2+ ions on the photo-Fenton type reaction. *Bulletin of the Chemical Society of Japan,* 1982. 55(3): 666–671.

43. Sato, S. and White, J.M., Photo-decomposition of water over P-TiO₂ catalysts. *Chemical Physics Letters,* 1980. 72(1):83–86.

44. Sato, S. and White, J.M., Photoassisted reaction of CO₂ with H₂O on Pt-TiO₂. *Abstracts of Papers of the American Chemical Society,* 1980. 179(MAR): 75-COLL.

45. Bockelmann, D., Goslich, R., and Bahnemann, D., In *Solar Thermal Energy Utilization.* Vol. 6. 1992, Heidelberg: Springer Verlag GmbH.

46. Pichat, P., et al., Pt content and temperature effects on the photocatalytic H-2 production from aliphatic-alcohols over Pt TiO₂. *Nouveau Journal De Chimie—New Journal of Chemistry,* 1982. 6(11): 559–564.

47. Sclafani, A., Mozzanega, M.N., and Pichat, P., Effect of silver deposits on the photocatalytic activity of titanium-dioxide samples for the dehydrogenation or oxidation of 2-propanol. *Journal of Photochemistry and Photobiology A—Chemistry,* 1991. 59(2): 181–189.

48. Bahnemann, D., Bockelmann, D., and Goslich, R., Mechanistic studies of water detoxification in illuminated TiO₂ suspensions. *Solar Energy Materials,* 1991. 24(1–4): 564–583.

49. Gopidas, K.R., Bohorquez, M., and Kamat, P.V., Photoelectrochemistry in semiconductor particulate systems. 16. Photophysical and photochemical aspects of coupled semiconductors—Charge-transfer processes in colloidal Cds-TiO₂ and Cds-Agi systems. *Journal of Physical Chemistry,* 1990. 94(16): 6435–6440.

50. Duonghong, D., Borgarello, E., and Gratzel, M., Dynamics of light-induced water cleavage in colloidal systems. *Journal of the American Chemical Society*, 1981. 103(16): 4685–4690.

51. Hurum, D.C., et al., Recombination pathways in the Degussa P25 formulation of TiO_2: Surface versus lattice mechanisms. *Journal of Physical Chemistry B*, 2005. 109(2): 977–980.

52. Sun, B. and Smirniotis, P.G., Interaction of anatase and rutile TiO_2 particles in aqueous photooxidation. *Catalysis Today*, 2003. 88(1–2): 49–59.

53. Davis, R.J., et al., Photocatalytic decolorization of waste-water dyes. *Water Environment Research*, 1994. 66(1): 50–53.

54. Dibble, L.A. and Raupp, G.B., Fluidized-bed photocatalytic oxidation of trichloroethylene in contaminated airstreams. *Environmental Science & Technology*, 1992. 26(3): 492–495.

55. Faust, B.C. and Hoffmann, M.R., Photoinduced reductive dissolution of alpha-Fe_2O_3 by bisulfite. *Environmental Science & Technology*, 1986. 20(9): 943–948.

56. Matsunaga, T., Namba, Y., and Nakajima, T., Electrochemical sterilization of microbial-cells. *Bioelectrochemistry and Bioenergetics*, 1984. 13(4–6): 393–400.

57. Matsunaga, T., et al., Photoelectrochemical sterilization of microbial-cells by semiconductor powders. *FEMS Microbiology Letters*, 1985. 29(1–2): 211–214.

58. Rincon, A.G. and Pulgarin, C., Bactericidal action of illuminated TiO_2 on pure *Escherichia coli* and natural bacterial consortia: Post-irradiation events in the dark and assessment of the effective disinfection time. *Applied Catalysis B—Environmental*, 2004. 49(2): 99–112.

59. Rincon, A.G., et al., Interaction between *E. coli* inactivation and DBP-precursors—dihydroxybenzene isomers—in the photocatalytic process of drinking-water disinfection with TiO_2. *Journal of Photochemistry and Photobiology A—Chemistry*, 2001. 139(2–3): 233–241.

60. Carraway, E.R., Hoffman, A.J., and Hoffmann, M.R., Photocatalytic oxidation of organic-acids on quantum-sized semiconductor colloids. *Environmental Science & Technology*, 1994. 28(5): 786–793.

61. Kormann, C., Bahnemann, D.W., and Hoffmann, M.R., Photolysis of chloroform and other organic-molecules in aqueous TiO_2 suspensions. *Environmental Science & Technology*, 1991. 25(3): 494–500.

62. Henderson, R.S., et al., A cylindrical drift chamber for radiative muon-capture experiments at Triumf. *Transactions on Nuclear Science*, 1990. 37(3): 1116–1119.

63. Doushita, K. and Kawahara, T., Evaluation of photocatalytic activity by dye decomposition. *Journal of Sol-Gel Science and Technology*, 2001. 22(1–2): 91–98.

64. Galindo, C., Jacques, P., and Kalt, A., Photodegradation of the aminoazobenzene acid Orange 52 by three advanced oxidation processes: UV/H_2O_2 UV/TiO_2 and VIS/TiO_2—Comparative mechanistic and kinetic investigations. *Journal of Photochemistry and Photobiology A—Chemistry*, 2000. 130(1): 35–47.

65. Guillard, C., et al., Influence of chemical structure of dyes, of pH and of inorganic salts on their photocatalytic degradation by TiO_2 comparison of the efficiency of powder and supported TiO_2. *Journal of Photochemistry and Photobiology A—Chemistry*, 2003. 158(1): 27–36.

66. Houas, A., et al., Photocatalytic degradation pathway of methylene blue in water. *Applied Catalysis B—Environmental*, 2001. 31(2): 145–157.

67. Hu, C., et al., Photocatalytic degradation of triazine-containing azo dyes in aqueous TiO_2 suspensions. *Applied Catalysis B—Environmental*, 2003. 42(1): 47–55.

68. Hu, C., et al., Effects of acidity and inorganic ions on the photocatalytic degradation of different azo dyes. *Applied Catalysis B—Environmental*, 2003. 46(1): 35–47.

69. Konstantinou, I.K. and Albanis, T.A., TiO₂-assisted photocatalytic degradation of azo dyes in aqueous solution: kinetic and mechanistic investigations—A Review. *Applied Catalysis B—Environmental,* 2004. 49(1): 1–14.

70. Lachheb, H., et al., Photocatalytic degradation of various types of dyes (Alizarin S, Crocein Orange G, Methyl Red, Congo Red, Methylene Blue) in water by UV-irradiated titania. *Applied Catalysis B—Environmental,* 2002. 39(1): 75–90.

71. Mora-Sero, I. and Bisquert, J., Fermi level of surface states in TiO₂ nanoparticles. *Nano Letters,* 2003. 3(7):945–949.

72. Sivalingam, G., et al., Photocatalytic degradation of various dyes by combustion synthesized nano anatase TiO₂. *Applied Catalysis B—Environmental,* 2003. 45(1): 23–38.

73. Neumann, B., et al., Electrochemical mass spectroscopic and surface photovoltage studies of catalytic water photooxidation by undoped and carbon-doped titania. *Journal of Physical Chemistry B,* 2005. 109: 16579–16586.

74. Wang, W., et al., Preparation and crystalline phase of a TiO₂ porous film by anodic oxidation. *Rare Metals,* 2005. 24(4): 330–335.

75. Tao, H.J., et al., Fabrication of self-organized TiO₂ nanotubes by anodic oxidation and their photocatalysis. *Transactions of Nonferrous Metals Society of China,* 2005. 15: 462–466.

76. Xie, L.J., et al., Isopropanol-assisted hydrothermal synthesis of bismuth titanate nanophotocatalysts. *Materials Letters,* 2006. 60(2): 284–286.

77. Wu, Z.Y., et al., Structural determination of titanium-oxide nanoparticles by x-ray absorption spectroscopy. *Applied Physics Letters,* 2002. 80(16): 2973–2975.

78. Fu, X.R., et al., Influence of annealing atmosphere on the structures and photocatalytic properties of Ag-doped TiO₂ thin film. *Acta Chimica Sinica,* 1998. 56(6): 521–526.

79. Yuan Gao and Huitao Liu, Preparation and catalytic property study of a novel kind of suspended photocatalyst of TiO₂-activated carbon immobilized on silicone rubber film, *Materials Chemistry and Physics,* 2005. 92(2–3): 604–608.

80. Sakthivel, S., Janczarek, M., and Kisch, H., Visible light activity and photoelectrochemical properties of nitrogen-doped TiO₂. *Journal of Physical Chemistry B,* 2004. 108(50): 19384–19387.

81. Kowalska, E., et al., H₂O₂/UV enhanced degradation of pesticides in wastewater. *Water Science and Technology,* 2004. 49(4): 261–266.

82. Dabrowski, B., et al., Photo-oxidation of dissolved cyanide using TiO₂ catalyst. *Journal of Photochemistry and Photobiology A—Chemistry,* 2002. 151(1–3): 201–205.

83. Lee, S.W., et al., Synthesis and characterization of hard magnetic composite photocatalyst (barium ferrite silica titania) and its photoactivity. *Abstracts of Papers of the American Chemical Society,* 2004. 227: U1246–U1246.

84. Khan, A.Y. and Mazyck, D.W., The effect of UV irradiation on adsorption by activated carbon/TiO₂ composites. *Carbon,* 2006. 44: 158–193.

85. Schwarzenbach, R.P., Gschwend, P.M., and Imboden, D.M., *Environmental Organic Chemistry* (2nd ed.). 2002, New Jersey: John Wiley & Sons.

86. Jitianu, A., et al., Synthesis and characterization of carbon nanotubes-TiO₂ nanocomposites. *Carbon,* 2004. 42(5–6): 1147–1151.

87. Lee, S.W. and Sigmund, W.M., Formation of anatase TiO₂ nanoparticles on carbon nanotubes. *Chemical Communications,* 2003(6): 780–781.

88. Pyrgiotakis, G., Titania carbon nanotubes interphase for advanced photocatalysis, in *Materials Science and Engineering.* 2006, University of Florida: Gainesville, FL. p. 212.

89. Pyrgiotakis, G., Lee, S.-H., and Sigmund, W., Advanced photocatalysis with anatase nanocoated multi-walled carbon nanotubes, in *MRS Spring Meeting 2005.* San Fransisco, CA: MRS.

14 Zero-valent Iron Nanoparticles for Abatement of Environmental Pollutants
Materials and Engineering Aspects

X.-Q. Li, D.W. Elliott, and W.-X. Zhang

CONTENTS

INTRODUCTION

Due to their subcolloidal size and unique molecular and/or atomic structures, many nanomaterials have been shown to possess distinctive mechanical, magnetic, optical, electronic, catalytic, and chemical properties, which contribute to promising applications in machinery, energy, optics, electronics, drug delivery, and medical

309

diagnostics.[1-6] The large surface-to-volume ratio of nanomaterials can lead to surprising surface and quantum size effects. As particle size decreases, the proportion of surface and near surface atoms increases. Surface atoms tend to have more unsatisfied or dangling bonds with concomitantly higher surface energy. Thus, the surface atoms have a stronger tendency to interact, adsorb, and react with other atoms or molecules in order to achieve surface stabilization.[7] For example, carbon nanotubes have been widely reported to exhibit an extremely high storage capacity for hydrogen and may serve as an ideal material for fuel cells, energy storage and transmission.[7]

New discoveries on nanomaterials are being reported at an ever-increasing rate. Generally speaking, gold is considered to be a rather stable and inert substance. However, nano-gold particles (<10 nm) exhibit very high affinity for many functionalized species including amino and nucleic acids and have been used as molecular delivery carriers. When functionalized with oligonucleotides, gold nanoparticles can act as intracellular gene regulation agents for the control of protein expression in cells.[8] Moreover, titania nanoparticles have far better chemical stability, optical properties, and higher photochemical reactivity than their conventional microscale counterparts.[9] It has been suggested that titania nanoparticles are more suitable for photoelectrical energy conversion and photocatalysis. At the nanoscale, the energy and electronic properties of materials are discrete rather than existent in a continuum as is the case for bulk materials.[5] The quantum size effect of nanomaterials has found many interesting applications, notably in electronics, information storage, and also in nanosensors. The quantum size effect even allows detection of single molecules and/or single biological cells.

Interestingly, nanoscale materials also exist ubiquitously in the natural environment. Biomolecules such as DNAs and proteins have characteristic dimension(s) in the nano domain (1–100 nm). The particle sizes of natural materials, such as viruses, smog aerosols, and weathered minerals such as iron oxides and silicates, are often on the nanoscale or have nanoscale components. Engineered nanomaterials have found increasing environmental applications.[10-13] For example, nanomaterials with promising sorptive and reactive properties have been used in water and air purification, hazardous waste treatment, and environmental remediation. This trend points to a considerable increase in the rate of usage of nanomaterials in environmental technologies. According to recent technical market research conducted by BCC Research, the total global market for nanotechnology in environmental remediation in 2003 was minuscule at $2.8 million. Nonetheless, it escalated rapidly to $4.8 million in 2004 and $11.2 million in 2005. With increasing concern for environmental quality and improvements on performance and public acceptance of nanotechnology, it is estimated that the market for nanomaterials in the environmental remediation marketplace alone could reach $2.4 billion by 2010.[14]

In this chapter, recent research in the utilization of zero-valent iron (ZVI) nanoparticles for treatment of contaminated soils and groundwater is highlighted. Moreover, the synthesis and characterization of ZVI nanoparticles are discussed and examples provided regarding the treatment of both organic and inorganic contaminants. Issues related to fate and transport of nanoparticles in the environment are also addressed.

IRON NANOPARTICLES FOR ENVIRONMENTAL REMEDIATION

The nanoparticles (<100 nm) discussed in this contribution are zero-valent iron (ZVI) particles and exhibit a typical core-shell structure as illustrated in Figure 14.1. The core consists primarily of zero-valent or metallic iron while the mixed valent (i.e., Fe[II] and Fe[III]) oxide shell is formed as a result of oxidation of the metallic iron. Iron typically exists in the environment as iron(II)- and iron(III)-oxides, and as such, ZVI is a manufactured material. Thus far, applications of ZVI have focused primarily on the electron-donating properties of ZVI. Under ambient conditions, ZVI is fairly reactive in water and can serve as an excellent electron donor, which makes it a versatile remediation material.[15]

The use of ZVI as a remediation agent in groundwater treatment started in the early 1990s when granular ZVI was first employed in permeable reactive barrier (PRBs) systems.[16–19] In a PRB structure, groundwater flows passively through an engineered iron wall while contaminants are precipitated, adsorbed, or transformed in contact with the ZVI surface. Over 100 such PRB structures have been constructed in the United States since early 1990s. While PRBs containing ZVI powders may serve as useful *in-situ* remedies for some sites, important challenges still exist for this technology, which may limit its practical application. For example, a large amount (e.g., tons) of iron powder is usually needed even for a modest PRB structure. Costs associated with the PRB construction, especially for deep aquifers, remains too high for many potential users of the technology. Another important limitation is

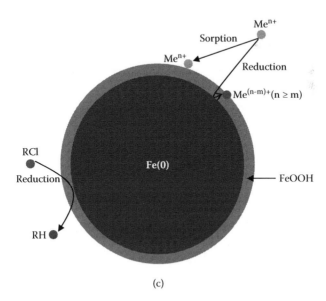

(c)

FIGURE 14.1 The core-shell model of zero-valent iron nanoparticles. The core consists of mainly zero-valent iron and provides the reducing power for reactions with environmental contaminants. The shell is largely iron oxides/hydroxides formed from the oxidation of zero-valent iron. The shell provides sites for chemical complex formation (e.g., chemosorption).

the relative lack of flexibility after a PRB is installed. Relocation or major modifications to the PRB infrastructure is often impractical.

The nanoscale iron (nZVI) technology discussed herein can be regarded as an extension of ZVI technology.[20–22] In some cases, it may serve as an alternative to the conventional ZVI PRBs. For other sites, the nZVI process can complement (or supplement) the fixed PRBs. For example, nZVI injections can be used to address the heavily contaminated source area or other "hot spots" while ZVI PRB functions as a barrier to contain the dispersion of contaminants. Because of their small size, nanoparticle slurries in water can be injected under pressure and/or even by gravity flow to the contaminated area and under certain conditions remain in suspension and flow with water for extended periods of time. An *in-situ* treatment zone may thus be formed. Alternatively, iron nanoparticles also can be used in some *ex-situ* applications.

TABLE 14.1
ZVI Nanoparticles for Site Remediation—A List of Recent Projects

Site	Location
Phoenix—Goodyear Airport (Unidynamics)	Phoenix, AZ
Defense contractor site	CA
Jacksonville dry cleaner sites, pilot tests using nZVI (several sites)	FL
State LEAD SITE	ID
Groveland Wells Superfund Site	Groveland, MA
Aberdeen site	MD
Sierra Army Depot	NV
Pharmaceutical plant. Pilot test using BNP	Research Triangle Park, NC
Industrial Site	Edison, NJ
Picattiny Arsenal	Dover, NJ
Shieldalloy plant	NJ
Manufacturing site	Passaic, NJ
Klockner Road site	Hamilton Township, NJ
Manufacturing plant	Trenton, NJ
Naval Air Engineering Station	Lakehurst, NJ
Confidential site. Pilot test using nZVI	Winslow Township, NJ
Confidential site. Pilot test using BNP	Rochester, NY
Nease superfund site, Pilot test using BNP	OH
Former Electronics Manufacturing Plant	PA
Rock Hill, pharmaceutical plant, full scale using nZVI	SC
Memphis Defense Depot	TN
Grand Plaza Drycleaning Site	Dallas, TX
Industrial plant, pilot tests using nZVI	Ontario, Canada
Public domain, pilot test using nZVI	Quebec, Canada
Solvent manufacturing plant, pilot test using nZVI	Czech Republic
Industrial plant, pilot test using nZVI	Czech Republic
Industrial plant, pilot test using nZVI	Germany
Industrial plant, pilot test using nZVI	Italy
Brownfields, pilot test using nZVI	Slovakia

Over the past decade, extensive studies have demonstrated that ZVI nanoparticles are effective for the treatment of many pollutants commonly identified in groundwater including perchloroethene (PCE) and trichloroethene (TCE), carbon tetrachloride (CT), nitrate, energetic munitions such as TNT and RDX, legacy organohalogen pesticides such as lindane and DDT, as well as heavy metals like chromium and lead[10,21–35]. Dozens of pilot and large-scale in-situ applications have also been conducted and demonstrated that rapid in-situ remediation with nZVI can be achieved.[21,36–37] Examples of some recent field applications using ZVI nanoparticles are presented in Table 14.1.

SYNTHESIS OF IRON NANOPARTICLES

Robust methods for large-scale and cost-effective production of nanomaterials are essential to the growth of nanotechnology. Environmental applications often require usage of considerable quantities of treatment reagents and/or amendments for the remediation of huge volumes of contaminated water and soil. Unlike many industrial applications, environmental technologies often exhibit relatively low market values. Therefore, the application of these technologies can be particularly sensitive to the costs of nanomaterials. This may have been the main factor for the relatively slow adaptation of some environmental nanotechnologies.

There are two general strategies in terms of nanoparticle synthesis: top-down and bottom-up approaches. The former starts with large-size (i.e., granular or microscale) materials with the generation of nanoparticles by mechanical and/or chemical steps including milling, etching, and/or machining. The latter approach entails the "growth" of nanostructures atom-by-atom or molecule-by-molecule via chemical synthesis, self-assembling, positional assembling, etc.

Both approaches have been successfully applied in the preparation of nZVI nanoparticles. For example, ZVI nanoparticles have been fabricated by vacuum sputtering,[38] synthesized from the reduction of goethite and hematite particles with hydrogen gas at elevated temperatures (e.g., 200–600°C),[39] by decomposition of iron pentacarbonyl ($Fe(CO)_5$) in organic solvents or in argon,[40–42] and by electrodeposition of ferrous salts. The generation of nZVI by the bottom-up reduction of ferric (Fe(III)) or ferrous (Fe(II)) salts with sodium borohydride has been used by many research groups:[10]

$$4Fe^{3+} + 3BH_4^- + 9H_2O \leftrightarrows 4Fe^0\downarrow + 3H_2BO_3^- + 12H^+ + 6H_2\uparrow \qquad (14.1)$$

A major advantage of this method is its relative simplicity with the need of only two common reagents and no need for any special equipment/instrument, it can been done in almost any wet chemistry lab. In our laboratory at Lehigh University, we have used this approach to produce high-quality iron nanoparticles for over ten years. Typically it is achieved by slowly adding 1:1 volume ratio of 0.25 M sodium borohydride into 0.045 M ferric chloride solution. Nevertheless, there are important health and safety considerations associated with the borohydride reduction approach. The synthesis needs to be conducted in a fume hood as the chemical reactions produce hydrogen gas as a byproduct. Moreover, explosion-resistant mixers should be used to

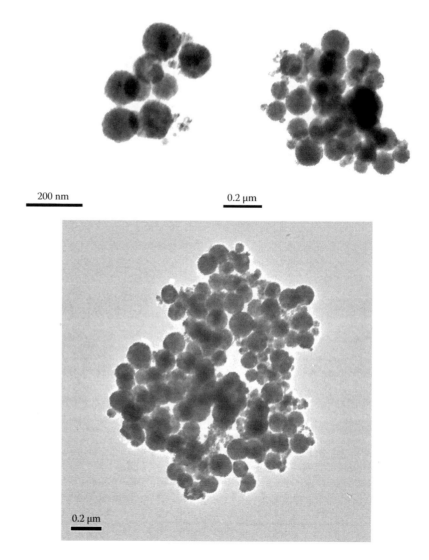

FIGURE 14.2 TEM images of zero-valent iron nanoparticles.

minimize the possibility of sparks, etc. The jet-black nanoparticle aggregates can be collected by vacuum filtration. Selected images of nZVI are given in Figure 14.2.

Bimetallic iron nanoparticles, in which a second and often less-reactive metal such as Pd, Ni, Pt, or Ag, can be prepared simply by soaking the freshly prepared nZVI in a solution of the second metal salt.[23] It is believed that the second noble metal promotes iron oxidation and may act as a catalyst for electron transfer and hydrogenation. Several studies have demonstrated that bimetallic iron nanoparticles (Pd-Fe, Pt-Fe, Ni-Fe, Ag-Fe) can achieve significantly higher degradation rates and prevent or reduce the formation of toxic byproducts.[20,43]

The iron nanoparticles are colloidal in nature and exhibit a strong tendency to aggregate as well as adhere to the surfaces of natural materials such soil and sediment. Increasing efforts have been directed on the dispersion of iron nanoparticles. For example, He and Zhao used water-soluble starch as a potential nanoparticle stabilizer.[27] The average size of iron nanoparticles was reduced to 14.1 nm in diameter. In another work, Mallouk and coworkers have synthesized carbon platelets with 50–200 nm in diameter and water soluble polyelectrolyte like poly(acrylic acid) (PAA) as supports and dispersants.[34] Saleh and others used poly(methacrylic acid)-block-ploy(methyl methacrylate)-block-poly(styrensulfonate) to modify the iron nanoparticle surface. Although the hydrodynamic diameter of the modified particles increased 30–50 nm, the colloidal stability of the modified nZVI was enhanced.[44]

Over the last decade, the costs for ZVI nanoparticles have been reduced dramatically (e.g., from >$500/kg to $50–100/kg). At the time of the initial field demonstration of the nZVI technology in 2001, there were no commercial suppliers of iron nanoparticles while presently multiple vendor options exist. Still, prices remain too high for many applications. Compared to laboratory-prepared ZVI particles, the quality and efficacy of commercially available nZVI products can be highly variable as inadequate quality assurance and quality control and characterization protocols exist to support the products.

CHARACTERIZATION OF IRON NANOPARTICLES

Only limited work has been published thus far on the surface characterization of ZVI nanoparticles. A detailed knowledge of the surface properties is vital for understanding the salient reaction mechanisms, kinetics and intermediate/product profiles. The transport, distribution and fate of nanoparticles in the environment also depend on these surface properties. However, it is often not practical to define an average or typical iron nanoparticle because ZVI nanoparticles produced with different methods may exhibit widely varying properties. Fundamentally, iron nanoparticles are reactive species and their surface properties change rapidly and profoundly over time and solution chemistry and with environmental conditions. In practice, the term "aging" has often been used to describe the observed changes of iron nanoparticles.

Figure 14.2 presents the Transmission Electron Microscopy (TEM) images of iron nanoparticles synthesized using the sodium borohydride method. The nanoparticles are mostly spherical in shape and exist as chain-like aggregates. An accumulative size distribution survey of over 400 nanoparticles from TEM images suggests that over 80% of the nanoparticles have diameters of less than 100 nm while 50% are less than 60 nm (Figure 14.3).[45] The average BET surface area is about 30,000–35,000 m^2/kg.

According to the core-shell model, the mixed valence iron oxide shell is largely insoluble under neutral pH conditions and may protect the ZVI core from rapid oxidation. The speciation of the actual oxide layer is complicated due to the lack of long-range order and its amorphous structures. The composition of the oxide shell may also depend on the fabrication processes and environmental conditions. For example, the oxide shell of α-Fe nanoparticles generated by sputtering consists mainly of maghaemite (γ-Fe_2O_3) or partially oxidized magnetite (Fe_3O_4). Nanoparticles formed

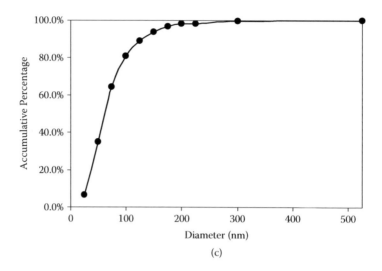

(c)

FIGURE 14.3 An accumulative size distribution of zero-valent iron nanoparticles.

via nucleation of metallic vapor also contain γ-Fe_2O_3 and Fe_3O_4, with richer γ-Fe_2O_3 for smaller particles due to the higher surface-to-volume ratio and rapid surface oxidation.[38] On the other hand, particles produced by hydrogen reduction of goethite and hematite particles reportedly have only Fe_3O_4 in the shell.[39,46] The presence of wustite (FeO) has also been noted.[38] It is not clear from the existing literature whether variations in the shell structure and composition have any effect on the iron nanoparticle reactivity, aggregation, and transport.

Detailed x-ray photoelectron spectroscopy (XPS) studies on the ZVI nanoparticles have been performed in our laboratory. Figure 14.4 shows the XPS survey on the $Fe2p_{3/2}$ and O1s regions. For the $Fe2p_{3/2}$ spectrum, the binding energy of the main peak was located at about 711 eV, which is attributable to ferric iron (Fe[III]). The smaller peak at about 707 eV suggests the presence of the elemental metallic iron. The photoelectron peak of O1s in Figure 14.4b can be decomposed into three separate peaks at 529.9 eV, 531.2 eV, and 532.5 eV, representing the binding energies of oxygen in O^{2-}, OH^-, and chemically or physically adsorbed water, respectively. The oxygen species are similar to those on the surface of iron oxides in water.

Further examination of the peak area ratios of Fe/OH and OH^-/O^{2-} suggest that the oxide shell is composed of mainly iron hydroxides or iron oxyhydroxide. As a result of iron oxidation, Fe^{2+} is first formed on the surface:

$$2Fe + O_2 + 2H_2O \rightleftarrows 2Fe^{2+} + 4OH^- \tag{14.2}$$

$$Fe + 2H_2O \rightleftarrows Fe^{2+} + H_2 + 2OH^- \tag{14.3}$$

Fe^{2+} can be further oxidized to Fe^{3+}:

$$4Fe^{2+} + 4H^+ + O_2 \rightleftarrows 4Fe^{3+} + 2H_2O \tag{14.4}$$

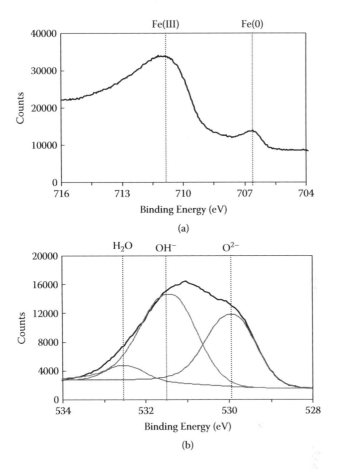

FIGURE 14.4 (a) XPS survey on Fe $2p_{3/2}$ and (b) XPS survey on O 1s of zero-valent iron nanoparticles. Results suggest that the iron on the particle's surface is extensively oxidized.

Fe^{3+} reacts with OH^- or H_2O and yields hydroxide or oxyhydroxide:

$$Fe^{3+} + 3OH^- \rightleftarrows Fe(OH)_3 \tag{14.5}$$

$$Fe^{3+} + 2H_2O \rightleftarrows FeOOH + 3H^+ \tag{14.6}$$

$Fe(OH)_3$ can also dehydrate to form $FeOOH$:

$$Fe(OH)_3 + 3H^+ \rightleftarrows FeOOH + H_2O \tag{14.7}$$

The above results constitute the foundation of the conceptual model of the ZVI nanoparticle as illustrated in Figure 14.1. In water, ZVI nanoparticles can exhibit metal-like or ligand-like coordination properties depending on the solution chemistry. At low pH ($<pH_{zpc} \cong 8$), iron oxides are positively charged and attract anionic ligands including key environmental species such as sulfate and phosphate.

When the solution pH is above the isoelectric point, the oxide surface becomes negatively charged and can form surface complexes with cations (e.g., metal ions). As described previously, zero-valent iron can serve as very effective electron donors (i.e., reductants). ZVI (Fe^{2+}/Fe) has a standard reduction potential ($E°$) of -0.44 V, which is lower than many metals such as Pb, Cd, Ni, and Cr, as well as many organic compounds such as chlorinated hydrocarbons. These compounds are thus susceptible to reduction by ZVI nanoparticles.

Reactions at the nZVI surface involve many steps, for example, mass transport of molecules to the surface and electron transfer (ET) from the ZVI to the surface-adsorbed molecules. Kinetic analysis suggests that for many organic compounds such as PCE and TCE, the surface reaction or electron transfer is the dominant factor or the rate-limiting step. There are several potential pathways for the electron transfer (ET) to occur: (1) direct ET from Fe(0) through defects such as pits or pinholes, where the oxide layer acts as a physical barrier; (2) indirect ET from Fe(0) through the oxide layer via the oxide conduction band, impurity bands, or localized bands; or (3) ET from sorbed or lattice Fe(II) surface site.[48–49] In this regard, the iron oxide shell can appropriately be considered an n-type semiconductor. On the other hand, we are not aware of direct evidence on the ET mechanisms at this time.

DEGRADATION OF ORGANIC CONTAMINANTS

A substantial portion of the peer-reviewed nZVI literature has been limited to the degradation of various organic contaminants such as chlorinated organic solvents, organochlorine pesticides, polychlorinated biphenyls (PCBs), and organic dyes. Selected compounds studied in our laboratory at Lehigh University are listed in Table 14.2.

The chemical principles underlying the transformation of halogenated hydrocarbons have been particularly well documented as those compounds are among the most commonly detected soil and groundwater pollutants. Metallic iron (Fe^0) serves effectively as an electron donor:

$$Fe^0 \rightleftarrows Fe^{2+} + 2e^- \tag{14.8}$$

Chlorinated hydrocarbons on the other hand accept the electrons and undergo reductive dechlorination:[50,51]

$$RCl + H^+ + 2e^- \rightleftarrows RH + Cl^- \tag{14.9}$$

From a thermodynamic perspective, the coupling of the reactions (14.8) and (14.9) is often energetically highly favorable:

$$RCl + Fe^0 + H^+ \rightleftarrows RCl + Fe^{2+} + X^- \tag{14.10}$$

For example, tetrachloroethene (C_2Cl_4), a common solvent, can be completely reduced to ethane by nZVI in accordance with the following overall equation:

$$C_2Cl_4 + 5Fe^0 + 6H^+ \rightleftarrows C_2H_6 + 5Fe^{2+} + 4Cl^- \tag{14.11}$$

TABLE 14.2

Common Contaminants That Can Be Remediated by Iron Nanoparticles (20)

Chlorinated Methanes

Carbon tetrachloride (CCl_4)

Chloroform ($CHCl_3$)

Dichloromethane (CH_2Cl_2)

Chloromethane (CH_3Cl)

Chlorinated Benzenes

Hexachlorobenzene (C_6Cl_6)

Pentachlorobenzene (C_6HCl_5)

Tetrachlorobenzenes ($C_6H_2Cl_4$)

Trichlorobenzenes ($C_6H_3Cl_3$)

Dichlorobenzenes ($C_6H_4Cl_2$)

Chlorobenzene (C_6H_5Cl)

Pesticides

DDT ($C_{14}H_9Cl_5$)

Lindane ($C_6H_6Cl_6$)

Organic Dyes

Orange II ($C_{16}H_{11}N_2NaO_4S$)

Chrysoidin ($C_{12}H_{13}ClN_4$)

Tropaeolin O ($C_{12}H_9N_2NaO_5S$)

Heavy Metals

Mercury (Hg^{2+})

Nickel (Ni^{2+})

Cadium (Cd^{2+})

Lead (Pb^{2+})

Chromium (Cr(VI))

Trihalomethanes

Bromoform ($CHBr_3$)

Dibromochloromethane ($CHBr_2Cl$)

Dichlorobromomethane ($CHBrCl_2$)

Chlorinated Ethenes

Tetrachloroethene (C_2Cl_4)

Trichloroethene (C_2HCl_3)

cis-Dichloroethene ($C_2H_2Cl_2$)

trans-Dichloroethene ($C_2H_2Cl_2$)

1,1-Dichloroethene ($C_2H_2Cl_2$)

Vinyl chloride (C_2H_3Cl)

Other Polychlorinated Hydrocarbons

PCBs

Pentachlorophenol

1,1,1-Trichloroethane

Other Organic Contaminants

N-nitrosodimethylamine (NDMA) ($C_4H_{10}N_2O$)

TNT ($C_7H_5N_3O_6$)

Inorganic Anions

Perchlorate (ClO_4^-)

Nitrate (NO_3^-)

Figure 14.5 presents an example of a laboratory study on a mixture of chlorinated hydrocarbons. Six common compounds including *trans*-dichloroethene (*t*-DCE), *cis*-dichloroethene (*c*-DCE), 1,1,1-trichloroethane (1,1,1-TCA), tetrachloroethylene (PCE), trichloroethylene (TCE), and tetrachloromethane were investigated. The initial concentration was 10 mg/L for each of the six compounds and the total nZVI loading was 5 g/L. As illustrated by the trend illustrated in the gas chromatograms over the reaction time course, all of the six compounds were reduced by the ZVI nanoparticles. Within 1 hour, over 99% removal of tetrachloroethylene was observed. Greater than 95% removal efficiency of the six compounds was achieved within 120 hours.

The importance of electron donors in the reduction of chlorinated hydrocarbons has received significant attention in the research and environmental remediation communities. Extensive research on both biological and chemical dechlorination has been published over the last two decades. In general, the rate of dechlorination

Peak #	Compound Name
1	*Trans*-dichloroethene (*trans*-C$_2$H$_2$Cl$_2$)
2	*Cis*-dichloroethene (*cis*-C$_2$H$_2$Cl$_2$)
3	1,1,1-trichloroethane (CCl$_3$CH$_3$)
4	Tetrachloroethylene (C$_2$Cl$_4$)
5	Trichloroethylene (C$_2$HCl$_3$)
6	Tetrachloromethane (CCl$_4$)

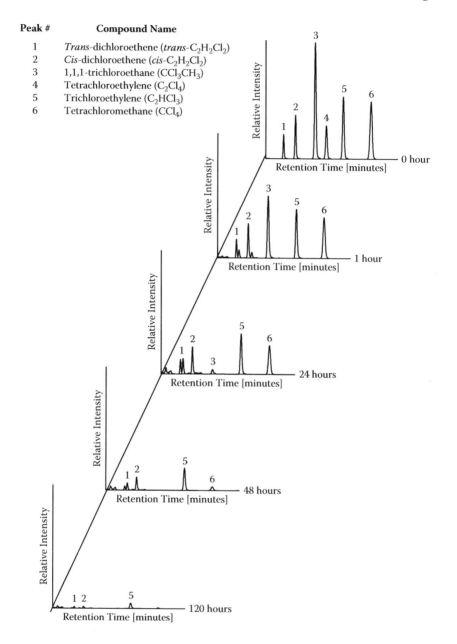

FIGURE 14.5 Reactions of iron nanoparticles (5 g/L) with a mixture of chlorinated aliphatic hydrocarbons. Gas chromatograms are shown in this figure. Six compounds with initial concentration at 10 mg/L are presented. *Trans*-dichloroethene (*t*-DCE), *cis*-dichloroethene (*c*-DCE), 1,1,1-trichloroethane (1,1,1-TCA), tetrachloroethylene (PCE), trichloroethylene (TCE), and tetrachloromethane.

increases with the number of chlorine substituents. For dechlorination with ZVI, three potential transformation mechanisms have been proposed: (1) direct reduction at the metal surface; (2) reduction by ferrous iron at the surface; and (3) reduction by hydrogen.[51] Ferrous iron, in concert with certain ligands, can slowly reduce the chlorinated hydrocarbons. However, electron transfer from ligand ferrous ions through the nZVI oxide shell is thought to be relatively slow and is probably not of great consequence. Dissolved hydrogen gas has revealed little reactivity in the absence of a suitable catalytic surface (e.g., Pd).

The incorporation of a noble or catalytic metal such as Pd, Ni, Pt, or Ag can substantially enhance the overall nZVI reaction rate. It has been observed that the surface area normalized rate constant (K_{SA}) of Fe/Pd nanoparticles for the degradation of tetrachloromethane is over two orders of magnitude higher than that of microscale iron particles. The activation energy of Fe/Pd nanoparticles in the transformation of tetrachloroethylene (PCE) was calculated to be 31.1 kJ/mole, compared to 44.9 kJ/mole for iron nanoparticles.[52] The fast reactions generated by the bimetallic nanoparticles also reduce possibility of toxic by-product formation and accumulation.

While chlorinated aliphatic compounds with one or two carbons have been studied extensively, chlorinated alicyclic and aromatic compounds have received far less attention. Without a doubt, chlorinated alicyclic and aromatic compounds feature more complicated chemical structures, are less aqueous soluble, react more slowly with nZVI, and often generate more intermediates and by-products. Limited research indicates that iron nanoparticles can exhibit fairly high reactivity toward these compounds even though some researches with micro- and millimeter iron particles reported little reactions. Xu and Zhang evaluated the degradability of hexachlorobenzene (HCB) by Fe/Ag bimetallic nanoparticles.[43] With an initial HCB concentration of 4 mg/L and an applied loading of 25 g/L nZVI, over 50% of HCB was reduced from aqueous solution within 30 minutes. HCB concentrations were reduced below the detection limit (<1 μg/L) within approximately 4 days. HCB was gradually transformed to a series of lesser chlorinated benzenes such as 1,2,4,5-tetrachlorobenzene, 1,2,4-trichlorobenzene, and 1,4-dichlorobenzene.

The degradation of lindane (γ-hexachlorocyclohexane, γ-HCH), one of the most widely used organochlorine pesticides over the timeframe from the 1940s through the 1990s, was investigated by Elliott.[53] Experiments conducted with groundwater contaminated with lindane (~700 μg/L of HCH, a summation of the four environmentally significant HCH isomers identified) showed that over 95% of the lindane was removed from solution within 48 hours by 2.2–27.0 g/L nZVI. In contrast, about 59% of HCH remained in solution after 24 hours with 49.0 g/L microscale iron particles (Figure 14.6). It is hypothesized that lindane is reduced via a dihaloelimination reaction to γ-TeCCH:

$$C_6H_6Cl_6 + Fe^0 \rightarrow C_6H_6Cl_4 + Fe^{2+} + 2Cl^- \qquad (14.12)$$

It was further determined that at least 35–65% of the chlorine initially presented in lindane was converted into chloride. The reaction followed pseudo first-order rate with the observed rate constant (k_{obs}) in the range of 0.04–0.65/hr.

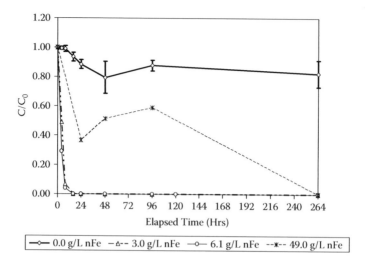

FIGURE 14.6 Removal of lindane by iron nanoparticles at various doses over 264 hours of reaction time. Contaminated groundwater sample from a site in Florida was used in this study. The total lindane concentration was approximately 1500 μg/L. Error bars represent the standard error at the 95% confidence interval.

REMEDIATION OF INORGANIC CONTAMINANTS

Several recent studies provided valuable insights into key nZVI properties associated with the potential to transform metal ions such as Cd, Ni, Zn, As, Cr, Ag, and Pb, as well as notorious inorganic anions like perchlorate and nitrate.[31,33–35,47] ZVI nanoparticles can rapidly remove and/or reduce these inorganic ions and also have relatively higher capacity than conventional sorptive media and granular iron particles. As an example, studies with chromium ore processing residual (COPR) containing highly concentrated hexavalent chromium show that 1 g of ZVI nanoparticles can reduce and immobilize 65–110 mg Cr(VI).[34,35] In comparison, the capacity for Cr(VI) removal by microscale ZVI is only 1–3 mg Cr(VI)/g Fe. Furthermore, the reaction rate with the ZVI nanoparticles is at least 25–30 times faster.

The work on reductive precipitation of Cr(VI) also confirmed that the nZVI reactions are surface-mediated. The initial reaction can be treated as a pseudo first-order reaction in which the rate constant is normalized to the total surface area of iron. Cr(VI) is reduced to Cr(III), which is then incorporated into the iron oxide layer as $(Cr_xFe_{1-x})(OH)_{3X}$ or $Cr_xFe_{1-x}(OOH)_x$.

While reduction is the predominant mechanism for Cr(VI) removal, both reduction and surface complex formation (sorption) are observed for Ni(II) removal. The capacity was experimentally determined to be 0.13 g Ni(II)/g Fe or 4.43 meq Ni(II)/g), which is well over 100% higher than the best inorganic sorbents (e.g., zeolites) available. High-resolution X-ray photoelectron spectroscopy (HR-XPS) reveals that the amount of reduced Ni(0) at the iron nanoparticle surface increases with time. At equilibrium, about 50% of Ni(II) is reduced as Ni(0) at the surface and 50% Ni(II)

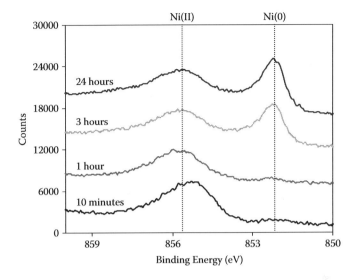

FIGURE 14.7 Reactions of Ni(II) with iron nanoparticles. This figure shows high-resolution XPS survey on Ni $2p_{3/2}$. Both Ni(II) and Ni(0) were observed on the nanoparticle surface. Ni concentration was 1000 mg/L. Iron nanoparticle loading was 5 g/L (47).

remains adsorbed at the iron nanoparticle surface (Figure 14.7). The surface complex is primarily nickel hydroxide.[47]

Iron nanoparticles can also reduce some relatively stable inorganic compounds such as perchlorate and nitrate. According to the several studies on the reduction of nitrate by granular and microscale ZVI particles published thus far, the reduction follows pseudo first-order kinetics with ammonia being the end product:[30,54]

$$NO_3^- + 4Fe^0 + 10H^+ \rightleftarrows 4\ Fe^{2+} + NH_4^+ + 3H_2O \qquad (14.13)$$

The reduction of nitrate by iron nanoparticles is fast even at relatively high pH (e.g., 8–10), a range that is known to be inhibitory to iron oxidation reactions. For example, Sohn and colleagues[33] designed repetitive experiments and observed that reaction rates remained about the same even after exposed to new nitrate solutions for six times. It is speculated that in alkaline solutions, the anionic hydroxo species (Fe(OH)$_y^{2-y}$ or Fe(OH)$_x^{3-x}$) are dissolved and then precipitated, yielding different phase of iron oxide, most likely magnetite (Fe$_3$O$_4$). The phase transformation may expose the fresh ZVI iron surface to nitrate, resulting in a relatively stable reaction rate.

Perchlorate is another high-profile contaminant in the United States, particularly in the western states. It is persistent in the environment due to its intrinsic chemical stability. Consequently, rates of natural attention rate under most cases are very slow. Because of its stability in water, only a few reductants, such as complexes of tin(II), vanadium(II, III), molybdenum(III), titanium(III), and ruthenium(III, IV), have been identified as being capable of reducing perchlorate. Still, the half-lives of perchlorate may range from 0.83 years to up to 11.3 years for these transition metal reductants.[55] In a study by Cao, Elliott, and Zhang,[31] ZVI nanoparticles were observed to reduce perchlorate to chloride with no observance of the sequential degradation products. The nZVI-mediated reduction of perchlorate proceeds according to equation (14.14) below:

$$ClO_4^- + 4Fe^0 + 8H^+ \rightleftarrows Cl^- + 4Fe^{2+} + 4H_2O \qquad (14.14)$$

In contrast, the reduction by microscale iron particles is negligibly slow. Not surprisingly, the activation energy was calculated to be 79 kJ/mole, indicative that the reduction is limited by the slow kinetics.

TRANSPORT OF THE IRON NANOPARTICLES

Because ZVI nanoparticles are increasingly being used in site remediation, a critical issue for the future development of the technology involves nZVI transport, dispersion, and fate in the subsurface environment. At present, little information is available on the transport and fate of ZVI nanoparticles in the environment. In laboratory soil column experiments, the mobility of pure iron nanoparticles has been observed to be limited due to the colloidal nature and efficient filtration mechanisms of aquifer materials. Data from some recent field tests indicate that the iron nanoparticles may migrate only a few inches to a few feet from the point of injection. The mobility of nanoparticles in the subsurface environment depends on many factors including the particle size, solution pH, ionic strength, soil composition, groundwater flow velocity, etc.[27,34,37,44,56–58] For example, groundwater usually has relatively high values of ionic strength, which results in a reduction in the electrostatic repulsion among particles and an increase in particle aggregation.

Recent research indicates that promising new synthetic methods are being developed to produce more mobile ZVI nanoparticles without sacrificing significant surface reactivity. Mallouk and his group at Penn State[57] have successfully synthesized ZVI nanoparticles on supports, which feature a high density of negative charges (dubbed as delivery vehicles) to aid in electrostatic repulsion between nZVI particles and with the predominantly negative surface aquifer materials. These nZVI delivery vehicles, including anionic hydrophilic carbon and poly(acrylic acid) (PAA), bind strongly to nZVI, create highly negative surfaces, thus effectively reducing the aggregation among ZVI particles, and reduce the filtration removal by aquifer materials. Laboratory soil column tests with Fe/C, Fe/PAA, and unsupported iron nanoparticles suggest that the anionic surface charges can enhance the transport of iron nanoparticle through soil- and sand-packed columns, while the unsupported iron nanoparticles aggregate and impede the flow of water through the column. Further analysis shows that the Fe/C and Fe/PAA nanoparticles have lower sticking coefficients.

Work by Y.P. Sun at Lehigh University on the use of poly(vinyl alcohol-co-vinyl acetate-co-itaconic acid) (PV3A) as a dispersant also helped to generate a new class of nZVI with substantially better subsurface mobility potential.[58] As illustrated in Figure 14.8, the iron nanoparticles were well dispersed with most of the resultant particles less than 20–30 nm. Further analysis using acoustic spectrometry indicated that the dispersed iron nanoparticles have a mean size at 15.5 nm with 90% of the

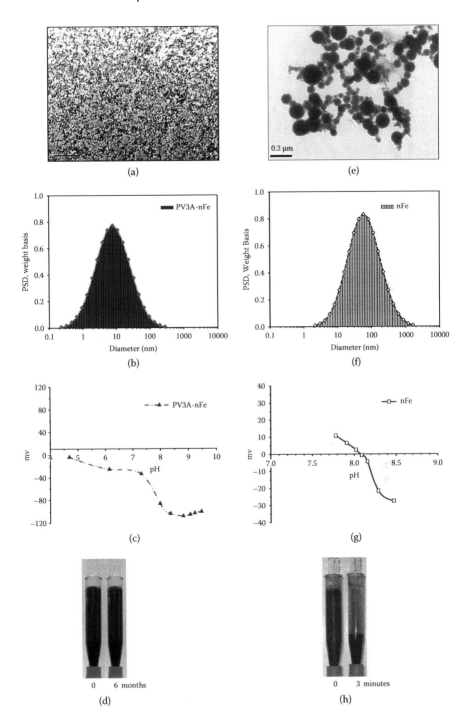

FIGURE 14.8 Characterization of iron nanoparticles (a–d) dispersed with poly(vinyl alcohol-co-vinyl acetate-co-itaconic acid) (PV3A), and (e) to (h) pure nanoparticles.

particles smaller than 33.2 nm. The stabilized particles also have high reactivity and high stability in terms of suspension (over 6 months). Measurements of ζ-potential suggest that dispersed iron nanoparticles have net negative charge at pH 4.5, likely the result of the dissociation of the carboxylic acid groups within the PV3A molecules. In comparison, the original or bare iron nanoparticles have much larger particle size and settle out of solution in less than 10 minutes. The bare iron actually has slight positive charges in the neutral pH range. This may help to explain the low mobility of ZVI nanoparticles.

CHALLENGES AHEAD

Since the initial field demonstration of nZVI technology in 2001, significant progress has been made in research and development of iron nanoparticles for soil and groundwater treatment. New research and development efforts should be directed toward enhancing real-world performance and minimizing potential economic and environmental risks. The following subsections present some questions, problems, and/or challenges we have experienced in our laboratory and fieldwork.

MATERIALS CHEMISTRY

The synthesis of iron nanoparticles represents the foundation of this technology. Several nZVI synthetic methods have been identified but the economics of each is considerably different. The challenge is to scale up the processes and produce nano-iron in large enough quantity such that further price reductions into the $10–25 per pound range become feasible. Methods for quality control and assurance (e.g., particle size, reactivity, surface charge, and mobility) should be established. Other issues requiring resolution might include optimization or customization of the ZVI surface properties (hydrophobicity, charge, functional group, etc.) for efficient subsurface transport under site-specific conditions and for degradation of targeted contaminants (e.g., perchlorate, PCBs, DNAPL).

ENVIRONMENTAL CHEMISTRY

This work includes the study of reaction rate, mechanism, and effect of environmental factors (e.g., pH, ionic strength, competing contaminants) on the transformation of targeted environmental contaminants. As discussed earlier, the published work so far has been largely limited to one- and two-carbon chlorinated aliphatic compounds. Details on the reaction mechanisms at the nanoparticle-water interface are still scarce. More work needs to be focused on exploring the feasibility of using nZVI for the treatment of halogenated aromatic compounds, PCBs, dioxins, pesticides, and other organic complex pollutants. Remediation of metal-contaminated sites also presents interesting opportunities and challenges.

GEOCHEMISTRY

This entails the nZVI reactions with both surface and groundwater, interactions (sorption, desorption) with soil and sediment, settling, aggregation, and transport phenomena in porous media. Evaluating the long-term fate of iron nanoparticles represents another topic worthy of more focused attention. To date, virtually no studies in this key area have been published.

ENVIRONMENTAL IMPACT

Thus far, no reports on the ecotoxicity of low-level ZVI in soil and water have been published in the peer-reviewed literature. However, it is the authors' view that systematic research on the environmental transport, fate, and ecotoxicity is needed to overcome increasing concerns and fear in the environmental use of nanomaterials, and minimize any unintended impact. To date, the overwhelming proportion of nZVI research deals with the applications of the technology and work is needed to adequately assess its implications.

Iron nanoparticles may actually provide a valuable opportunity to demonstrate the positive effect on environmental quality. Iron is the fifth most-used element; only hydrogen, carbon, oxygen, and calcium are consumed in greater quantities. It has been found at the active center of many biological molecules and likely plays an important role in the chemistry of living organisms. It is well documented that iron is an essential constituent of the blood and tissues. Iron in the body is mostly present as iron porphyrin or heme proteins, which include hemoglobin in the blood, myoglobin, and the heme enzymes. The challenge is to determine the eco- and human toxicity of highly reactive ZVI nanoparticles.

ACKNOWLEDGMENTS

Research described in this work has been supported by the Pennsylvania Infrastructure Technology Alliance (PITA), by U.S. National Science Foundation (NSF), and by U.S. Environmental Protection Agency (EPA). The authors thank Mr. Sisheng Zhong for his able assistance in the preparation of this manuscript.

REFERENCES

1. Roco, M.C., Williams, R.S., and Alivasatos, P., Eds., *Nanotechnology Research Directions: IWGN Workshop Report*, Kluwer Academic Publishers, Norwell, MA (1999).
2. Chan, W.C.W., Maxwell, D.J., Gao, X., Bailey, R.E., Han, M., and Nie, S., *Curr. Opin. Biotechnol.*, **13**, 40 (2002).
3. Thess, A., Lee, R., Nikolaev, P., Dai, H., Petit, P., Robert, J., Xu, C., Lee, Y.H., Kim, S.G., and Rinzler, A.G., *Science*, **273**, 483 (1996).
4. Zhan, B.Z., White, M.A., Sham, T.K., Pincock, J.A., Doucet, R.J., Rao, K.V.R., Robertson, K.N., and Cameron, T.S., *J. Am. Chem. Soc.,* **125**, 2195 (2003).
5. Linsebigler, A.L., Lu, G., and Yates, J.T., *Chem. Rev.*, **95**, 735 (1995).
6. Zhang, H., Penn, R.L., Hamers, R.J. ,and Banfield, J.F., *J. Phys. Chem. B*, **103**, 4656 (1999).
7. Service, R.F., *Science*, **281**, 940 (1998).
8. Rosi, N.L., Giljohann, D.A., Thaxton, C.S., Lytton-Jean, A.K.R., Han, M.S., and Mirkin, C.A., *Science*, **312**, 1027 (2006).

9. Shchukin, D.G., Schattka, J.H., Antonietti, M., and Caruso, R.A., *J. Phys. Chem. B*, **107**, 952 (2003).
10. Wang C. and Zhang, W., *Environ. Sci. Technol.*, **35**, 4922 (1997).
11. Fujishima, A., Rao, T.N., and Tryk, D.A., *J. Photochem. Photobiol. C: Photochem. Rev.*, **1**, 1 (2000).
12. Kamat, P.V., Huehn, R., and Nicolaescu, R., *J. Phys. Chem. B.*, **106**, 788 (2002).
13. Masciangioli T. and Zhang, W.X., *Environ. Sci. Technol.*, **36**, 102A (2003).
14. BCC Research, *Nanotechnology in Environmental Applications* (RNAN039A).
15. Stumm W. and Morgan, J.J., *Aquatic Chemistry*, 3rd ed., John Wiley & Sons, Inc., New York (1996).
16. Reynolds, G.W., Hoff, J.T., and Gillham, R.W., *Environ. Sci. Technol.*, **24**, 135 (1990).
17. Gillham R.W. and O'Hannesin, S.F., *Groundwater*, **32**, 958 (1994).
18. Gavaskar, A.R., Gupta, N., Sass, N.M., Janoy, R.J., and O'Sullivan, D., *Permeable Barriers for Goundwater Remediation—Design, Construction, and Monitoring*, Battelle Memorial Institute, Columbus, OH (1998).
19. Gu, B., Phelps, T.J., Liang, L., Dickey, M.J., Roh, Y., Kinsall, B.L., Palumbo, A.V., and Jacobs, G.K., *Environ. Sci. Technol.*, **33**, 2170 (1999).
20. Zhang, W., *J. Nanopart. Res.*, **5**, 323 (2003).
21. Elliott, D.W. and Zhang, W., *Environ. Sci. Technol.*, **35**, 4922 (2001).
22. Zhang, W. and Elliott, D.W., *Remediation Journal*, **16**, 7 (2006).
23. Zhang, W., Wang, C., and Lien, H., *Catal. Today*, **40**, 387 (1998).
24. Nutt, M., Hughes, J., and Wong, M., *Environ. Sci. Technol.*, **39**, 1346 (2005).
25. Doyle, J., Miles, T., Parker, E., and Cheng, I., *Microchem. J.* **60**, 290 (1998).
26. Elliot, D., Cao, J., Zhang, W., and Spear, S. The 225th ACS National Meeting, New Orleans, LA, **43**, 564 (2003).
27. He F. and Zhao, D.Y., *Environ. Sci. Technol.*, **39**, 3314 (2005).
28. Liu, Y.Q., Majetich, S.A., and Tilton, R.D., *Environ. Sci. Technol.*, **39**, 1338 (2005).
29. Xu, J., Dozier, A., and Bhattacharyya, D., *J. Nanopart. Res.*, **7**, 449 (2005).
30. Alowitz, M. and Scherer, M., *Environ. Sci. Technol.*, **36**, 299 (2002).
31. Cao, J., Elliott, D.W., and Zhang, W., *J. Nanopart. Res.*, **7**, 499 (2005).
32. Kanel, S., Manning, B., Charlet, L., and Choi, H., *Environ. Sci. Technol.*, **39**, 1291 (2005).
33. Sohn, K., Kang, S.W., Ahn, S., Woo, M., and Yang, S.K., *Environ. Sci. Technol.*, **40**, 5514 (2006).
34. Ponder, S.M., Darab, J.G., and Mallouk, T.E., *Environ. Sci. Technol.*, **34**, 2564 (2000).
35. Cao, J., and Zhang, W., *J. Hazardous Materials*, **132**, 213 (2006).
36. Glazier, R., Venkatakrishnan, R., Gheorghiu, F., Walata, L., Nash, R., and Zhang, W., *Civ. Eng.* **73**, 64 (2003).
37. Quinn, J., Geiger, C., Clausen, C., Brooks, K., Coon, C., and O'Hara, S., *Environ. Sci. Technol.*, **39**, 1309 (2005).
38. Kuhn, L.T., Bojesen, A., Timmermann, L., Nielsen, M.M., and Mørup, S., *J. Phys. Condens. Matter*, **14**, 13551 (2002).
39. Uegami, M., Kawano, J., Okita, T., Fujii, Y., Okinaka, K., Kakuya, K., and Yatagai, S., Toda Kogyo Corp., U.S. Patent Application (2003).
40. Karlsson, A., Deppert, K., Wacaser, A., Karlsson, S., and Malm, O., *Appl. Phys. A Mater*, **A80**, 1579 (2005).
41. Choi, Ch., Dong, X., and Kim, B., *Mater. Trans.* **42**, 2046 (2001).
42. Elihn, K., Otten, F., Boman, M., Kruis, F., Fissan, J., and Carlsson, J. *Nanostruct. Mater.*, **12**, 79 (1999).
43. Xu, Y. and Zhang, W., *Indus. Eng. Chem. Res.*, **39** (2000) 2238.

44. Saleh, N., Phenrat, T., Sirk, K., Dufour, B., Ok, J., Sarby, T., Matyjaszewski, K., Tilton, R.D., and Lowry, G.V., *Nano. Lett.*, **5**, 2489 (2005).
45. Sun, Y.P., Li, X.Q., Cao, J., Zhang, W., and Wang, H.P., *Adv. Colloid Interfac.*, **120**, 47 (2006).
46. Nurmi, J.T., Tratnyek, P.G., Sarathy, V., Bear, D.R., Amonette, J.E., Pecher, K., Wang, C., Linehan, J.C., Matson, D.W., Penn, R.L., and Driessen, M.D., *Environ. Sci. Technol.*, **39**, 1221 (2005).
47. Li, X.Q. and Zhang, W., *Langmuir*, **22**, 4638 (2006).
48. Scherer, M.M., Richter, S., Valentine, R.L., and Alvarez, P.J.J., *Crit. Rev. Microbiol.*, **26**, 221 (2000).
49. Balko, B. and Tratnyek, P.G., *J. Phys. Chem. B*, **102**, 1459 (1998).
50. Vogel, T.M., Criddle, C.S., and McCarty, P.L., *Environ. Sci. Technol.*, **21**, 722 (1987).
51. Matheson, L.J. and Tratnyek, P.G., *Environ. Sci. Technol.*, **28**, 2045 (1994).
52. Lien, H.L. and Zhang, W.X., *J. Environ. Eng.*, **125**, 1042 (1999).
53. Elliott, D.W., Iron nanoparticles: Reaction with lindane and the hexachlorocyclo-hexanes. Doctoral dissertation, Department of Civil and Environmental Engineering, Lehigh University, Bethlehem, PA, 2005.
54. Huang, C.P., Wang, H.W., and Chiu, P.C., *Water Res.*, **32**, 2257 (1998).
55. Kallen, T.W. and Earley, J.E., *Inorg. Chem.*, **10**, 1152 (1971).
56. Zhang, P., Tao, X., Li, Z., and Rowman, R.S., *Environ. Sci. Technol.*, **36**, 3597 (2002).
57. Schrick, B., Hydutsky, B.W., Blough, J.L., and Mallouk, T.E., *Chem. Mater.*, **16**, 2187 (2004).
58. Sun, Y.P., Dispersion of nanoscale iron particles. Doctoral dissertation, Department of Civil and Environmental Engineering, Lehigh University, Bethlehem, PA, 2006.

15 Functionalized Magnetite Nanoparticles— Synthesis, Properties, and Bioapplications

P. Majewski and B. Thierry

CONTENTS

INTRODUCTION

In recent years, functional nanostructures such as superparamagnetic nanoparticles, quantum dots, noble metal nanorods and nanoshells have come to the fore to potentially revolutionize the biomedical field. These nanotechnology constructs may indeed overcome the limitation of conventional diagnostic systems and offer promising alternatives to conventional therapeutic agents. Their small size and subsequent high surface-to-volume ratio translates into unique physical properties such as superparamagnetism, high yield and photostable luminescence, or shape and size-tuneable optical properties. There is not to date a unanimously accepted definition of what is a nanoparticle; this review will, however, consider a nanoparticle as a nanometer-sized solid structure inheriting unique physico-chemical properties, different from those of the bulk materials, from their small dimension. In the case of magnetite nanoparticles, this usually occurs for nanoparticles less than 30 nm in diameter, which exhibit superparamagnetism at room temperature, a property related to the large magnetic moment resulting from the coupling of the atomic spins within the nano-sized magnetite nanoparticles. Small single-domain superparamagnetic magnetite nanoparticles are often referred as SPIONs (superparamagnetic iron oxide nanoparticles) or USPIONs (ultrasmall superparamagnetic iron oxide nanoparticles) depending on their hydrodynamic diameter. Several magnetic nanoparticles are currently commercially available or under advanced clinical investigations, for instance, Ferimoxides (Endorem/Feridex) and Ferumoxtran-10 from Guerbet Advanced Magnetics.[1]

In vitro and *in vivo* biomedical applications of magnetite nanoparticles have flourished in the last decade, exploiting the magnetic properties of the nanoparticles to achieve magnetic separation of cells and biomolecules, as well as to enhance contrast in magnetic resonance imaging (MRI) and enable intracellular hyperthermia procedures. In addition, magnetite nanoparticles are generally considered biocompatible materials and could receive FDA approval in the near future. The question of nanomaterial biocompatibility and safety is, however, a complex one, and has received much attention recently.[2,3] While recent advances in the synthesis of magnetic nanoparticles enable fine-tuning of their physical properties, a major challenge remains to engineer the nanoparticle surfaces to perform optimally in complex *in vitro* or *in vivo* biological environments. This is especially critical in the field of molecular imaging, where functional nanoparticles are used as contrast agents in noninvasive medical imaging procedures. Successful integration of nanoparticulate constructs into the clinical practice will require efficient molecular targeting of high-quality functional nanostructures. The objective of this chapter is to review the critical properties of magnetite nanoparticles and discuss their current status in nanotechnology with an emphasis on biomedical applications. It is, however, out of the scope of this chapter to provide an exhaustive review of the tremendous works that have been done in this area recently. More exhaustive reviews can be found elsewhere.[4–8] We intend rather to provide the readers with key concepts and issues and refer them to more detailed reviews when available.

MAGNETITE

Magnetite is a ferrimagnetic mineral form of iron(II,III) oxide with the chemical formula Fe_3O_4, one of several iron oxides and a member of the spinel group ($MgAl_2O_4$). Magnetite is a common iron oxide mineral, named for an ancient region of Greece where metal production was prominent. It is the only mineral that exhibits strong magnetism, whereas others, such as ilmenite ($FeTiO_3$) and hematite (Fe_2O_3), have weakly magnetic properties. Most natural magnetite occurs in very small grains. A chunk of crystallized magnetite is called a lodestone, which was the earliest form of the sailor's compass. A lodestone was mounted to a rod on cork and floated in a bowl of water. When the rod aligns with the earth's magnetic field, it points roughly north–south, which provides a useful but rather limited way of positioning.

Small grains of magnetite occur in almost all crystalline rocks as well as many sedimentary rocks. With an iron content of about 70 wt%, magnetite is a valuable source of iron ore and, consequently, was explored for the production of iron metal since ancient times. Its mineral hardness is 5.5 to 6.5, and it has a density of 5.2 g/cm^3 and dissolves slowly in hydrochloric acid. Crystals of magnetite have also been found in some bacteria, such as *Magnetospirillum magnetotacticum*, and in the brains of bees, of termites, of some birds, for example, the pigeon, and even of humans. These crystals are thought to be involved in magnetoreception, the ability to sense the polarity or the inclination of the earth's magnetic field, and to be involved in navigation—an ability that may be lost by humans. Also, some snails have teeth made of magnetite on their radula, making them unique among animals. This means they have an exceptionally abrasive tongue with which they scrape food from rocks.

CRYSTAL STRUCTURE

Magnetite has the spinel ($MgAl_2O_4$) structure, with a cubic close-packed oxygen array, and iron in both fourfold (tetrahedral sites) and sixfold (octahedral sites) coordination. The tetrahedral and octahedral sites form the two magnetic sublattices, A and B, respectively. The spins on the A sublattice are antiparallel to those on the B sublattice, which is defined as ferrimagnetism. The two crystal sites are very different and result in complex forms of exchange interactions of the iron ions between and within the two types of sites.

MAGNETIC PROPERTIES

Magnetite has recently attracted attention because bulk Fe_3O_4 has a high Curie temperature of 850K and nearly full spin polarization at room temperature. Both properties are considered of great potential for applications in giant magnetoelectronic and spin-valve devices based on magnetite films. Although magnetite is perhaps the oldest magnetic material known, some aspects on the basic mechanisms related to the Verwey transition are still being investigated, especially in nanosized magnetite.[10,11] In addition, the magnetic properties of magnetite nanocrystals and also nanostructured films based on magnetite have been reported to strongly depend on their synthesis route as well as on their nonmagnetic matrix or substrate.[12–14] Moreover, for

---

TABLE 15.1
Crystal Structure Data of Magnetite[9]

Structure:	Cubic
Space group:	Fd-3m (No. 227)
Lattice parameter:	a = 8.3941 Å
Lattice angle:	a = 90.00
Number of formulas per unit cell:	Z = 8
Atomic positional parameters:	Fe1: 8a, 0.1250, 0.1250, 0.1250
	Fe2: 16d, 0.5000, 0.5000, 0.5000
	O: 32e, 0.2549, 0.2549, 0.2549

---

a given synthesis route, the resulting magnetic properties of nano-sized magnetite appear to strongly depend on nano-sized changes in the crystal morphology, like antiphase boundaries, and crystal structure, such as oxygen deficiency and local ionic disorder.[14–16]

Above the Verwey temperature (T_V = 120K), which represents the transition from an electric conductor to an insulator, magnetite has a cubic spinel structure as described above. Bulk magnetite has cubic magnetic anisotropy, with the {111} and {100} directions being the easy and hard axes of magnetization, respectively. At room temperature, the first-order magnetocrystalline anisotropy constant, K_1, has a negative value of $-1.35 \cdot 10^5$ erg/cm^3. The value changes sign at low temperature, passing through an isotropic point at few degrees above the Verwey transition.[17] On cooling below the Verwey temperature, the magnetic structure changes from cubic to triclinic structure, yielding a change to uniaxial anisotropy with {001} easy axis.[18]

In recent studies, the magnetic properties of magnetite nanoparticles of sizes between 5 and 150 nm have been investigated closely[19] (Figure 15.1), and a gradual evolution from bulk-like magnetite to single-domain behavior has been observed with decreasing grain size. Bulk-like properties, such as saturation magnetization, hyperfine parameters, coercive field, and Verwey transition were observed in 150-nm particles. With decreasing particle size, the Verwey temperature was reported to shift down to 20K for particles with a size of about 50 nm. At smaller particle sizes, the Verwey temperature is no longer observable. Magnetite crystals with a particles size of 5 nm were observed to display superparamagnetic behavior at room temperature, with transition to a blocked state at T_B of about 45K, which depends on the

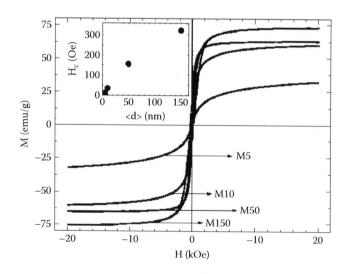

FIGURE 15.1 Magnetization hysteresis curves measured at 296K for Fe$_3$O$_4$ with different particles size (M5: 5 nm, M10: 10 nm; M50: 50 nm, M150: 150 nm). (Reused with permission from Goya, G.F. et al., *J Appl Phys* **2003**, 94, (5), 3520–3528. Copyright 2003, American Institute of Physics.)

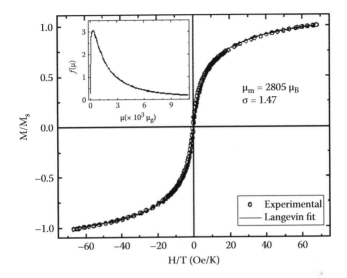

FIGURE 15.2 Superparamagnetic contribution M^η of the sample with Fe_3O_4 nanoparticles with 5 nm grain size at 296K. (Reused with permission from Goya, G.F. et al., *J Appl Phys* **2003,** 94, (5), 3520–3528. Copyright 2003, American Institute of Physics.)

applied field (Figure 15.1 and Figure 15.2). The existence of surface spin disorder could be inferred from the decrease in saturation magnetization at low temperatures, as the average particle size is reduced. This disordered surface did not show effects of exchange coupling to the particle core, as observed from hysteresis loops after field cooling in a 7 T magnetic field.[19] For particles with an average size of 5 nm, dynamic ac susceptibility measurements showed a thermally activated Arrhenius–Neel dependence of the blocking temperature with applied frequency. Analysis of the magnetic anisotropy indicated that there is no structural transition from cubic to triclinic symmetry for magnetite crystals of 5 nm particle size, which is in agreement with the absence of the Verwey transition.[19]

SYNTHESIS OF MAGNETITE NANOPARTICLES

Considering the high number of potential applications for high-quality magnetite nanoparticles, it is not surprising that numerous synthetic routes have been described with different levels of control on the size, polydispersity, shape, and crystallinity. Similar to other functional nanoparticle systems such as noble metal nanoparticles, monodisperse magnetite nanoparticles can now be prepared in good yield with good control over size and shape.[20–26] The influence of the experimental environment during synthesis of nanoparticles has been recently demonstrated.[27] In this study, the contamination of gold nanoparticle suspension with lipopolysaccharides drastically affected dendritic cells.

The most widely used synthesis routes for iron oxide nanoparticles are based on precipitation from solution. In these processes, a nucleation phase is followed by a

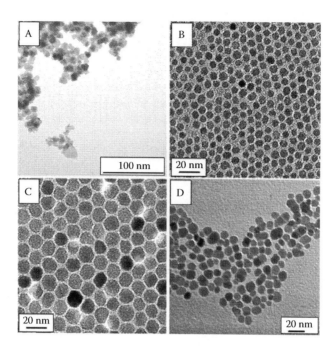

FIGURE 15.3 TEM of magnetite nanoparticles prepared using different synthetic routes offering more or less control on the size, shape, polydispersity, and crystallinity. A: Coprecipitation from an aqueous solution. B. and C: Thermal decomposition of iron(III) acetylacetonate in benzyl ether. (Reprinted from Sun, S. et al., *J Am Chem Soc* **2004,** 126, (1), 273–279. With permission from the American Chemical Society.) D: Surfactant-free benzyl alcohol route described in Ref. 42.

growth phase, affording fairly good control over the particle size and polydispersity. Most of the nanoparticles available to date have been prepared using a variation of the aqueous coprecipitation technique. Typically, magnetite is precipitated from basic aqueous solutions of ferric and ferrous salts. While some control over the size and composition of the particles can be achieved by changing the nature and ratio of ferric/ferrous salts (e.g., chlorides) as well as by controlling the reaction conditions (e.g., pH, temperature), coprecipitation processes usually result in polydisperse nanoparticle suspensions due to significant aggregation (Figure 15.3A). Controlled oxidation of the magnetite particles to maghemite is often performed.[28,29] The nanoparticle preparation can be achieved in presence of stabilizing agents (e.g., dextran).[30] Alternatively, the synthesized nanocrystals have been modified with various molecules such as dextran, starch,[31] polyvinyl alcohol,[29,32] citrate,[33] polyethyleneimine,[34] block copolymers,[28,35,36] and using silane-based chemistry.[31,37–39] Surface modification of nanoparticles will be discussed in more detail later in this chapter. Better control over size, monodispersity, and shape can be achieved using emulsions (water-in-oil or oil-in-water) that provide a confined environment during nucleation and growth of the iron oxide nanoparticles.[40]

Thermal decomposition processes have been recently developed to produce high-quality monodisperse and monocrystalline iron oxide nanoparticles. In these procedures, iron precursors are decomposed in hot organic solvents in the presence of stabilizing surfactants such as oleylamine, oleic acid, and steric acid. Iron precursors include iron acetylacetonate, iron cupferronates, and iron carbonyls. Thermal decomposition of iron(III) acetylacetonate in phenyl/benzyl ether and 2-pyrrolidone have been used, for instance, to synthesize high-quality magnetite nanoparticles with size ranging from 3 to 20 nm[22,23,41] (Figure 15.3B and Figure 15.3C). The modest contrast enhancement created on MR images by magnetic nanoparticles is a limitation to their clinical use but the excellent quality of the magnetite nanoparticles prepared by these novel synthetic schemes may contribute to their more widespread integration into clinical practice.[25] One potential issue is the possible presence of residual surfactants, which may hamper efficient subsequent surface modifications of the nanoparticles. In addition, the use of toxic solvents and surfactants may reduce the biocompatibility of the produced nanoparticles, although iron oxide itself is considered a biocompatible material.

An intermediate approach has been proposed by Niederberger and coworkers. Using a one-pot surfactant-free procedure, where benzyl alcohol acts as both solvent and ligand, high-quality monocrystalline magnetite nanoparticles could be prepared in good yield.[42] Phase transfer resulting from ligand exchange at the surface of the iron oxide particles can then be performed to create a stable aqueous suspension of these high-quality nanocrystals. For instance, dopamine and citric acid can be used to introduce reactive amino and carboxylic acid groups on the magnetite particles. The ligand exchange/phase transfer can be achieved without significant aggregation, as shown by dynamic light scattering (Figure 15.4; see also Figure 15.3D for TEM image).

Magnetite nanoparticles have been also been integrated into functional hybrid, and potentially multifunctional, structures. Many examples of such hybrid (nano) structures have been reported recently, including magnetic liposomes developed for *in vivo* MRI imaging,[43] magnetic nanotubes for separation and drug delivery,[44] and magnetite-coated fluorescent-silica particles used as a dual probe for magnetic resonance and fluorescent imaging of neuroblastoma[45] (Figure 15.4). Numerous procedures have also been reported for the surface engineering of iron oxide nanoparticles with inorganic nonfunctional (e.g., gold, silica)[46,47] and functional shells/coatings (e.g., fluorescent loaded silica, mesoporous silica).[48–52] Adding another level of complexity, Prasad and coworkers have recently described sophisticated approaches towards binary and ternary nanocrystals displaying magnetic, plasmonic, and semiconducting properties.[53]

COLLOIDAL STABILITY

As a result of the Brownian motion, colloidal nanostructures dispersed in a medium collide with each other frequently and the overall colloidal stability of the dispersion, critical to most potential nanoparticle applications, is dictated by the fate of the individual particles after each collision.[54] When attractive interactions (e.g., van der Waals forces) dominate, Brownian motion leads to irreversible aggregation of the nanoparticles. In the case of magnetite particles, magnetic dipole-dipole interactions

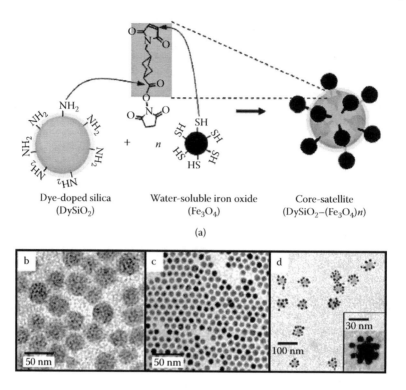

Dye-doped silica (DySiO₂)   Water-soluble iron oxide (Fe₃O₄)   Core-satellite (DySiO₂-(Fe₃O₄)n)

(a)

FIGURE 15.4 (a) Schematic diagram for the synthesis of core-satellite $DySiO_2$-$(Fe_3O_4)_n$ nanoparticles. (b–d) TEM images of (b) rhodamine-doped silica ($DySiO_2$), (c) iron oxide (Fe_3O_4), and (d) core-satellite $DySiO_2$-$(Fe_3O_4)_n$ nanoparticles. (Reprinted from Lee, J.H. et al., *Angew Chem Int Ed Engl* **2006,** 45, (48), 8160–8162. With permission from Wiley Interscience.)

resulting from the residual magnetic moment can provide a substantial attractive force in addition to van der Waals forces. A critical requirement is, therefore, to surface-engineer the nanoparticles with (macro)molecules that provide repulsive forces large enough to counter the attractive ones in the collision processes. Repulsive forces can be achieved in the presence of an electrical double layer on the particles (electrostatic stabilization) or in presence of polymeric chains providing steric stabilization. Attractive and repulsive forces are strongly influenced by the dispersion medium properties, for instance, temperature, ionic strength, and polarity. Colloidal stability of nanoparticle dispersion must therefore be considered for a specific system, especially in the setting of bioapplications, which require colloidal stability in complex biological medium such as blood or plasma. Steric stabilization provided by adsorption or grafting of polymers on the nanoparticles is the most efficient way to prevent aggregation in biosystems. Steric stabilization is indeed less sensitive to the ionic strength of the dispersion medium, can be achieved in both polar and nonpolar media, and is relatively insensitive to the total solid content. Steric stabilization results from the configurational entropic loss associated with compression of the polymer chains when two particles approach at a distance less than twice the thickness of the polymeric coating. The thickness of the polymeric coating, which needs to be equal or greater than the attractive forces, as well as the polymer conformation and density, which dictate the entropic cost of chain compression, must be therefore appropriately engineered to meet the colloidal stability requirements.

BIOMEDICAL APPLICATIONS OF MAGNETITE NANOPARTICLES

Magnetite nanoparticles have attracted a great deal of attention in the biomedical field owing to their magnetic properties, but also to their biocompatibility. A major area of application has been the field of bioassays where the magnetic properties of the nanoparticles are exploited in vitro to manipulate the nanoparticles with an external magnetic field. The use of magnetite in vivo has also been reported by many groups, targeting applications such as MRI contrast agents, thermoablation agents, and magnetically guided drug delivery.

SURFACE MODIFICATION OF MAGNETITE NANOPARTICLES FOR BIOAPPLICATIONS

Biomedical applications often require stringent control of the biointerfaces of the nanoparticles. Along with the need for colloidal stability in complex biological environments, a major requirement for the successful integration of magnetic nanoparticles in biomedical application is indeed to minimize biologically nonspecific adsorption events, for example, the adsorption of plasma proteins on the nanoparticle surfaces. Such nonspecific events can drastically hamper molecular recognition processes at the surface of the nanoparticles, therefore reducing the efficiency of magnetic nanoparticle-based bioassays. A prerequisite to the widespread use of nanoparticles *in vivo* is also their ability to resist nonspecific adsorption of opsonins. Opsonization of nanoparticles by plasma proteins results in rapid elimination from the blood by the mononuclear phagocyte system (MPS) with consequent accumulation in organs of the reticuloendothelial system (RES) such as spleen and liver. The nature (e.g., complement proteins and immunoglobulins) and amount of plasma proteins adsorbing on nanoparticles is directly related to the physicochemical characteristics of the surfaces of the nanoparticles. Adsorbed opsonins potentially lead to specific interactions with receptors on the surface of macrophages and hepatocytes and the subsequent elimination of the nanoparticles. It is out of the scope of the present review to describe in detail the sequences of biological events leading to clearance of nanoparticles from blood circulation but this topic has been reviewed recently.[55,56]

Control of biointerfacial interactions and prevention/reduction of nonspecific adsorption events have therefore been the object of much work toward biomedical applications of magnetite nanoparticles. Procedures to achieve high-quality biointerfaces able to resist nonspecific interaction have been implemented on macroscopic surfaces. It is commonly admitted that nonfouling surfaces should possess the following characteristics: (1) hydrophilic, (2) hydrogen bond acceptors, (3) no hydrogen bond donors, (4) neutral. PEGylation, that is, the introduction of the ethylene glycol unit on the surface, has been to date the most successful strategy to design nonfouling interfaces. Tethered PEG layers can greatly reduce the adsorption of proteins and interaction with cells if their properties are appropriately tailored. Especially, dense

PEG molecules extend in a good solvent (e.g., water) to form overlapping brush, following the concept introduced by de Gennes,[57] can almost completely eliminate protein adsorption. For instance, PEG layers immobilized under low solubility conditions, for example, just below the cloud-point condition of PEG in solution, can be used to create nonfouling surfaces.[58] The unique property of ethylene oxide rich surfaces, although fundamental mechanisms are still under debate, is related to the large hydration volume and unique interaction of PEG with water as well as to osmotic repulsion by polymer chains associated with protein adsorption. Other strategies to engineer surface resistance to biologically nonspecific adsorption include immobilization of polysaccharides such as dextran and hyaluoran and immobilization of zwitterionic polymers such as phosphorylcholine. Excellent reviews on this topic can be found elsewhere.[59,60]

Surface engineering of nanoparticles is, however, more challenging to perform than macroscopic surfaces. A major issue is to avoid/minimize irreversible aggregation of the nanoparticles during the coating procedure. For instance, conditions of high salt and temperature that can be used to achieve dense PEG brushes on macroscopic surfaces are not compatible with the colloidal stability of most aqueous nanoparticle suspensions. Routes to nanoparticle PEGylation have been published, but the data suggest that they are not as effective as on macroscopic material surfaces, and significant improvements may still be required to optimize *in-vivo* use of magnetite nanoparticles. PEGylated liposomes, or "Stealth liposome," have been a success story in the field of colloidal drug delivery. Stealth liposomes can indeed circulate in the blood for an extended time in comparison to unpegylated ones, although there is significant evidence that PEG layers on the surface of liposomes are not fully efficient to prevent protein adsorption.[61] To a lesser extent, PEGylation of magnetic nanoparticles has been shown to increase their blood circulation time and reduce uptake by phagocytic cells like macrophages.[36] The discrepancy between PEGylated solid nanoparticulates and liposomes in terms of circulation time remains to be elucidated but may be related to the difference in the PEG layers' conformation.

One can differentiate two general strategies toward surface modification of nanoparticles for bioapplications. First, "one pot" approaches have been described where the nanoparticles are simultaneously synthesized and coated with suitable macromolecules. Polysaccharides such as dextran and starch have been used widely to stabilize magnetite nanoparticles prepared in an aqueous coprecipitation reaction, and commercial products such as ferumodextran are available.[30] While thermal decomposition routes are usually achieved in the presence of surfactants such as oleic acid and oleylamine, Li and coworkers recently reported that thermal decomposition of ferric triacetylacetonate in 2-pyrrolidone in presence of mPEG-COOH could be used to produced crystalline PEGylated nanoparticles.[41,62] The second set of strategies used to surface-engineer magnetite nanoparticles relies on processing the particles postsynthesis to physically adsorb or chemically graft (macro)molecules. A widely used approach has been to introduce reactive functional groups on the nanocrystals, and to further conjugate them with stabilizing molecules such as PEG. Reactive groups can be introduced by ligand-exchange reaction, with or without phase transfer, using ligands with a strong affinity for iron oxide, such as dopamine and 2,3-dimercaptosuccinic acid.[63,64] Immobilization of a maleimide-activated Herceptin antibody onto

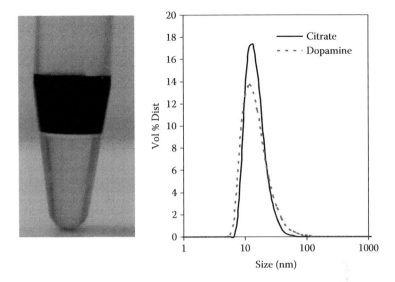

FIGURE 15.5 Ligand exchange procedure: Magnetite nanoparticles synthesized using the surfactant-free benzyl alcohol route described in Ref. 42 are functionalized in a ligand exchange/phase transfer procedure *(left panel)* with dopamine and citrate. The right panel displays the dynamic light scattering volume distribution of the magnetite nanoparticles in water (see also Figure 15.3B). The dopamine and citrate functionalized nanoparticles are well dispersed even after removal of excess ligands.

2,3-dimercaptosuccinic acid-protected magnetite nanoparticles enabled efficient immunotargeting and MR imaging *in vitro* and *in vivo* of tumor cells expressing the HER2/neu receptor. Alternatively, stabilizing molecules such as PEG have been chemically conjugated to a good iron oxide ligand and used in ligand-exchange procedures to coat the nanoparticles.[38,63,65]

Although not as efficient as PEG to control nonspecific adsorption, dextran has been widely used to prepare magnetite nanoparticles. As mentioned before, dextran coatings can be achieved directly during the preparation of magnetite nanoparticles in the coprecipitation technique but also by conjugating reactive dextran, for instance, carboxymethyl dextran and partially oxidized dextran (via formation of Schiff's bases linkages), onto functionalized nanoparticles.[66–68] The dextran shell can be further used to equip the nanoparticles with monoclonal antibodies or peptides for targeting purposes.[69–71] The importance of polysaccharide conformations on their blood compatibility has been long recognized. The nature of the dextran derivative (e.g., carboxymethy dextran), the number of reactive groups, and immobilization conditions (e.g., ionic strength, pH) can be used to control the final conformation on the surfaces.[72–74] As mentioned above, an issue when working with nanoparticles is to avoid irreversible aggregation during coating procedures subsequent to synthesis. Optimizing experimental conjugation conditions, monodisperse dextran-coated monocrystalline magnetite nanoparticles have been prepared without detectable

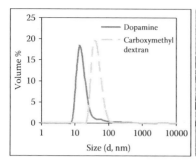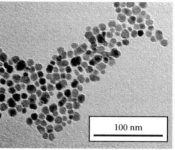

FIGURE 15.6 Monodisperse dextran-coated magnetite nanoparticles are prepared by covalent conjugation of carboxymethyl dextran to dopamine-functionalized nanoparticles. *Left panel*: Dynamic light scattering volume distribution of the magnetite nanoparticles before and after the coating. *Middle panel*: TEM of the dextran-coated nanoparticles. *Right panel*: Dextran-coated particles are stable in PBS buffer even in the presence of high salt concentration (1 M NaCl).

aggregation (Figure 15.6). The derivatization ratio of the carboxymethyl dextran had a substantial influence on the conjugation procedure (B. Thierry, unpublished data).

In Vitro Applications: Magnetic Separation and Bioassays

Albeit not limited to magnetite nanoparticles, magnetic separation processes have been widely investigated and used in many biotechnology applications such as purification, cell sorting, and immunoassays. Magnetic nanoparticle-based separation processes are attractive as they can be efficiently operated with small quantity of target entities (e.g., protein markers). Commercially available products can be found for various applications such as DNA, RNA, and protein extraction and purification.[75,76] Streptavidin conjugated magnetic beads are also available for specific capture of biotinylated targets.

Magnetic nanoparticles represent an intriguing alternative to conventional biolabels, which rely on fluorescence and chemiluminescence.[77–79] Using ferromagnetic "spin valve" sensors, extremely sensitive magnetic microarrays are, for instance, being developed by Wang and coworkers to detect binding of target DNA and proteins. The electrical resistance of these sensors changes in a predictable way in the presence of single domain magnetic nanoparticles captured on the surface of the chip in a sandwich assay strategy.[80]

As mentioned elsewhere, a key challenge in the use of magnetic nanoparticles in bioassay is the optimization of the molecular recognition processes. Lin and colleagues recently reported the development of ethylene glycol protected magnetite nanoparticles for a Matrix-Assisted Laser Desorption/Ionization Mass Spectroscopy (MALDI-MS) based immunoassay.[81] The presence of ethylene glycol groups on the nanoparticles significantly reduced nonspecific binding and consequently substantially improved the sensitivity of the proposed immunoassay.

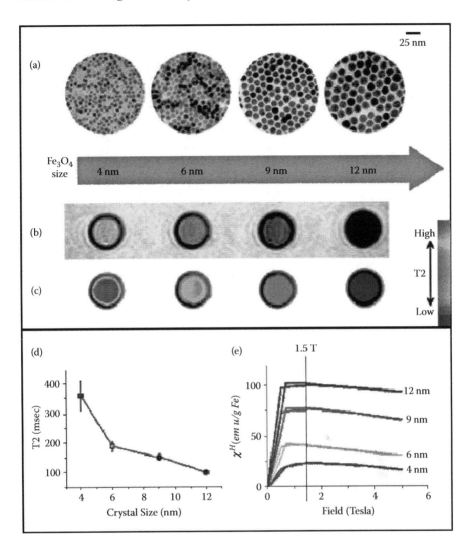

FIGURE 15.7 Nanoscale size effect of WSIO nanocrystals on magnetism and induced MR signals. (a) TEM images of Fe_3O_4 nanocrystals of 4 to 6, 9, and 12 nm. (b) Size-dependent $T2$-weighted MR images of WSIO nanocrystals in aqueous solution at 1.5 T. (c) Size-dependent changes from red to blue in color-coded MR images based on $T2$ values. (d) Graph of $T2$ value versus WSIO nanocrystal size. (e) Magnetization of WSIO nanocrystals measured by a SQUID magnetometer. (Reprinted from Y.W. Jun, Y.M. Huh, J.S. Choi, J.H. Lee, H.T. Song, S. Kim, S. Yoon, K.S. Kim, J.S. Shin, J.S. Suh, and J. Cheon, *Journal of the American Chemical Society.* 127(16), 5732 (2005) with permission from ACS Publications.)

In clinical application, magnetite nanoparticles such as the CELLection Dynabeads coated with the monoclonal antibody toward the human Epithelial Cell Adhesion Molecule (EpCam) have been used to specifically enrich epithelial tumor cells from biological samples (e.g., whole blood or bone marrow).[82]

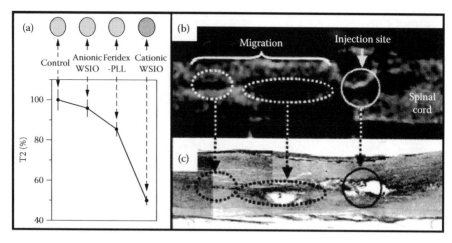

FIGURE 15.8 *In vitro* MR images (a) of anoinic WSIO-, Feridex-PLL-, cationic WSIO-treated neural stem cells. *In vivo* MR tracking (b) and histological examination by using Prussian blue (c) of cationic WSIO nanocrystal-labeled neural stem cells in an SD rat spinal cord. (Reprint from H.T. Song, J.S. Choi, Y.M. Huh, S. Kim, Y.W. Jun, J.S. Suh, and J. Cheon, *Journal of the American Chemical Society*, 127(28), 9992 (2005) with permission from ACS Publications.)

In Vivo Diagnostic Applications: MRI Contrast Agents for Molecular and Cell Imaging

MRI enables noninvasive high spatial-resolution anatomical and functional imaging. A prerequisite to the development of functional MRI has been the development of contrast agents with nanomolar sensitivity. Magnetite nanoparticles exhibit strong T1 and T2/T2* relaxation properties, significantly larger than that of paramagnetic molecules used as MRI contrast agents such as Gd-chelates. The dramatic shortening of the T2/T2* relaxation time associated with accumulation of magnetite nanoparticles in tissues or cells can therefore be used to create a negative contrast on T2-weighted MRI images, with sensitivity down to the nanomolar range. The ability of magnetite nanoparticles to generate MR contrast is influenced by particle size, crystallographic structure, and the presence of contaminants. The influence of the size and composition of magnetite nanoparticles and other Fe-based spinels on the MR contrast has been studied recently.[24,25] These studies demonstrated that with increasing size of the nanoparticles within the size range of 6 to 12 nm, the MR contrasts increases. This observation agrees with the findings of Goya and others[19] who observed an increase in the magnetization of magnetite nanoparticles with increasing size as shown in Figure 15.1. A higher MR contrast of larger magnetite nanoparticles is reasonable, because faster spin-spin relaxation processes of the water molecules are induced by materials with a larger magnetization.[83] However, it has yet to be determined what size range is optimum for highest MR contrast without losing the superparamagnetic properties of the nanoparticles.

The use of magnetic nanoparticles as MRI contrast agents can be divided into molecularly targeted approaches, which rely on equipping the nanoparticles with specific ligands to recognize specifically the target markers and the passive approaches. The passive approaches are based on the nonspecific accumulation of the nanoparticles into the tissues to be imaged, for instance, the organs of the RES system such as liver and spleen following intravenous injection of the nanoparticles. Dextran-coated magnetite nanoparticles have been clinically used as MRI contrast agents for more than a decade and commercially available products are available with different sizes and surface chemistry. Imaging of the lymph node system has also been achieved through either intravenous or subcutaneous injection of dextran-coated nanoparticles. A major development in the field has been the demonstration of the ability of small lymphotropic superparamagnetic nanoparticles to enable MRI detection of clinically occult lymph-node metastases in patients with prostate cancer.[84]

Molecularly targeted approaches are more challenging but promising *in vivo* results have been reported, enabling MR imaging of rare molecular events at the cellular and subcellular level. The onset of the proteomics era has made available a wealth of information about molecular processes at work in diseases such as cancer and an increasing numbers of novel target markers for molecular imaging are being identified. Monoclonal antibody and peptide sequences have been conjugated to magnetite nanoparticles and used to target them to pathologic areas such as tumors, myocardial infarction, or beta-amyloid plaques.[85,86] Magnetite nanoparticles conjugated with the synaptotagmin I protein that bind to anionic phospholipids expressed in apoptotic cells have, for instance, been used to detect MRI apoptotic tumor cells following treatment with chemotherapeutics in a murine lymphoma model.[87] Herceptin-conjugated nanoparticles could also be used to detect cancer cells implanted in mice *in vivo* in T2 weighted MRI sequences.[64] Recent years have witnessed an explosion of work in this molecular imaging and interested readers are referred to recent reviews.[86,88–90]

In-vivo MRI cell tracking has also been a successfully performed using magnetite nanocrystals, enabling better monitoring of cell transplantation therapy. Cell labeling can be performed either through passive or active loading with the magnetic nanocrystals. The biointerfacial properties of the nanoparticles have a strong influence on their ability to be nonspecifically internalized by target cells. Song and coworkers have, for instance, demonstrated substantial increase in neural stem cell labeling efficiency for cationic magnetite nanocrystals prepared by thermal decomposition of $Fe(CO)_5$ in dioctyl ether.[26] Silica-coated magnetic nanoparticles could also be used to achieve efficient stem cell labeling.[48] To enable labeling of cells lacking phagocytic abilities, it is necessary to equip the particles with biomacromolecules such as monoclonal antibodies or active peptide sequences. HIV-Tat peptide conjugated dextran-coated nanoparticles have, for instance, been used to achieve *in vivo* tracking and recovery of progenitor cells.[70] The use of MRI-cell tracking in cell therapy applications has been recently reviewed.[91,92]

In Vivo Therapeutic Applications: Drug Delivery and Hyperthermia

The surface of magnetite nanoparticles have been functionalized with drugs, proteins, and genetic materials to achieve localized delivery of therapeutic agents.[4,37,40,93,94] The need for localized delivery of bioactive agents stems from the systemic toxicity often associated with nontargeted administration of these agents. Potential elimination of systemic side-effects could potentially be achieved through efficient targeting of nanoparticles carrying a high payload of active agents. For example, gene delivery procedures achieved using magnetic nanoparticles (magnetofection) have been investigated *in vivo*, building on successful *in-vitro* studies.[34,95] Although still the object of much debate, the ability to magnetically target *in vivo* drug-loaded magnetic nanoparticles has attracted considerable attention.[4,95,96] Whether an external magnetic field can be used to capture magnetic nanoparticles injected intravenously depends on the strength of the applied magnetic field, as well as on the magnetic properties and volume of the nanoparticles. Alternatively, targeting moieties on the surface of the nanoparticles functionalized with bioactive agents have been used to target tumors *in vivo*. The recent work from Medarova and colleagues illustrates well the tremendous potential of advanced functional magnetic nanoparticles in the medical field. Simultaneous MR/near-infrared optical imaging and tumor delivery of silencing RNA could be achieved using nanoparticles coated with aminodextran and further functionalized with a membrane translocation peptide for intracellular delivery.[93]

Hyperthermia is a promising cancer therapy, which has been known for decades. Cancer cells are indeed sensitive to heat and when exposed to temperatures around 42 to 46°C, significant cell death can be obtained. Hyperthermia can be differentiated from thermoablation procedures where tissues heated above 46°C undergo significant necrosis. The major technical difficulty associated with heat-based cancer therapy is to heat only the tumor tissues without damaging the healthy ones. Considerable interest has therefore been drawn to magnetic nanoparticles to achieve intracellular hyperthermia procedures, for instance, by targeting magnetic nanoparticles conjugated with a monoclonal antibody such as anti-HER2. Magnetic nanoparticles exposed to an alternating magnetic field can indeed generate heat by hysteresis loss (multidomain particles) or energy dissipation (superparamagnetic nanoparticles). Proof-of-principle has been demonstrated *in vivo* by a few groups.[68,71,97–100] In addition, the synergistic combination of hyperthermia procedure and local delivery of chemotherapeutics holds promise of improved therapeutic activity. As for other *in-vivo* diagnostic and therapeutic applications, the major challenge is to accumulate with high specificity the magnetic nanoparticles in the target tissues.

METAL IONS REMOVAL

Most metal ions display substantial toxicity above critical concentrations. There is therefore a need to develop efficient procedures to remove metal ions from water and physiological fluids. Iron nanostructures have been investigated in many environmental decontamination procedures.[101] Iron oxides are, for instance, known to bind arsenic in aqueous solutions. Colvin and others recently studied the effect of magnetite nanoparticles functionalized with oleic acid on the removal of arsenic from water.[102] They found that Fe_3O_4 particles 12 nm in diameter removed nearly all the arsenic from the solution containing up to 45 µmol/L of arsenic. However, the same mass of 300 nm Fe_3O_4 particles eliminated less than 30% of the poison. After arsenic remediation, the particles can easily removed from water by a permanent magnet. It is easily conceivable that by having tailored functionalities for the remediation of targeted pollution in water, magnetite nanoparticles may have very promising opportunities for applications in water treatment.

CONCLUDING REMARKS

Recent advances in synthetic procedures enable the preparation of magnetite nanoparticles with narrow size distribution and well-controlled shape and crystallinity. Further research should aim toward greener synthetic routes reducing the use of toxic solvents and surfactants. The availability of such high-quality magnetic nanoparticles will continue to drive an intense interdisciplinary research activity toward their application in the biotechnology and medical fields. Integration of magnetite nanoparticles into hybrid multifunctional structures will likely create novel opportunities and broaden the fields of application. A major challenge with the clinical use of magnetite nanoparticles (and other magnetic nanostructures) is the design of effective biointerfacial coating procedures imparting the magnetic suspensions with longer residence time in blood and improved ability to engage in specific molecular recognition binding events. With these requirements met, it is likely that the next years will see an increase in the number of magnetite nanoparticle products on the market.

REFERENCES

1. Jung, C. W., Surface-properties of superparamagnetic iron-oxide MR contrast agents—Ferumoxides, ferumoxtran, ferumoxsil. *Mag Res Imaging* **1995,** 13, (5), 675–691.
2. Tsuji, J.S.; Maynard, A.D.; Howard, P.C.; James, J.T.; Lam, C.W.; Warheit, D.B.; Santamaria, A.B., Research strategies for safety evaluation of nanomaterials, Part IV: Risk assessment of nanoparticles. *Toxicol Sci* **2006,** 89, (1), 42–50.
3. Maynard, A.D., Nanotechnology: The next big thing, or much ado about nothing? *Ann Occup Hyg* **2007,** 51, (1), 1–12.
4. Neuberger, T.; Schopf, B.; Hofmann, H.; Hofmann, M.; von Rechenberg, B., Superparamagnetic nanoparticles for biomedical applications: Possibilities and limitations of a new drug delivery system. *J Mag Mag Mater* **2005,** 293, (1), 483–496.
5. Duguet, E.; Vasseur, S.; Mornet, S.; Devoisselle, J.M., Magnetic nanoparticles and their applications in medicine. *Nanomedicine* **2006,** 1, (2), 157–168.
6. Bulte, J.W.; Kraitchman, D.L., Iron oxide MR contrast agents for molecular and cellular imaging. *NMR Biomed* **2004,** 17, (7), 484–499.
7. Gupta, A.K.; Gupta, M., Synthesis and surface engineering of iron oxide nanoparticles for biomedical applications. *Biomaterials* **2005,** 26, (18), 3995–4021.
8. Lu, A.H.; Salabas, E.L.; Schuth, F., Magnetic nanoparticles: synthesis, protection, functionalization, and application. *Angew Chem Int Ed Engl* **2007,** 46, (8), 1222–1244.
9. Fleet, M., The structure of magnetite. *Acta Crystallogr B* **1981,** 37, 917–920.
10. Poddar, P.; Fried, T.; Markovich, G., First-order metal-insulator transition and spin-polarized tunneling in Fe_3O_4 nanocrystals. *Physical Review B* **2002,** 65, (17).
11. Seo, H.; Ogata, M.; Fukuyama, H., Aspects of the Verwey transition in magnetite. *Phys Rev B* **2002,** 65, (8).
12. Sena, S.P.; Lindley, R.A.; Blythe, H.J.; Sauer, C.; Al-Kafarji, M.; Gehring, G.A., Investigation of magnetite thin films produced by pulsed laser deposition. *J Mag Mag Mater* **1997,** 176, (2–3), 111–126.
13. Margulies, D.T.; Parker, F.T.; Spada, F.E.; Goldman, R.S.; Li, J.; Sinclair, R.; Berkowitz, A.E., Anomalous moment and anisotropy behavior in Fe_3O_4 films. *Phys Rev B* **1996,** 53, (14), 9175–9187.

14. Voogt, F.C.; Palstra, T.T.M.; Niesen, L.; Rogojanu, O.C.; James, M.A.; Hibma, T., Superparamagnetic behavior of structural domains in epitaxial ultrathin magnetite films. *Phys Rev B* **1998**, 57, (14), R8107–R8110.

15. Margulies, D.T.; Parker, F.T.; Rudee, M.L.; Spada, F.E.; Chapman, J.N.; Aitchison, P.R.; Berkowitz, A.E., Origin of the anomalous magnetic behaviour in single crystal Fe_3O_4 films. *Phys Rev Lett* **1997**, 79, (25), 5162–5165.

16. Hibma, T.; Voogt, F.C.; Niesen, L.; van der Heijden, P.A.A.; de Jonge, W.J.M.; Donkers, J.; van der Zaag, P.J., Anti-phase domains and magnetism in epitaxial magnetite layers. *J Appl Phys* **1999**, 85, (8), 5291–5293.

17. Bickford, L.; Brownlow, J.; Penoyer, F.R., Magneto-crystalline anisotropy in cobalt-substituted magnetite single crystals. *Proc. IEEE* **1957**, 104, 238–244.

18. Medrano, C.; Schlenker, M.; Baruchel, J.; Espeso, J.; Miyamoto, Y., Domains in the low-temperature phase of magnetite from synchrotron-radiation x-ray topographs. *Phys Rev B* **1999**, 59, (2), 1185–1195.

19. Goya, G.F.; Berquo, T.S.; Fonseca, F.C.; Morales, M.P., Static and dynamic magnetic properties of spherical magnetite nanoparticles. *J Appl Phys* **2003**, 94, (5), 3520–3528.

20. Yin, Y.; Alivisatos, A.P., Colloidal nanocrystal synthesis and the organic-inorganic interface. *Nature* **2005**, 437, (7059), 664–670.

21. Liang, X.; Wang, X.; Zhuang, J.; Chen, Y.T.; Wang, D.S.; Li, Y.D., Synthesis of nearly monodisperse iron oxide and oxyhydroxide nanocrystals. *Adv Functional Mater* **2006**, 16, (14), 1805–1813.

22. Sun, S.; Zeng, H., Size-controlled synthesis of magnetite nanoparticles. *J Am Chem Soc* **2002**, 124, (28), 8204–8205.

23. Sun, S.; Zeng, H.; Robinson, D.B.; Raoux, S.; Rice, P.M.; Wang, S.X.; Li, G., Monodisperse MFe_2O_4 (M = Fe, Co, Mn) nanoparticles. *J Am Chem Soc* **2004**, 126, (1), 273–279.

24. Jun, Y.W.; Huh, Y.M.; Choi, J.S.; Lee, J.H.; Song, H.T.; Kim, S.; Yoon, S.; Kim, K.S.; Shin, J.S.; Suh, J.S.; Cheon, J., Nanoscale size effect of magnetic nanocrystals and their utilization for cancer diagnosis via magnetic resonance imaging. *J Amer Chem Soc* **2005**, 127, (16), 5732–5733.

25. Lee, J.H.; Huh, Y.M.; Jun, Y.W.; Seo, J.W.; Jang, J.T.; Song, H.T.; Kim, S.; Cho, E.J.; Yoon, H.G.; Suh, J.S.; Cheon, J., Artificially engineered magnetic nanoparticles for ultra-sensitive molecular imaging. *Nat Med* **2007**, 13, (1), 95–99.

26. Song, H.T.; Choi, J.S.; Huh, Y.M.; Kim, S.; Jun, Y.W.; Suh, J.S.; Cheon, J., Surface modulation of magnetic nanocrystals in the development of highly efficient magnetic resonance probes for intracellular labeling. *J Amer Chem Soc* **2005**, 127, (28), 9992–9993.

27. Vallhov, H.; Qin, J.; Johansson, S.M.; Ahlborg, N.; Muhammed, M.A.; Scheynius, A.; Gabrielsson, S., The importance of an endotoxin-free environment during the production of nanoparticles used in medical applications. *Nano Lett* **2006**, 6, (8), 1682–1686.

28. Thunemann, A.F.; Schutt, D.; Kaufner, L.; Pison, U.; Mohwald, H., Maghemite nanoparticles protectively coated with poly(ethylene imine) and poly(ethylene oxide)-block-poly(glutamic acid). *Langmuir* **2006**, 22, (5), 2351–2357.

29. Schulze, K.; Koch, A.; Petri-Fink, A.; Steitz, B.; Kamau, S.; Hottiger, M.; Hilbe, M.; Vaughan, L.; Hofmann, M.; Hofmann, H.; von Rechenberg, B., Uptake and biocompatibility of functionalized poly(vinylalcohol) coated superparamagnetic maghemite nanoparticles by synoviocytes *in vitro*. *J Nanosci Nanotechnol* **2006**, 6, (9–10), 2829–2840.

30. Palmacci, S.; Josephson, L. Synthesis of polysaccharide covered superparamagntic oxide colloids. 1993, U.S. Patent 5262176.

31. Mikhaylova, M.; Kim, D.K.; Bobrysheva, N.; Osmolowsky, M.; Semenov, V.; Tsakalakos, T.; Muhammed, M., Superparamagnetism of magnetite nanoparticles: dependence on surface modification. *Langmuir* **2004**, 20, (6), 2472–2477.

32. Petri-Fink, A.; Chastellain, M.; Juillerat-Jeanneret, L.; Ferrari, A.; Hofmann, H., Development of functionalized superparamagnetic iron oxide nanoparticles for interaction with human cancer cells. *Biomaterials* **2005,** 26, (15), 2685–2694.

33. Taupitz, M.; Schnorr, J.; Abramjuk, C.; Wagner, S.; Pilgrimm, H.; Hunigen, H.; Hamm, B., New generation of monomer—stabilized very small superparamagnetic iron oxide particles (VSOP) as contrast medium for MR angiography: preclinical results in rats and rabbits. *J Magn Reson Imaging* **2000,** 12, (6), 905–911.

34. Huth, S.; Lausier, J.; Gersting, S.W.; Rudolph, C.; Plank, C.; Welsch, U.; Rosenecker, J., Insights into the mechanism of magnetofection using PEI-based magnetofectins for gene transfer. *J Gene Med* **2004,** 6, (8), 923–936.

35. Berret, J.F.; Schonbeck, N.; Gazeau, F.; El Kharrat, D.; Sandre, O.; Vacher, A.; Airiau, M., Controlled clustering of superparamagnetic nanoparticles using block copolymers: design of new contrast agents for magnetic resonance imaging. *J Am Chem Soc* **2006,** 128, (5), 1755–1761.

36. Lee, H.; Lee, E.; Kim Do, K.; Jang, N.K.; Jeong, Y.Y.; Jon, S., Antibiofouling polymer-coated superparamagnetic iron oxide nanoparticles as potential magnetic resonance contrast agents for *in vivo* cancer imaging. *J Am Chem Soc* **2006,** 128, (22), 7383–7389.

37. Kohler, N.; Sun, C.; Wang, J.; Zhang, M., Methotrexate-modified superparamagnetic nanoparticles and their intracellular uptake into human cancer cells. *Langmuir* **2005,** 21, (19), 8858–8864.

38. Kohler, N.; Fryxell, G.E.; Zhang, M., A bifunctional poly(ethylene glycol) silane immobilized on metallic oxide-based nanoparticles for conjugation with cell targeting agents. *J Am Chem Soc* **2004,** 126, (23), 7206–7211.

39. Sun, C.; Sze, R.; Zhang, M., Folic acid-PEG conjugated superparamagnetic nanoparticles for targeted cellular uptake and detection by MRI. *J Biomed Mater Res A* **2006,** 78, (3), 550–557.

40. Gupta, A.K.; Wells, S., Surface-modified superparamagnetic nanoparticles for drug delivery: Preparation, characterization, and cytotoxicity studies. *IEEE Trans Nanobiosci* **2004,** 3, (1), 66–73.

41. Li, Z.; Wei, L.; Gao, M.Y.; Lei, H., One-pot reaction to synthesize biocompatible magnetite nanoparticles. *Advanced Materials* **2005,** 17, (8), 1001–1005.

42. Pinna, N.; Grancharov, S.; Beato, P.; Bonville, P.; Antonietti, M.; Niederberger, M., Magnetite nanocrystals: Nonaqueous synthesis, characterization, and solubility. *Chem Mater* **2005,** 17, (11), 3044–3049.

43. Martina, M.S.; Fortin, J.P.; Menager, C.; Clement, O.; Barratt, G.; Grabielle-Madelmont, C.; Gazeau, F.; Cabuil, V.; Lesieur, S., Generation of superparamagnetic liposomes revealed as highly efficient MRI contrast agents for *in vivo* imaging. *J Am Chem Soc* **2005,** 127, (30), 10676–10685.

44. Son, S.J.; Reichel, J.; He, B.; Schuchman, M.; Lee, S.B., Magnetic nanotubes for magnetic-field-assisted bioseparation, biointeraction, and drug delivery. *J Am Chem Soc* **2005,** 127, (20), 7316–7317.

45. Lee, J.H.; Jun, Y.W.; Yeon, S.I.; Shin, J.S.; Cheon, J., Dual-mode nanoparticle probes for high-performance magnetic resonance and fluorescence imaging of neuroblastoma. *Angew Chem Int Ed Engl* **2006,** 45, (48), 8160–8162.

46. Wang, L.Y.; Luo, J.; Fan, Q.; Suzuki, M.; Suzuki, I.S.; Engelhard, M.H.; Lin, Y.H.; Kim, N.; Wang, J.Q.; Zhong, C.J., Monodispersed core-shell $Fe_3O_4@Au$ nanoparticles. *J Phys Chem B* **2005,** 109, (46), 21593–21601.

47. Lyon, J.L.; Fleming, D.A.; Stone, M.B.; Schiffer, P.; Williams, M.E., Synthesis of Fe oxide core/Au shell nanoparticles by iterative hydroxylamine seeding. *Nano Letters* **2004,** 4, (4), 719–723.

48. Lu, C.W.; Hung, Y.; Hsiao, J.K.; Yao, M.; Chung, T.H.; Lin, Y.S.; Wu, S.H.; Hsu, S.C.; Liu, H.M.; Mou, C.Y.; Yang, C.S.; Huang, D.M.; Chen, Y.C., Bifunctional magnetic

silica nanoparticles for highly efficient human stem cell labeling. *Nano Lett* **2007,** 7, (1), 149–154.

49. Ma, D.L.; Jakubek, Z.J.; Simard, B., A new approach towards controlled synthesis of multifunctional core-shell nano-architectures: Luminescent and superparamagnetic. *J Nanosci Nanotechnol* **2006,** 6, (12), 3677–3684.

50. Zhao, W.; Gu, J.; Zhang, L.; Chen, H.; Shi, J., Fabrication of uniform magnetic nano-composite spheres with a magnetic core/mesoporous silica shell structure. *J Am Chem Soc* **2005,** 127, (25), 8916–8917.

51. Kim, J.; Lee, J.E.; Lee, J.; Yu, J.H.; Kim, B.C.; An, K.; Hwang, Y.; Shin, C.H.; Park, J.G.; Kim, J.; Hyeon, T., Magnetic fluorescent delivery vehicle using uniform mesoporous silica spheres embedded with monodisperse magnetic and semiconductor nanocrystals. *J Am Chem Soc* **2006,** 128, (3), 688–689.

52. Deng, Y.; Wang, C.; Shen, X.; Yang, W.; Jin, L.; Gao, H.; Fu, S., Preparation, characterization, and application of multistimuli-responsive microspheres with fluorescence-labeled magnetic cores and thermoresponsive shells. *Chemistry* **2005,** 11, (20), 6006–6013.

53. Shi, W.; Zeng, H.; Sahoo, Y.; Ohulchanskyy, T.Y.; Ding, Y.; Wang, Z.L.; Swihart, M.; Prasad, P.N., A general approach to binary and ternary hybrid nanocrystals. *Nano Lett* **2006,** 6, (4), 875–881.

54. Hunter, J.R., *Foundations of Colloid Science.* Oxford University Press: 2001.

55. Vonarbourg, A.; Passirani, C.; Saulnier, P.; Benoit, J.P., Parameters influencing the stealthiness of colloidal drug delivery systems. *Biomaterials* **2006,** 27, (24), 4356–4373.

56. Yan, X.; Scherphof, G.L.; Kamps, J.A., Liposome opsonization. *J Liposome Res* **2005,** 15, (1–2), 109–139.

57. de Gennes, P.G., Conformations of polymers a attached to an interface. *Macromolecules* **1980,** 13, 1069–1075.

58. Kingshott, P.; Thissen, H.; Griesser, H.J., Effects of cloud-point grafting, chain length, and density of PEG layers on competitive adsorption of ocular proteins. *Biomaterials* **2002,** 23, (9), 2043–2056.

59. Nath, N.; Hyun, J.; Ma, H.; Chilkoti, A., Surface engineering strategies for control of protein and cell interactions. *Surf Sci* **2004,** 570, (1–2), 98–110.

60. Kingshott, P.; Griesser, H. J., Surfaces that resist bioadhesion. *Curr Opin Solid State & Mater Sci* **1999,** 4, (4), 403–412.

61. Allen, C.; Dos Santos, N.; Gallagher, R.; Chiu, G.N.; Shu, Y.; Li, W.M.; Johnstone, S.A.; Janoff, A.S.; Mayer, L.D.; Webb, M.S.; Bally, M.B., Controlling the physical behavior and biological performance of liposome formulations through use of surface grafted poly(ethylene glycol). *Biosci Rep* **2002,** 22, (2), 225–250.

62. Hu, F.Q.; Wei, L.; Zhou, Z.; Ran, Y.L.; Li, Z.; Gao, M.Y., Preparation of biocompatible magnetite nanocrystals for *in vivo* magnetic resonance detection of cancer. *Advanced Materials* **2006,** 18, (19), 2553–2556.

63. Xu, C.; Xu, K.; Gu, H.; Zheng, R.; Liu, H.; Zhang, X.; Guo, Z.; Xu, B., Dopamine as a robust anchor to immobilize functional molecules on the iron oxide shell of magnetic nanoparticles. *J Am Chem Soc* **2004,** 126, (32), 9938–9939.

64. Huh, Y.M.; Jun, Y.W.; Song, H.T.; Kim, S.; Choi, J.S.; Lee, J.H.; Yoon, S.; Kim, K.S.; Shin, J.S.; Suh, J.S.; Cheon, J., *In vivo* magnetic resonance detection of cancer by using multifunctional magnetic nanocrystals. *J Amer Chem Soc* **2005,** 127, (35), 12387–12391.

65. Hong, R.; Fischer, N.O.; Emrick, T.; Rotello, V.M., Surface PEGylation and ligand exchange chemistry of FePt nanoparticles for biological applications. *Chem Mater* **2005,** 17, (18), 4617–4621.

66. Mornet, S.; Portier, J.; Duguet, E., A method for synthesis and functionalization of ultrasmall superparamagnetic covalent carriers based on maghemite and dextran. *J Magnet Magnet Mater* **2005,** 293, (1), 127–134.

67. Schwalbe, M.; Jorke, C.; Buske, N.; Hoffken, K.; Pachmann, K.; Clement, J.H., Selective reduction of the interaction of magnetic nanoparticles with leukocytes and tumor cells by human plasma. *J Magnetism Magnetic Materials* **2005,** 293, (1), 433–437.

68. Sonvico, F.; Mornet, S.; Vasseur, S.; Dubernet, C.; Jaillard, D.; Degrouard, J.; Hoebeke, J.; Duguet, E.; Colombo, P.; Couvreur, P., Folate-conjugated iron oxide nanoparticles for solid tumor targeting as potential specific magnetic hyperthermia mediators: synthesis, physicochemical characterization, and *in vitro* experiments. *Bioconjug Chem* **2005,** 16, (5), 1181–1188.

69. Chen, J.; Wu, H.; Han, D.Y.; Xie, C.S., Using anti-VEGF McAb and magnetic nanoparticles as double-targeting vector for the radioimmunotherapy of liver cancer. *Cancer Lett* **2006,** 231, (2), 169–175.

70. Lewin, M.; Carlesso, N.; Tung, C.H.; Tang, X.W.; Cory, D.; Scadden, D.T.; Weissleder, R., Tat peptide-derivatized magnetic nanoparticles allow *in vivo* tracking and recovery of progenitor cells. *Nat Biotechnol* **2000,** 18, (4), 410–414.

71. Denardo, S.J.; Denardo, G.L.; Miers, L.A.; Natarajan, A.; Foreman, A.R.; Gruettner, C.; Adamson, G.N.; Ivkov, R., Development of tumor targeting bioprobes (in-111-chimeric L6 monoclonal antibody nanoparticles) for alternating magnetic field cancer therapy. *Clin Cancer Res* **2005,** 11, (19), 7087S–7092S.

72. Chupa, J.M.; Foster, A.M.; Sumner, S.R.; Madihally, S.V.; Matthew, H.W., Vascular cell responses to polysaccharide materials: *in vitro* and *in vivo* evaluations. *Biomaterials* **2000,** 21, (22), 2315–2322.

73. McLean, K.M.; Johnson, G.; Chatelier, R.C.; Beumer, G.J.; Steele, J.G.; Griesser, H.J., Method of immobilization of carboxymethyl-dextran affects resistance to tissue and cell colonization. *Colloids Surf B Biointerf* **2000,** 18, (3–4), 221–234.

74. Lemarchand, C.; Gref, R.; Passirani, C.; Garcion, E.; Petri, B.; Muller, R.; Costantini, D.; Couvreur, P., Influence of polysaccharide coating on the interactions of nanoparticles with biological systems. *Biomaterials* **2006,** 27, (1), 108–118.

75. Franzreb, M.; Siemann-Herzberg, M.; Hobley, T.J.; Thomas, O.R., Protein purification using magnetic adsorbent particles. *Appl Microbiol Biotechnol* **2006,** 70, (5), 505–516.

76. Berensmeier, S., Magnetic particles for the separation and purification of nucleic acids. *Appl Microbiol Biotechnol* **2006,** 73, (3), 495–504.

77. Osaka, T.; Matsunaga, T.; Nakanishi, T.; Arakaki, A.; Niwa, D.; Iida, H., Synthesis of magnetic nanoparticles and their application to bioassays. *Anal Bioanal Chem* **2006,** 384, (3), 593–600.

78. Chemla, Y.R.; Grossman, H.L.; Poon, Y.; McDermott, R.; Stevens, R.; Alper, M.D.; Clarke, J., Ultrasensitive magnetic biosensor for homogeneous immunoassay. *Proc Natl Acad Sci USA* **2000,** 97, (26), 14268–14272.

79. Perez, J.M.; Josephson, L.; O'Loughlin, T.; Hogemann, D.; Weissleder, R., Magnetic relaxation switches capable of sensing molecular interactions. *Nat Biotechnol* **2002,** 20, (8), 816–820.

80. Wang, S.X.; Bae, S.Y.; Li, G.X.; Sun, S.H.; White, R.L.; Kemp, J.T.; Webb, C.D., Towards a magnetic microarray for sensitive diagnostics. *J Magnet Magnet Mater* **2005,** 293, (1), 731–736.

81. Lin, P.C.; Chou, P.H.; Chen, S.H.; Liao, H.K.; Wang, K.Y.; Chen, Y.J.; Lin, C.C., Ethylene glycol-protected magnetic nanoparticles for a multiplexed immunoassay in human plasma. *Small* **2006,** 2, (4), 485–489.

82. Kielhorn, E.; Schofield, K.; Rimm, D.L., Use of magnetic enrichment for detection of carcinoma cells in fluid specimens. *Cancer* **2002,** 94, (1), 205–211.

83. Koenig, S.H.; Kellar, K.E., Theory of 1/T1 and 1/T2 NMRD profiles of solutions of magnetic nanoparticles. *Magn Reson Med* **1995,** 34, (2), 227–233.

84. Harisinghani, M.G.; Barentsz, J.; Hahn, P.F.; Deserno, W.M.; Tabatabaei, S.; Van De Kaa, C.H.; De La Rosette, J.; Weissleder, R., Noninvasive detection of clinically occult lymph-node metastases in prostate cancer. *New Eng J Med* **2003**, 348, (25), 2491–2499.

85. Corot, C.; Robert, P.; Idee, J.M.; Port, M., Recent advances in iron oxide nanocrystal technology for medical imaging. *Adv Drug Deliv Rev* **2006**, 58, (14), 1471–1504.

86. Sosnovik, D.E.; Weissleder, R., Emerging concepts in molecular MRI. *Curr Opin Biotechnol* **2007**, 18, (1), 4–10.

87. Zhao, M.; Beauregard, D.A.; Loizou, L.; Davletov, B.; Brindle, K.M., Non-invasive detection of apoptosis using magnetic resonance imaging and a targeted contrast agent. *Nature Medicine* **2001**, 7, (11), 1241–1244.

88. Wickline, S.A.; Neubauer, A.M.; Winter, P.M.; Caruthers, S.D.; Lanza, G.M., Molecular imaging and therapy of atherosclerosis with targeted nanoparticles. *J Magn Reson Imaging* **2007**, 25, (4), 667–680.

89. Jasanoff, A., Functional MRI using molecular imaging agents. *Trends Neurosci* **2005**, 28, (3), 120–126.

90. Winter, P.M.; Caruthers, S.D.; Wickline, S.A.; Lanza, G.M., Molecular imaging by MRI. *Curr Cardiol Rep* **2006**, 8, (1), 65–69.

91. Rogers, W.J.; Meyer, C.H.; Kramer, C.M., Technology insight: *in vivo* cell tracking by use of MRI. *Nat Clin Pract Cardiovasc Med* **2006**, 3, (10), 554–562.

92. Modo, M.; Hoehn, M.; Bulte, J.W., Cellular MR imaging. *Mol Imaging* **2005**, 4, (3), 143–164.

93. Medarova, Z.; Pham, W.; Farrar, C.; Petkova, V.; Moore, A., *In vivo* imaging of siRNA delivery and silencing in tumors. *Nat Med* **2007**, 13, (3), 372–377.

94. Wang, X.; Zhang, R.; Wu, C.; Dai, Y.; Song, M.; Gutmann, S.; Gao, F.; Lv, G.; Li, J.; Li, X.; Guan, Z.; Fu, D.; Chen, B., The application of Fe(3)O(4) nanoparticles in cancer research: A new strategy to inhibit drug resistance. *J Biomed Mater Res A* **2007**, 80, (4), 852–860.

95. Plank, C.; Anton, M.; Rudolph, C.; Rosenecker, J.; Krotz, F., Enhancing and targeting nucleic acid delivery by magnetic force. *Expert Opin Biol Ther* **2003**, 3, (5), 745–758.

96. Dobson, J., Magnetic nanoparticles for drug delivery. *Drug Dev Res* **2006**, 67, (1), 55–60.

97. Ito, A.; Kuga, Y.; Honda, H.; Kikkawa, H.; Horiuchi, A.; Watanabe, Y.; Kobayashi, T., Magnetite nanoparticle-loaded anti-Her2 immunoliposomes for combination of anti-body therapy with hyperthermia. *Cancer Lett* **2004**, 212, (2), 167–175.

98. Wust, P.; Gneveckow, U.; Johannsen, M.; Bohmer, D.; Henkel, T.; Kahmann, F.; Sehouli, J.; Felix, R.; Ricke, J.; Jordan, A., Magnetic nanoparticles for interstitial thermother-apy—feasibility, tolerance and achieved temperatures. *Int J Hyperther* **2006**, 22, (8), 673–685.

99. Ito, A.; Fujioka, M.; Yoshida, T.; Wakamatsu, K.; Ito, S.; Yamashita, T.; Jimbow, K.; Honda, H., 4-S-Cysteaminylphenol-loaded magnetite cationic liposomes for combina-tion therapy of hyperthermia with chemotherapy against malignant melanoma. *Cancer Sci* **2007**, 98, (3), 424–430.

100. Ivkov, R.; Denardo, S.J.; Daum, W.; Foreman, A.R.; Goldstein, R.C.; Nemkov, V.S.; Denardo, G.L., Application of high amplitude alternating magnetic fields for heat induction of nanoparticies localized in cancer. *Clin Cancer Res* **2005**, 11, (19), 7093S–7103S.

101. Li, L.; Fan, M.H.; Brown, R.C.; Van Leeuwen, J.H.; Wang, J.J.; Wang, W.H.; Song, Y.H.; Zhang, P.Y., Synthesis, properties, and environmental applications of nanoscale iron-based materials: A review. *Crit Rev Environ Sci Technol* **2006**, 36, (5), 405–431.

102. Yavuz, C.; Mayo, J.T., Yu, W., Prakash, A., Falkner, S. Yean, S., et al. Low-field magnetic separation of mondisperse Fe 304 nanocrystals. *Science* 2006, 314, 964–967.

16 The Emergence of "Magnetic and Fluorescent" Multimodal Nanoparticles as Contrast Agents in Bioimaging

P. Sharma, A. Singh, S.C. Brown,
G.A. Walter, S. Santra, S.R. Grobmyer,
E.W. Scott, and B.M. Moudgil

CONTENTS

INTRODUCTION

Bioimaging technologies have seen significant growth over the last two decades and are now a mainstay in research and diagnosis. Conventional imaging methods such as computed x-ray tomography (CT), ultrasound (US), magnetic resonance imaging (MRI), and positron emission tomography (PET) are continually being advanced to enable improved diagnosis by enhancing spatial resolution and providing more reliable structural information. Although many existing imaging methods can generate contrast by exploiting anatomical heterogeneities, the obtained images are usually difficult to decipher. The demand for more sensitive imagery with improved signal-to-noise ratios has led to the rapid development of contrast agents. Today, these agents are abundantly available for nearly all imaging modalities.

Modern contrast agents are frequently used to enhance gross structural changes through nonspecific labeling. Although this is an effective method for the diagnosis of disease, it is usually only feasible in the late stages of pathogenesis when treatments are less effective. To enhance prognosis there has been a desire for imaging methods that enable early and accurate detection of illness. Research in this area has led to the emergence of molecular imaging techniques, where the underlying pathways that underpin disease are targeted for detection at the cellular and molecular levels.[1,2] In these methods, contrast agents are modified to specifically label molecular targets or chemical sequences—in contrast to gross anatomical and structural features that are labeled by nonspecific contrast agents. Because they offer improved sensitivity and selectivity, molecular imaging approaches have and will continue to play an important role in genomic and proteomic technologies, as well as in the diagnosis and treatment of disease.

In recent years, advances in the field of nanoscience and nanotechnology have further fueled the research and development of contrast agents. Nanotechnology-based contrast agents have impacted medicinal biology by enabling novel methods of imaging and detection, and are anticipated to bring new opportunities in clinical diagnosis and therapy. The ability to manipulate the physical, chemical, and biological properties of nanomaterials allows one to design tailored nanoparticles with high sensitivity and selectivity for probing complex biological systems *in vivo* and in real time. An emerging trend in this field is the development of constructs that have multimodal capabilities, that is, the ability to generate contrast from more than one imaging modality simultaneously from a single entity. The combination of simultaneous imaging with different techniques in a single construct integrates the strengths of each modality and offers the possibility of improved diagnostics, preclinical research, and therapeutic monitoring. The development of future technologies is underway and has spurred research for the development of multifunctional particles that unify targeting, imaging, therapeutic, and reporting capabilities into a single entity.

This chapter focuses on the synthesis and application of optical and magnetic nanoparticulate probes and culminates with their integration into single multimodal nanoparticulate entities. It begins with a brief review of existing imaging techniques and provides rationale for the development of both nanoparticulate and multimodal probes. Luminescent and MRI contrast agents are then discussed with a focus on their composition, synthesis, and application.

IMAGING TECHNIQUES

A number of noninvasive bioimaging methods are available for clinical diagnostic applications that rely on different waveforms (electromagnetic or mechanical) to produce imaging signals from the specimen or patient of interest. Each applied waveform has a characteristic interaction that is used to provide information about the biological system. Common methods include CT-x-rays, MRI-radio waves, PET, single photon emission computed tomography (SPECT)-gamma rays, and ultrasound-ultrasonic waves. The imaging modalities differ from one another primarily in terms of penetration depth, detection sensitivity, spatial and temporal resolution, signal-to-noise ratio, and quantitative accuracy—in addition to different exposure/data acquisition times, cost of imaging hardware, and maintenance as shown in Table 16.1. The details of these interactions are outside the scope of the current discussion and are extensively reviewed elsewhere.[1,3–6]

As evident in Table 16.1, a motley of advantages and disadvantages of modern imaging techniques exist. For instance, although CT is a sensitive three-dimensional technique that is used routinely for visualizing internal anatomy with rapid data acquisition, it exposes subjects to potentially harmful ionizing radiation. Ultrasound, on the other hand, is a safer and relatively inexpensive real-time imaging tool that has low resolution and poor quantification—however, it is commonly used for monitoring fetus development, cardiac wall function, tissue structure, and motion. MRI has higher resolution capabilities (down to 25 μm), has unlimited depth penetration, and can be used to simultaneously obtain anatomical and molecular information; however, it lacks the portability of ultrasound and is relatively high in cost. Additionally, the high magnetic fields required prohibits the use of MRIs with patients who have pacemakers or certain metallic entities (implants or debris) in their body. PET and SPECT, employ high- and low-energy gamma rays and short-lived radioisotopes to provide excellent sensitivity and unlimited depth penetration. The major drawbacks

TABLE 16.1

Comparison of Imaging Modalities (Adapted from Ref [44])

Modality	Radiation, Resolution and Time	Typical Reagents	Principle Advantages	Principle Disadvantages
Positron Emission Tomography (PET)	high energy γ rays; 1–2 mm; minutes	^{18}F, ^{11}C, ^{13}N, 15O labeled probes or substrates for Reporter transgenes	High sensitivity; variety of probes and strategies enables a high degree of versatility	Cyclotron required; low resolution;
Single Photon Emission Computed Tomography (SPECT)	low energy γ–rays; 1–2 mm; minutes	^{99m}Tc, ^{111}In, ^{125}I labeled Probes	Multiplexing possible; radioisotopes used have longer half-lives than those in PET	Between 10 to 100-fold less sensitive than PET
Magnetic Resonance Imaging (MRI)	Radiowaves; 25–100 μm; minutes to hours	Paramagnetic cation probes*	High spatial resolution; provides both anatomical detail and functional information	Low sensitivity, long acquisition and image process times, lead to relatively low throughput
Computed Tomograpy (CT)	X-rays; 50 μm; minutes	Iodine*	Morphological information; Gold standard in the diagnosis of a large number of disease entities.	Relatively poor soft-tissue contrast
Bioluminescence Imaging (BMI)	visible light; 1–10 mm— dependent on tissue depth; seconds to minutes	Luciferase and substrate	High sensitivity; provides relative measure of cell viability or cell function; high throughput; transgene-based approach enhances versatility	Low anatomic resolution; light prone to attenuation with increased tissue depth
Whole-body fluorescence imaging	visible and near-infrared light; 1–10 mm— dependent on tissue depth; seconds to minutes	Fluorescent proteins, fluorescent dyes, and quantum dots	Multiplexing possible; highly compatible with a range of *ex vivo* analysis methodologies; transgene-based approach confers versatility	Excitation and emission light <600 nm prone to attenuation with increased tissue depth; autofluorescence artifacts/noise

TABLE 16.1 (CONTINUED)
Comparison of Imaging Modalities (Adapted from Ref [44])

Modality	Radiation, Resolution and Time	Typical Reagents	Principle Advantages	Principle Disadvantages
Intra-vital microscopy	visible and near-infrared light; Single cell; minutes	Fluorescent proteins, fluorescent dyes, and quantum dots	Microscopic resolution; multiplexing possible; enables real-time imaging and tracking of labeled cell populations	Surgery required to implant tissue window; small field of view; limited to relatively superficial tissues
Ultrasound	high-frequency sound; 50 μm; minutes	Microbubbles*	Images morphology and physiology of tissue relatively close to the surface of the mouse in real time	Limited ability to image through bone or lungs

* Reagents used for contrast enhancement

of these forms of imaging are low resolutions and the need for an in-house cyclotron to generate short-lived radioisotopes, which are used as probes. Optical imaging, on the other hand, is one of the most sensitive imaging modalities; however, its clinical application has been limited due to poor tissue penetration depth and scattering issues. However, the method is frequently used in developmental research where small animals play a critical role in disease models.

SMALL ANIMAL IMAGING

The most advanced imaging techniques have been routinely developed with the use of small animals.[7] Indeed, techniques are commercially available for small animals that have had little to no clinical applications. A prime example is optical imaging (OI). OI is largely used for *in vitro* and *ex vivo* experiments in cellular and molecular biology. OI is an easy, quick, and relatively low-cost imaging tool with very high sensitivity. It provides resolution of down to a couple of microns and can detect femtomolar concentrations of fluorescent probes.[8,9] Various modifications of OI are employed for *in vivo* investigations currently limited to small animals such as optical coherence tomography, bioluminescence, fluorescence reflectance imaging, fluorescence-mediated tomography, bioluminescence, and intravital microscopy.[10,11] The main restriction of OI is poor penetration of the visible light in the biological tissue due to high absorption and scattering from tissue and blood components. This has led to the development of fluorescence methods based on near-infrared (NIR) probes for *in vivo* investigations[12] that one day will enable a larger presence of OI in the clinical setting.

Advanced CT methods[2,13,14] are available for small animals enabling full three-dimensional images of small rodents to be obtained in a few minutes[15] at 50-μm resolution. This technology has been used to evaluate bone architecture dynamics,[16] murine myocardial infarction,[17] and thoracic imaging.[18] However, ionizing radiation exposure is still a problem limiting exposure duration and long-term monitoring. Ultrasound instruments have been specifically designed for small animals using a 30-MHz transducer,[19] and are used for evaluation of cardiac structure and function[20] and providing ~100-μm resolution. However, the field of view, depth penetration, and reproducibility (operator dependent) remain important hurdles to overcome. Significant advances in *in vivo* imaging, particularly in animal models, have been made using PET.[21–23] PET imaging now offers high volumetric spatial resolution of ~1 mm^3 and maximum sensitivity of ~7%;[24] however, the need for an in-house cyclotron remains. Scaled-down MR instruments having smaller magnets and coils but higher magnetic fields (up to 11 T), which generate higher signal-to-noise ratios have been successfully applied in small animals. Variations of the basic MRI technique such as MR microscopy, MR spectroscopy, and diffusional MR are being used to infer additional information about the tissue microenvironment and morphology pheotyping[25–28] in animals as well.

The imaging modalities discussed above also differ in their ability to generate a contrast from endogenous differences in the biological structure. In MRI, intrinsic contrast can be generated from proton density and tissue relaxation times; likewise, differences in the opacity of neighboring tissues can be imaged by x-rays and the reflection of the acoustic waves from the fluid/soft tissue and soft tissue/bone can be used to create image in US. PET and SPECT, on the other hand, rely on interaction with exogenously administered radioactive tracers. Exogenous contrast agents are also important in MRI, OI, CT, and US, as alluded to earlier. Magnetic contrast agents like iron, manganese, and lanthanide-based compounds are used in MRI, fluorophores and Luciferin substrates are employed in fluorescence and bioluminescence, iodine compounds in CT, and microbubbles in ultrasound as contrast enhancement agents. These agents enable improved imagery as well as provide a means to target specific information.

THE EMERGENCE OF MULTIMODAL IMAGING

Each imaging technique has inherent strengths as well as limitations in the type of information that it can furnish. By appropriately combining methods, imaging strengths can be built upon and weaknesses can be minimized. For example, CT is an established modality for imaging bone and other skeletal structures, yet it produces poor soft tissue contrast. However, by combining CT with MRI, the ability to obtain detailed information about both dense and soft tissues is enabled. Alone, CT would only provide information with regard to density contrast in a specimen, and MRI would be limited to soft tissue information. Because of the promising opportunities that can be enabled via the synergistic combination of techniques, there has been a movement toward the development of multimodal contrast agents, or single entities that can be employed to independently generate contrast from two or more imaging methods.

Some examples of multimodal imaging agents are found on the market and many more are expected to become available in the near future. Commercially available Gadolinium (Gd) chelates,[29,30] gadolinium oxide microspheres,[31,32] and iron nanoparticles[33] have been evaluated as contrast agents for MRI/CT. From our research group, Gd-doped fluorescent core nanoparticles have been demonstrated as a contrast agent for MRI/CT/OI.[34,35] Others have combined CT with PET via the use of multimodal agents. PET provides functional and metabolic information, whereas CT provides information about anatomical details with high spatial resolution.[36] PET/CT imaging is being increasingly used for clinical diagnosis, monitoring cancer staging and treatment.[37,38] μCT and bioluminescence imaging have been used together in a murine model to detect early tumor-bone destruction, demonstrating an effective combination of high sensitivity and quantitative morphological estimation.[39]

Although there are many forms of contrast agents and multimodal entities, the current chapter will focus on aspects of optical and magnetic resonance imaging and combinations thereof through the use of multimodal agents. The combination of MR and OI is beneficial as it integrates the advantages of high sensitivity (from optical method of detection, e.g., fluorescence) with the potential of true three-dimensional imaging of biological structures and processes at cellular resolution *in vivo* (using MRI). Magnetic and fluorescent probes have been used for tracking different cells,[40] including stem cells,[41] while simultaneously monitoring the presence of phenotypic enzymes and their activity.[42] In the following sections, nanoparticulate platforms currently being used for optical (fluorescent) and MR (magnetic and paramagnetic) imaging will be described.

LUMINESCENT CONSTITUENTS

In OI, light is used as the incident radiation and images are generated by the exploitation of different parameters such wavelength, intensity, polarization, coherence, or interference. Contrast can be created via a variety of mechanisms based on differential transmission, reflection, and fluorescence between the sample of interest and its background. The desire to apply these basic parameters for imaging has led to the emergence of many different optical techniques.[6] Although there are many techniques available, the selection of the best optical imaging technique for use will ultimately depend on the nature of information desired and the depth of investigation. While confocal and multiphoton microscopic techniques have primarily been applied for *in vitro* studies generating contrast for up to 800 μm, diffuse optical tomography and fluorescence molecular tomography can penetrate up to 20 cm with less resolution. In optical imaging, the use of both endogenous and exogenous fluorescent optical probes is frequently applied. For instance, bioluminescence is an indirect endogenous route that generates light as a result of chemiluminescence reaction from within the tissues. Luciferase is a common bioluminescence agent that has been frequently integrated in living cells (typically by gene transfection) for labeling applications *in vitro* and *in vivo*. The reaction between an externally administered substrate (typically a luciferin) and luciferase in the presence of ATP and oxygen leads to formation of the oxidized form of luciferin and the emission of visible light.

Various codon-optimized luciferases with emission wavelengths ranging from green to red are being used for *in vivo* tracking applications.[43–45] Bioluminescence imaging is a promising *in vivo* imaging technique because there is no background fluorescence and incident light penetration is not an issue; however, nonhomogeneous scattering and the need for a stable luciferase expression limit its applicability. Other OI techniques make use of fluorescent agents such as proteins and dyes. The advent of nanotechnology has also led to new nanoparticle-based platforms such as dye-doped silica, phosphors, and quantum dots. The selection of fluorescent agent for imaging is dependent on issues such as background autofluorescence and light penetration in tissues. These issues are briefly discussed.

(a) Autofluorescence: Autofluorescence is common in many biological tissues and can vary in excitation/emission properties depending on the area of the body investigated. It results from low amounts of endogenous fluorophores present in all tissues, which absorb visible light and subsequently emit photons at a higher wavelength. Some commonly known tissue fluorophores and their associated emission wavelengths are listed in Table 16.2.[46] From the emission wavelengths of the fluorophores such as nicotinamide (440–460 nm), flavins (520–560 nm), and collagen and elastin (470–520 nm), it is evident that these overlap with the commonly applied exogenous fluorescent agents (e.g., fluorescent proteins, dyes, quantum dots) leading to a diminution in contrast and the production of artifacts.[46] Since almost all the endogenous fluorophores emit in the visible region up to 530 nm, fluorescent agents emitting in the far red to near-infrared (NIR) region are being increasingly employed to overcome this issue.

TABLE 16.2

Common Sources of Autofluorescence in Biological Systems (Modified from Ref[46])

Autofluorescence Source	Organism/Tissue	Emission Wavelength (nm)	Excitation Wavelength (nm)
Collagen and elastin	Human aorta and coronary artery	>515	476
	Human skin	470–520	442
AGEs	Chinese hamster and human lenses	434	365
	Human cornea	385	320
	Diabetic human skin	440	365
	In vitro	440–450	370
Flavins	CHO cells	520	380, 460
	Rat hepatocytes	525	468
	Bovine and rat neural cells	540–560	488
	Goldfish inner ear	540	450
NAD(P)H	Rat cardiomyocytes	~509	~395
	S. cerevisiae, rat hepatocytes	440–470	366
	CHO cells	440–450	360
Chlorophyll	Green algae thylakoid	488	685(740)

(b) Light Penetration: One of principle limitations for *in vivo* applications of OI (in comparison to MRI, PET, and US) is the limited penetration of incident radiation in biological systems. Water, tissue chromophores, melanin, lipid, deoxy, and oxy hemoglobin strongly absorb in the UV-visible spectral range, thus limiting photon penetration to the first few micrometers of the tissue. In addition, visible light emitted from the fluorescent centers in the tissues additionally suffer from absorption and scattering from the constituents of the surrounding environment, providing further attenuation of the signal with increasing depth. The decreased absorption of hemoglobin, oxy hemoglobin, lipids, and water in the spectral region of 650–900 nm permits deeper penetration of light. This region is referred to as the "NIR window" for bioimaging applications. Fluorophores that emit in the NIR region have the additional advantage of very low noise from background fluorescence.

The luminescent agents commonly used in fluorescent imaging are briefly described.

FLUORESCENT PROTEINS

Fluorescent endogenous proteins have been leveraged for cell labeling and other purposes. Green fluorescent protein (GFP), in particular, has been used extensively for live cell imaging due to its high *in vivo* stability, negligible toxicity, and intrinsic fluorescence. The expression of the gene coding for nominal GFP leads to the appearance of a fluorescent signal at 510 nm when excited with 490-nm light (excitation) indicating the location of the mature protein. The yellow-green emission from nominal GFP, which makes it susceptible to issues associated with autofluorescence, absorption, and scattering, are major drawbacks that limit its application. The identification of the complete class of GFP family of proteins with emissions going into far red and near-IR regions has somewhat overcome this limitation.[47,48] Additional methods for eliminating autofluorescence have been systematically and critically reviewed by Billinton and Knight.[46]

DYES

Organic colorimetric and fluorescent dyes have been used extensively by cell biologists for labeling applications. Dyes like FITC, DAPI, Texas Red, Rhodamine, and their reactive derivatives have been used in numerous reports for *in vitro* labeling of cellular and subcellular organelles frequently due to their high quantum efficiency, high absorptivity, nontoxic biological profiles, and availability in a wide spectral range of excitation and emission wavelengths. Organometallic complexes such as lanthanide chelates and metal-ligand-complexes are used for highly fluorescent probes (e.g., ruthenium complexes such as $[Ru(bpy)_2 (dppz)]^{2+}$). These fluorophores have very sharp emission spectra in contrast to organic fluorophores and hence can be used for multicolor imaging. Antibody conjugation to dyes and peptides have enabled *in vivo* detection of tumors, tissues[49–54] in small animals. The overall sensitivity of optical detection in the conjugated dye-targeting ligand approach is limited by the number of fluorescent centers reaching the targeted site; however, other limitations are also present, such as photobleaching and broad emission wavelengths.

Ultrasensitive detection of dyes for *in vitro* and *in vivo* applications employs strong illumination sources such as lasers. Continuous exposure to incident radiation causes photobleaching of the dyes—diminishing their fluorescence. This phenomena can inhibit the use of certain fluorescent molecules for long-term experiments (e.g., cell tracking). In addition, the spectral profile of the dye is also vulnerable to its local cellular conditions such as changes in pH, presence of ions, etc. The broad emission spectrum of organic fluorescent dyes also provides an additional limitation, making multicolor imaging studies difficult. Dyes with their emission falling in the visible region of 400–600 nm also suffer from interferences from the background fluorescence in tissue imaging. However, dyes having NIR emission are being increasingly used to overcome this limitation, although photostability and low quantum yield (brightness) remain challenging issues. Some of the commonly used NIR dyes are cyanine-based dyes, Alexa fluor 750, Indocyanine green, and their reactive derivatives. These dyes are highly water soluble, stable, and can be coupled covalently with biological molecules.

The use of NIR dyes has enabled deep-tissue imaging with advanced imaging technologies such as fluorescence imaging tomography[55,56] and fluorescence reflectance imaging.[57,58] Weissleder's group has demonstrated that NIR dyes can be applied for quantitative *in vivo* visualization of biological processes.[59] Frangioni's group has described an NIR fluorescence-based imaging system that permits noninvasive, real-time monitoring of surgical anatomy with high spatial resolution.[60] Advances in development of instrumentation based on NIR fluorescence optical imaging for early detection of cancer have been reported.[61,62]

DYE-DOPED SILICA NANOPARTICLES

The advent of nanotechnology has led to new opportunities for overcoming the limitations of conventional organic dyes. Numerous research groups have investigated the encapsulation of organic dyes in nanoparticulate matrices, which are often metal oxides (e.g., silica) or polymeric. The use of silica as an encapsulating matrix will be focused upon in the following sections.

The Silica Advantage

Silica is optically transparent,[63] water dispersible,[64] biocompatible,[65] and in its amorphous form, a nontoxic material.[66] These properties have also led to the use of silica matrix for therapeutic applications such as drug delivery.[67–70] Silica is also resistant to microbial attack and remains stable in physiological conditions. In addition, for targeted delivery of the nanoparticles, several biomolecules such as proteins, peptides, antibodies, and oligonucleotides can be easily conjugated to organically modified silica using silane-based chemistry.[71–75] Numerous fluorescent dyes have been encapsulated in silica matrix by our research group.[64,76–78] The synthesis of silica nanoparticles encapsulating organic/inorganic fluorescent dyes can be performed readily using either Stober's method or microemulsions.[64,79,80] Surface modification of the silica nanoparticles can subsequently be achieved using suitable organically modified silica substrates as starting materials.

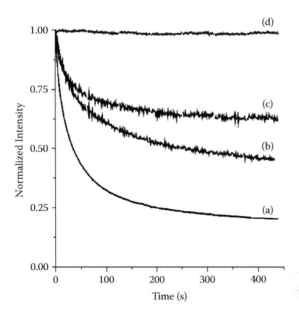

FIGURE 16.1 Luminescent intensity of (a) Rhodamine 6G, (b) Rubpy, (c) dye-doped silica, and (d) postcoated dye-doped silica nanoparticles upon excitation using a 0.503-W laser (Reprinted with permission from Analytical Chemistry, 73, 4988, 2001. Copyright 2001 American Chemical Society).

Dye Photostability

One of the key limitations of using dyes for long-term applications is their poor photostability. Silica nanoparticles have been widely used to encapsulate dyes in order to enhance their stability for *in vitro* and *in vivo* applications.[73,75–77,79–82] Silica provides a protective layer around the dye molecules that minimizes interaction between the dye and oxygen molecules, which can cause photodegradation. Hence, via encapsulation in silica the rapid photobleaching of fluorescent dyes can be inhibited, enabling longer term imaging. Furthermore, the placement of an additional shell of silica over an existing dye-doped silica particle leads to further photostability as indicated in Figure 16.1.

Increased Sensitivity

The sensitivity of fluorescent labeling techniques can be increased by increasing number of fluorescent centers in a particular region. Hence it is more favorable to have multiple fluorophores attached to each labeling entity. For fluorescent dyes there is usually only one fluorescent label per targeting group. In contrast, it has been demonstrated by the Tan group that tens of thousands of dye molecules can be encapsulated in a single silica nanoparticle of ~100 nm diameter (Figure 16.2).[83] Although some quenching of fluorescence is expected to occur due to the close proximity of the dyes, the fluorescence remains significantly intense for ultrasensitive bioanalysis.[84–86]

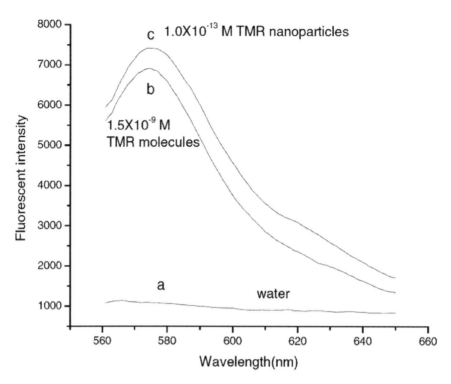

FIGURE 16.2 Fluorescent signal from (a) pure water; (b) tetramethylrhodamine dye, 1.5×10^{-9} M, and (c) TMR-doped NPs, 1×10^{-13} M. (Reprinted with permission from Advanced Materials, 16(2), 173, 2004. Copyright 2004 Wiley-VCH.)

Applications

Dye-doped silica (DDS) nanoparticles have been used for numerous applications including specific targeting, labeling cells, and immunoassays. Our group has demonstrated the selective labeling of mononuclear lymphoid target cells using DDS nanoparticles conjugated to antihuman CD10 antibodies.[71] Zhao and colleagues[86] have also developed a bioassay for precise and accurate determination of a single bacteria (*Escherichia coli*) cell using antibody-conjugated DDS nanoparticles. NIR dye labeled and antibody conjugated silica nanoparticles have recently been used for the detection of cancer markers.[87] Dual-luminophore-doped silica nanoparticles have also been demonstrated for simultaneous and multiplex signaling and bioanalysis.[88,89] DDS nanoparticles modified with TAT, a cell-penetrating peptide, have additionally been used for selectively labeling dysfunctional portions of rat brains.[75] Further applications of DSS nanoparticles have been covered in detail in several review articles.[76,83,90]

QUANTUM DOTS

Quantum dots (QDs) are semiconductor nanocrystals that possess several unique optical and electronic features due to quantum confinement. QDs are superior to dyes in many ways and are now used widely for many bioimaging applications.

Many reviews have described the synthesis, properties, and applications of QDs.[91-95] Semiconductor QDs are synthesized from elements combined of the II and VI groups, for example, CdS, CdSe, CdTe , III and V (e.g., InP, GaAs, InAs) and IV and VI (such as PbS, and PbSe). Alloyed QDs, those containing a mixture of two elements forming the optical core, are also employed as fluorescent particles. Two prominent examples of such QDs are those prepared from elements Zn and Mn and the other from chalcogenides, Te and Se. The unique properties of QDs in nanocrystals is the result of quantum confinement. Although the exhaustive explanation is covered in detail elsewhere, a brief simplified version is presented as follows.

Quantum Confinement and the Optical Properties of Quantum Dots

In semiconductors, conductivity is explained on the basis of "band theory" wherein the overlap of atomic orbitals from the constituent atoms leads to the formation of the valence band and conduction band. The energy difference between the highest energy level of the valence band and the lowest level of the conduction band is known as the band gap. When a semiconductor is optically excited with a photon of sufficient energy, electrons move from the valence band to the conduction band, leaving vacancies called holes in the valence band. This spatial distance between the hole and the electron is known as Bohr's radius. Quantum confinement arises when the dimensions of the particles are less than the Bohr's radius. Since the band gap in this regime is affected by the number of atoms constituting the particle, it connects size with emission. The tailoring of the band gap gives rise to the emission of photons of different energy (color). In QDs, the excitons—the electron-hole pairs formed upon excitation—are confined in a way similar to well-known quantum mechanics' particle in a box problem, leading to a finite band gap and discretization of energy levels in contrast to a bulk material with continuum energy.

Core–Shell Structure

The electron-hole pairs formed upon optical excitation (excitons) combine after a certain period of time, termed "lifetime", and produce light. For a single exciton, this lifetime is on the order of tens of nanoseconds. Some of these recombination processes could cause QDs to generate photons, depending on its quantum yield (QY). The product of the extinction coefficient at the excitation wavelength and QY is a primary indicator of the brightness of the QDs. QY is strongly affected by the presence of defects in the semiconductor nanocrystal. These defects lead to the formation of trap sites for the excited electron, nonradiative recombination, and ultimately a decrease in QY. Due to the extremely small size of the QDs, typically below 10 nm, the number of surface defects in the crystal can be significant, severely compromising QY. The common route to prevent the surface defects from affecting the QY of the QDs is to coat the QDs with another crystalline material. The coating material should have a larger band gap and similar lattice match with the QD core material. Lattice-matched shells mitigate surface defect formation and their higher band gap acts to confine excited electrons to the core. For instance, the encapsulation of a CdSe QD core by a ZnS shell reduces photochemical bleaching and drastically increases its quantum yield.[96] The shell also prevents the photooxidation of the QD by minimizing its interaction with environmental oxygen.

Epitaxially covered defect-free QDs have very large molar extinction coefficients and high QYs, resulting in bright fluorescent probes.[97] The QYs of QDs have been reported as high as 80%,[98,99] which is significantly greater than the QY of conventional organic fluorophores, which is about 15% in aqueous environments.[100] It is important to note that the QY of high-quality QDs has been shown to be quite stable and remains largely unaffected by bioconjugation,[101] making them attractive for imaging applications.

Optical Properties

QDs generally have broad excitation spectra and narrow emission spectra due to the discretization of energy levels. This is in contrast to the fluorescent dyes, which have a relatively sharp excitation and a broad emission typically with a red tail. The absorption spectrum of QDs typically increases in intensity toward the blue end. This implies that although the emission wavelength of the QDs is fixed, its excitation can be achieved over a broad range. The intensity of emission, however, decreases with increasing excitation wavelength. For instance, the extinction coefficient of QDs 605 streptavidin conjugates decreases from 3,500,000 M^{-1} cm^{-1} at 400 nm, to approximately 650,000 M^{-1} cm^{-1} at 600 nm.[102] This is still larger than extinction coefficients of the commonly used visible dyes that normally fall in the 70,000–250,000 M^{-1} cm^{-1} regime.[103]

The broad excitation and sharp emission of QDs permits the use of a single excitation source to excite QDs of different colors. Their narrow emission spectra reduce the probability of emission overlap, further enabling multiplexing application. Another important difference from fluorescent dyes is that QDs have a large Stokes shift, providing good separation of excitation and emission wavelengths. Collectively, these phenomena indicate that QDs are advantageous for multicolor imaging.[104,105]

Resistance to Photobleaching

One of the primary limitations of using dyes and fluorescent proteins for long-term bioimaging applications is their rapid photobleaching under strong illumination. This shortcoming can be overcome through the use of QDs. In dyes, photodegradation results in an irreversible loss in fluorescence; the inorganic crystalline semiconductor materials in QDs, however, are resistant to this phenomena. Yang and Holloway[106] have shown that the core/shell structured nanocrystals have remarkable photostability. The change in photoluminescence emission intensity was compared between organically (n-dodecanethiol) and inorganically (ZnS) passivated CdS:Mn upon UV irradiation (400 nm), as shown in Figure 16.3. A strong reduction in the photoluminescence intensity was observed for organically passivated QDs, while the crystals with ZnS shell showed an increase.[107]

A number of studies have compared the photobleaching of dyes with QDs. In a comparison with an organic dye, >90% photobleaching was observed for the dye, whereas the QDs were found to be reasonably stable for ~30 min duration under

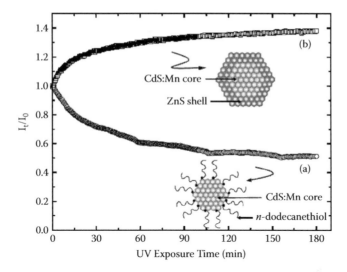

FIGURE 16.3 Change in PL intensity of (a) n-dodecanethiol and (b) ZnS- passivated CdS:Mn nanocrystals upon irradiation with 400-nm photons. (Reprinted with permission from Advanced Materials, 14(2), 152, 2004. Copyright 2004 Wiley-VCH.)

similar experimental conditions.[108] Ness and others have compared loss in fluorescence intensity between the frequently used dyes such as FITC, Cy3, and Alexa 546 dyes with QD 585 and QD 605.[109] It was found that in case of FITC and Alexa dyes, there was a loss in signal intensity of ~85% while for Cy3 it was ~50% after 5 minutes of illumination. In comparison, there was no loss in signal intensity for QDs even after 1 h of exposure. However, it is suggested in various reports that QDs underperform organic fluorophores as a function of molecular volume.[110–112] In a comparison to GFP and QDs for labeling of liver cells, only QDs could be detected under *in vivo* conditions, probably because of background autofluorescence.[112]

QDs Synthesis

QD synthesis for bioimaging applications can be fragmented into three segments, namely core shell synthesis, surface modification for aqueous dispersability, and bioconjugation. The hot-phase synthesis and microemulsion-mediated synthesis methods are the two main routes used for monodispersed quantum dot production. Murray, Norris, and Bawendi[113] introduced a hot-phase synthesis technique as a simple method for the synthesis of highly crystalline and monodisperse QDs. This method is based on the production of discrete homogenous nucleation via rapid injection of precursor reagents in a hot solvent maintained at ~290–350°C. The coordinating solvent commonly consists of a mixture of trioctylphosphine oxide and trioctylphosphine, both of which adsorb to the surface of the as-formed QDs from the basic end. This results in bulky alkyl groups on the outer surface of the QDs, making them nonpolar and readily dispersible in organic solvents. In the hot-phase synthesis method, control over particle size is achieved by temperature regulation—particle growth is severely inhibited in the highly viscous solvent at low temperatures. Modifications of the original technique have been made to include new coordinating ligands based on alkanoic acids and alkylamines,[114,115] noncoordinating solvents,[116] and alternative heat transfer fluids[117] for the improvement of quantum yields as well as the narrowing of size distributions.

The microemulsion-mediated methods are also frequently used for QD synthesis in which water-in-oil (w/o) microemulsions have been universally applied. In w/o microemulsions, water forms nano-sized droplets separated from the bulk organic phase by a surfactant layer. During QD synthesis, these droplets act as nanoreactors and their size can be controlled by manipulation of the water-to-surfactant molar ratio, Wo. In general, the water-soluble precursors for QDs are solubilized in separate microemulsions at the desired Wo value. Upon controlled mixing of the microemulsions, the nanoreactors collide continuously due to thermal Brownian motion and exchange their payloads, resulting in a precipitation reaction between the metal and the chalcogenide and formation of QDs. This yields surfactant-capped nanocrystals. Our group has used this method to produce CdS QDs doped with Mn and covered with a ZnS shell having a quantum yield of ~18%.[35,118,119] In comparison to the hot-phase synthesis route, QDs prepared via microemulsion routes have poor crystallinity and require longer reaction times. However, the advantage of the microemulsion-based method is that a wide variety of surface modifications can be carried out within the microemulsion environment (e.g., silica coatings on the QDs for improving their dispersion in aqueous medium and the introduction of functional groups to enable facile bioconjugation). These processes can be performed in one pot using the same reverse micellar system used for QDs synthesis. For example, CdS:Mn/ZnS QDs prepared in the sodium dioctyl sulfosuccinate (AOT)/heptane/water microemulsion have been consecutively overcoated with a 2.5-nm silica layer[119] in the same microemulsion system. Multimodal QDs generated by our group using microemulsions will be discussed in the multimodal nanoparticles section.

Attempts are being made to avoid using harsh reaction conditions and hazardous chemicals by making the QDs via aqueous routes. However, low-temperature aqueous methods lead to poor QYs.[120] Reports of the synthesis of bright QDs using aqueous-based methods are now being reported. For instance, CdTe QDs synthesized by a hydrothermal method at 180°C in the aqueous phase have shown high photoluminescence.[121] High QY QDs have been prepared by following postillumination procedures[98,122] and through microwave-assisted synthesis.[123,124] Microwave synthesis provides a fast (~5–45 min) reaction time, reproducibility, and an inexpensive route to generate QDs. High-quality QDs of CdSe, CdTe with 25–60% QY have been reported using a microwave route.[125,126]

Surface Modification and Aqueous Dispersion

QDs prepared from the hot-phase route are coordinated with hydrophobic groups as capping agents. For most applications in the biological systems, the particles are required to have good dispersion properties in aqueous medium. Indeed, poor aqueous dispersion was one of the primary limiting factors for the introduction of QDs for biological applications for a significant period of time. Multiple routes have now been developed to modify QDs surfaces to render them water "soluble" or highly hydrophilic. QD solubilization and functionalization has been covered in many review articles. Ligand exchange is one of the methods through which bifunctional ligands are exchanged with the hydrophobic groups on the QD surface. Using thiol affinity for the CdS and ZnS surfaces, TOPO-capped QDs are treated with bifunctional ligands such as mercaptoacetic acid, which binds to the QDs from the thiol end and the free acid group helps in the dispersion in polar medium.[127,128] Another commonly used method involves amphiphilic polymers to make QDs stable in aqueous medium. This involves polymers like polydimethylaminoethyl methacrylate,[129] polyethylenimine (PEI),[130] and modified polyacrylic acid[104] whose alkyl chains interact with the TOP/TOPO on the QD surface through hydrophobic interactions while the hydrophilic groups are projected from the surface making the QD dispersible in water. mphiphilic saccharides[131] and amphiphilic triblock copolymer[112] also interact with the QDs in a similar way. Another common way is to coat the QD surface with silica. As mentioned in the dye-doped silica section, silica surfaces offer numerous advantages. In our group, silica postcoating on QDs has been performed via a one-pot microemulsion route. CdS:Mn/ZnS QDs prepared in the AOT/heptane/water microemulsion was consecutively coated with a 2.5-nm silica layer by controlled hydrolysis and condensation of a silica precursor, tetraethyl orthosilicate with NH_4OH, as catalyst.[119] However, the coating increases the overall particle size of QDs to ~20 nm, which could limit their intracellular mobility to a certain extent.

Bioconjugation and Applications

In a number of applications it is desirable to attach homing ligands such as antibodies, antibody fragments, and peptides onto the QD surface to actively target specific cell phenotypes, tissues, or cellular components. The subject of bioconjugation strategies for QDs has been covered in detail in many reviews.[104,132] A number of strategies are used to attach the targeting ligands on QD surfaces. One of the common pathways is to use QDs with a carboxylic acid group on its surface, which can be introduced using the surface modification strategies discussed above. The proteins, peptides, and antibodies containing amino groups are conveniently covalently bound to these groups, leading to an amide bond using well-known 1-Ethyl-[3-dinethylaminopropyl] carbodiimide (EDC) chemistry. Aminated QDs can similarly be bound to the carboxy terminal of proteins/peptides. Another way is to electrostatically hold the proteins/peptides containing positively charged domains[133–135] on the negatively charged (in neutral and basic pH) carboxylated QDs. The biotin-avidin strong coupling interaction has also been used frequently to attach biomolecules such as antibodies. The conjugation has been performed by attaching either biotin/avidin to the QDs and the corresponding counter molecule on the antibodies.[133,136] This method is being used widely due to the ready availability of avidin-bound or biotinylated proteins/peptides[137] and established conjugation protocols. Making use of the thiol/amine binding affinity to the QD surface, particularly those with a ZnS or CdS shell, biomolecules containing thiol or amine groups are directly coupled to the QD surface.[138–140]

QDs have been used for labeling fixed/live cells and *in vivo* bioimaging applications. Both active and passive targeting approaches have been employed. It is found that uncoated QDs are endocytosed into the cell where they end up in the vesicles/

lysosomes in aggregated form.[141–145] QD-antibody conjugates have been used to label membrane proteins and nuclear antigens/fixed cells.[136,146] Similarly, QD conjugated with antibody fragments and proteins have been used to effectively label glycine receptors on neurons and[147,148] and human cancer cells.[149] For cellular labeling, the uptake of fluorescent particles in amounts that would suffice easy detection has always been challenging. Different ways such as microinjection,[144,150,151] electroporation,[144,152] endocytosis pathways mediated by receptor,[127,149] cationic lipids,[144,153] membrane-permeating proteins or peptides,[154–158] etc. are used to deliver the payloads inside the cytoplasm. Jaiswal and colleagues have used the QD antibody conjugate to selectively label the cell surface proteins and demonstrated the ability to perform long-term tracking of the labeled cells.[152] Similar approaches have been used for immunofluorescence labeling of mitochondria [144,159,160] and nucleus.[144,161] Antibody-conjugated QDs were used to detect human prostate cancer tumors in nude mice.[112] In a modified approach, the cells labeled with QDs were injected into mice and their distribution in the body was determined by fluorescence imaging.[153,162] We have successfully used QDs conjugated to TAT, a cell-penetrating peptide to effectively label the rat brain tissue.[118] There has been an increasing trend to employ NIR QDs for *in vivo* biomedical applications due to the advantage of higher penetration and low background fluorescence in the NIR window.[153,163–167] CdTe-based QDs coated with protein have been shown as an angiographic contrast agent for vessels surrounding and penetrating a murine squamous cell carcinoma in a mouse model.[168]

In general, the following have been suggested as desirable properties of a fluorescence-based optical imaging agent:[64]

1. *In vitro* and *in vivo* stability
2. Resistance to photobleaching
3. High quantum yield and high absorbency
4. Resistance to metabolic disintegration and nontoxicity
5. Emission in the NIR 700–900 nm window
6. Adequate dispersability in the biological environment

MAGNETIC RESONANCE IMAGING

MRI is one of the most used noninvasive imaging modalities in the clinical setting. It provides remarkable anatomical contrast with excellent spatial resolution and the ability to obtain 3-D images up to single-cell resolution. Technical advancements such as the availability of MR instruments with magnetic field strengths up to 9 T for whole-body imagers and up to 21 T for animal imagers promise further improvement of MR sensitivity and enhanced image contrast. Techniques such as high-resolution MR microscopy can be used to acquire images with resolutions below 100 μm. The high-resolution imaging offered by MRI has been used to detect as little as 2 cells/voxel in *in vitro* and ~3 cells/voxel *in vivo* using labeled cells.[169]

To understand the application of contrast agents in MRI it is important to know the basis of obtaining images in MRI. MRI works on the principles of nuclear magnetic resonance (NMR). Certain atomic nuclei, for example, H, C, N, F, P, etc., possess a nonzero spin resulting in a net magnetic moment and making them MRI active.

Essentially the NMR signals received from hydrogen nuclei located in different physiological environments are used to generate tissue contrast in most clinical cases. When the sample to be imaged is placed in a strong static magnetic field, the spins of the nuclei orient either parallel or anti-parallel to the magnetic field.[170–172] At any given field strength there are about 10^{-6} to 10^{-7} spin up states than spin down states producing a net magnetic moment. Although this difference is very small, the presence of large number of protons (from water) in biological samples makes it perceptible. The thermal equilibrium distribution in the sample is then perturbed by the application of a radiofrequency pulse at the appropriate resonant frequency. This results in the reorientation of the net nuclear magnetization in a new direction. After this perturbation, the transition from the excited state to the ground state (initial equilibrium orientation) occurs by the net loss of energy in a numbers of ways and this is collectively referred to as "relaxation." Generally the relaxation times[173,174] are described by (a) T_1: longitudinal relaxation time or spin lattice relaxation, denoting dissipation of energy to the surrounding biomolecules; (b) T_2: intrinsic transverse relaxation or spin-spin relaxation, denoting the interaction between the adjacent nuclei, and (c) T_2^*: apparent transverse relaxation. The inverse of each of the relaxation times, that is, $1/T_1$, $1/T_2$, are referred to as the relaxation rates.[175] Specific pulse sequences are used in MRI that are sensitive to changes in T_1 or T_2 relaxation times to generate an image.

The image contrast in MR is dependent on the distinguished relaxation rates of protons in proteins, tissue, and lipids. Inadequate image contrast is, however, an issue for imaging healthy and diseased tissues with small structural differences. Transition metals such as Fe, Mn, and lanthanide metals such as Gd and Dy have been employed for increasing the MR contrast due to their magnetic properties. MRI contrast agents are divided into two distinct groups, namely superparamagnetic and paramagnetic.[176] Although both types of contrast agents influence both T_1 and T_2, superparamagnetic agents have a greater influence on the T_2 and T_2^* relaxation time, producing a negative (dark) MR contrast; whereas paramagnetic ions shorten T_1 more and generate a positive (bright) contrast. Examples of magnetic contrast agent nanoparticles and their properties are described in the following sections.

SUPERPARAMAGNETIC CONTRAST AGENTS

Small crystalline iron oxide is the most important example of superparamagnetic contrast agents and will be the focus of this section. Although bulk iron exhibits ferromagnetism, below approximately 15 nm the cooperative phenomenon of ferromagnetism is no longer observed and no permanent magnetization is exhibited after the removal of magnetic field. In the presence of a magnetic field, the very small crystals of the iron oxide align completely with the field and acts as a single magnetic domain. The net magnetic moment causes rapid dephasing of the surrounding proton nuclei and alteration of their relaxation times.[177] Bulte and Kraitchman[178] have described additional advantages of superparamagnetic iron oxide (γFe_2O_3 and maghemite), which have led to its extensive use in MR. For instance, since it is composed of biocompatible metal it can be metabolized by cells using normal biochemical pathways. Additionally they can be detected in cells using electron microscopy techniques via atomic number contrast.

Superparamagnetic iron oxide (SPIO) particles are often hydrophilized for use in *in vivo* applications by coating the particles with dextran, which also enables further bioconjugation. Larger iron oxide nanoparticles are rapidly (less than 10 min) eliminated from blood by uptake into the RES in liver, spleen, and bone marrow.[179] The dextran coating, however, on the ultra-small particles promotes higher retention in blood, enabling long circulation times and wider biodistribution.[180,181] A variety of surface modification techniques and their influence on the magnetic properties have been described in review articles.[182,183] The effect of the coating on the particle surface leads to an overall increase in the particle size, characterized by "hydrodynamic" diameter.[179] Based on the final diameter, the iron oxide particles have been further classified into (a) SPIO, with more than 50 nm particle diameter; (b) ultrasmall iron oxide particles (USPIO), mean diameter between 20–50 nm; and (c) monocrystalline iron oxide nanoparticles (MION) that are composed of a single crystal, often less than 20 nm in diameter.

Another interesting feature of the SPIOs is that they perturb protons at distances that span multiple particle diameters from their surface. For example, it is shown that SPIO can influence media up to ~50 times the size of the core.[184] The comparison of relaxivity values between the superparamagnetic (iron oxide) and paramagnetic (commercial Gd chelates) have shown much larger relaxivities for iron oxide.[185] This permits use of very small amounts of these particles to generate contrast. Weissleder and others[186] have shown that as low as 1 mgFe/kg tissue can be detected *in vivo* in MR studies. Smirnov and colleagues have shown that individual cells can be imaged containing as little as ~1 pg/cell on a 9.4 T magnet.[187]

PARAMAGNETIC CONTRAST AGENTS

One of the principal applications of the lanthanides has emerged in the medical field of magnetic resonance imaging (MRI) contrast agents (CAs).[188] While many contrast agents are being increasingly used in MRI, Gd-based agents are the most frequently applied. Paramagnetic Gd, [Xe] $4f\ ^7 5d^1\ 6s^2$, with seven unpaired electrons having a large magnetic moment with a long electron spin relaxation time[189] has made it a favorite choice as contrast agent. Primarily, the proton relaxation results from the changes in the dipole coupling between the paramagnetic ion and that of water.[190] In comparison to SPIOs, which generate a dark contrast, the Gd chelates, with relatively larger shortening of T_1 than T_2, produce a more marked increase in the signal on T_1-weighted sequences, giving rise to bright contrast. In general, the low molecular weight contrast agents have relaxivities of about 3–5 mM^{-1} s^{-1}.[191]

Free Gd ions are toxic at as low as 10–20 μmol kg^{-1} concentration.[192] The LD$_{50}$ for Gd ions has been reported to be about 0.4 mmol kg^{-1} in rats. It has been suggested that the close proximity of the ionic radii of Gd (8 coordinate, ~119 pm) to calcium (8 coordinate, ~126 pm) leads to it inhibiting Ca metabolism and functioning in the body, giving rise to toxic side effects. Hence, it is necessary that Gd is appropriately chelated when applied to living systems to avoid toxic side effects. Gd contrast agents should satisfy certain prerequisites on issues related to its *in vivo* stability, stability in the presence of other competing bivalent and trivalent ions of relevance in body (conditional stability constant), biodistribution and elimination, and long-term toxicity.

The proton relaxation ability of the Gd chelates is also dependent on the design of the contrast agent. Factors such as number of water molecules present in the coordination sphere (directly bound to Gd), those present as hydrogen bonded, and those that diffuse close to the Gd but are not chemically bound, molecular size, pH, and temperature have a strong influence on the relaxivity of Gd contrast agents. A variety of factors involved in the design of Gd contrast agents that have an influence on their relaxivities have been discussed in detail in review articles.[192,193] A number of Gd chelates are now available commercially as extracellular and nonspecific contrast agents for MRI diagnostics. It is estimated that Gd-based contrast agents are used in more than 10 million studies per annum worldwide[194] and ~200 million patients have been administered Gd-based contrast agents.[195] The commonly used contrast agents are described briefly in the following section.

Derivatives of polyaminocarboxylic acids such as diethylenetriaminepentaacetic acid (DTPA) were among the first commercially approved chelating agent for contrast enhancement.[196] Gd forms an eight-coordinate water soluble complex with high thermodynamic and kinetic stability with DTPA. [Gd (DTPA)]$^{2-}$ has been used extensively as an extracellular contrast agent for contrast enhancement in cerebral perfusions,[197] renal functions, MR angiography,[198] cardiac MRI, etc. To lower the osmality of the aqueous formulations, neutral derivatives (diamides) of DTPA were later developed for clinical use. However, recent reports about the development of nephrogenic systemic fibrosis in patients with impaired kidney function from the use of nonionic forms (Omniscan) has raised concerns about the possible long-term effects.[195]

Macrocyclic Gd-chelates like DOTA (1,4,7,10-tetraazacyclododecane-1,4,7,10-tetraacetic acid), are known to provide high kinetic stability to the complex, have also been approved for clinical use. The macrocyclic ligand provides a higher *in vivo* stability to the Gd chelate as compared to the DTPA ligand. The Gd release/trans-metallation has been found to be significantly lower as compared to Gd-DTPA in a number of investigations.[199,200] It is suggested[201] that Gd is held strongly in the rigid ring framework of the macrocyclic chelate and its release would be affected by a simultaneous breakup of the coordinated network as compared to the possibility of successive breakup in an acyclic chelate. In a comparison of relative safety profiles for macrocyclic and acyclic chelates in mice, the median lethal dose of Gd-DOTA was found to be ~95% higher than that of Gd-DTPA.[202]

Gd chelates have been used intensively for cellular and molecular imaging with MRI.[203,204] Due to the lower sensitivity of Gd in comparison to SPIO, larger amounts of chelates are required for image enhancement. Arbab, Liu, and Frank have estimated ~100–1000 chelates per cell for effective enhancement.[177] This is a relatively high amount for cellular labeling and is not easily achieved without the help of transfection agents. Research activities have been directed toward developing new contrast agents with increased sensitivity as well as methods to concentrate the Gd concentration effectively without compromising the relaxation. While sensitivity of Gd chelates is achieved by suitable modification of the ligand design, which permits efficient water exchange, the formulation of contrast agents in various nanoplatforms enables higher concentrations of Gd and better signal enhancement. Various nanoplatforms such as liposomes, dendrimers, silica, and perfluorocarbon have been explored to achieve higher Gd payloads.

MAGNETIC AND LUMINESCENT NANOPARTICLES

One of the advantages in using nanoparticulate platforms as contrast agents is the ability to incorporate multiple contrast agents into a single nanoparticle, thus allowing for multimodal imagery. Pertaining to the focus of this chapter, the multimodal nanoparticles, which amalgamate MRI and optical contrast capabilities together, will be discussed in the following sections. These multimodal nanoparticles have been classified into two broad classes: (a) luminescent core—magnetically labeled, and (b) magnetic core—luminescent label particles. The various strategies used for integrating MR and OI as well as the size, surface modification, bioconjugation, and applications of the synthesized particles are briefly addressed in the following sections.

LUMINESCENT CORE AND MAGNETICALLY LABELED MULTIMODAL NANOPARTICLES

Dye-Doped Silica Core

One of the ways in which multimodal nanoparticles have been developed is through the use of a dye-doped silica matrix as the core. The magnetic component has been introduced by either attaching the Gd or iron oxide nanoparticles on the surface of the silica. We have reported the synthesis and characterization of ~100 nm multifunctional core-shell silica particle, which can be detected by fluorescence, MRI, and fluoroscopy. Narrowly dispersed nanoparticles have been prepared in one pot using the microemulsion-mediated approach. Rubpy dye-doped particles have been modified with a silane ligand to which the paramagnetic Gd species has been attached.[34] The number of Gd ions attached to each nanoparticle has been estimated by ICP to be ~16,000, and MR measurements reveal R_1 and R_2 to be 9.0 and 116 s^{-1} per mM of Gd^{3+} (4.7T). These values are up to 3 and 24 times that of Gadoteridol, a commercially used MR contrast agent, demonstrating that these nanoparticles can be used to generate contrast on both T_1- and T_2-weighted images. In addition, the particles have been demonstrated to produce x-ray contrast too because of the presence of heavy elements like Ru and Gd. The multimodal imaging ability of the particles has been demonstrated *in vitro* by labeling lung carcinoma A549 cells (Figure 16.4). [205]

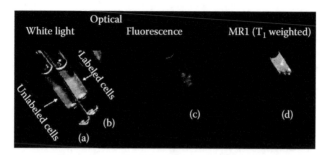

FIGURE 16.4 *In vitro* labeling of adenocarcinoma A549 cells using multimodal nanoparticles. (Published with permission from: Technology in Cancer Research and Treatment, 4, 593, 2005, Adenine Press, http://www.tcrt.org.)

Rieter and others have similarly prepared multimodal nanoparticles using ~40 nm Rubpy silica nanoparticles and condensing either Gd-(trimethoxysilylpropyl) diethylenetriamine tetraacetate complex (GdDTTA) or bis(silylated)-Gd-DTPA complex (GdDTPA) on the silica surface.[206] In these particles, Gd is coordinated from seven coordination sites in the silane complex, providing high stability and significantly reduced toxicity concerns. The number of Gd^{3+} ions has been estimated to be ~10,200 and 63,200 per nanoparticle in the GdDTTA and GdDTPA nanoparticles, respectively. It is suggested by the authors that the bis(silylated)-Gd-DTPA complex forms multilayers on the nanoparticle, forming a thick coating with ~63200 Gd ions per nanoparticle. The R_1 and R_2 values have been estimated to be 19.7 and 60.0 s^{-1} per mM of Gd^{3+} for GdDTTA and 7.8 and 12.3 s^{-1} per mM of Gd^{3+} (3T) for the GdDTPA complex. Despite spite having a higher loading of the Gd, lower relaxivity values for the GdDTPA nanoparticles have been explained by the poor accessibility of water to the inner layers of the Gd in the nanoparticle. This suggests that in addition to higher loading of the Gd, the design of the particle for maximizing the water exchange plays a crucial role in enhancing its ability as an MR contrast agent.

Iron oxide nanoparticles have also been used to introduce magnetic character into dye-doped silica nanoparticles.[207] 9 nm of water-dispersible iron oxide have been covalently attached to 30-nm Rhodamine dye-doped silica nanoparticles in what is described as the "core-satellite" structure. Iron oxide coated with dimercaptosuccinic acid was attached to the aminated silica nanoparticles using sulfo maleimide chemistry using sulfosuccinimidyl-(4-N-maleimidomethyl)cyclohexane-1-carboxylate as cross-linkers. The core-satellite nanoparticle having ~10 iron oxide nanoparticles per each fluorescent silica nanoparticle were further conjugated to HmemB1 antibodies and used to label neuroblastoma cell lines overexpressing polysialic acids. Comparison of MR measurements of the multimodal particles with the free iron oxide nanoparticles revealed significant improvement in MR signal, which was assigned to the synergestic effect of the individual iron oxide nanoparticles on the construct.[208]

Mesoporous silica nanoparticles exhibiting a high surface area, large pore volume, and uniform pore size have been used as a matrix in making multimodal nanoparticles.[209] Gd incorporated into the mesoporous silica matrix has been reported to have a higher relaxivity value when compared to commercially used $[Gd(DTPA)(H_2O)]^{2-}$. The authors have suggested that the use of such particles as multimodal imaging agents via additional doping of europium and terbium as luminescent lanthanides. However, the strength of binding Gd to silica and its stability in biological media have not been sufficiently addressed. Iron oxide has also been incorporated in mesoporous silica. In a multistep procedure, iron oxide nanoparticles prepared by the thermal decomposition method were made water dispersible by coating with a layer of silica. These together with organic dyes were co-incorporated in the mesoporous silica *in situ* during silica synthesis. The composite particles had tumblerlike morphology with average dimensions of 154 nm (long axis) by 115 nm (short axis). Phantom images and *in vitro* labeling of fibroblast cells were carried out to demonstrate the multimodal nature of these particles. Cytotoxicity studies demonstrate no effect on cell viability up to 200 μg/mL nanoparticle doses.

Quantum Dot-Based Multimodal Probes

The advantageous optical properties of the QDs over fluorescent dyes have led to an increasing number of reports that incorporate their use in multimodal probes. Gd has been the magnetic contrast agent mated with the QD in most examples. Various nanoparticulate platforms comprising QD and Gd, such as liposomes and dendrimers, are described later.

Mulder and colleagues have used CdSe/ZnS shell QDs prepared by the hot solvent method as the luminescent portion of OI/MRI multimodal probes.[210] The oil-soluble QDs were made water dispersible by coating them with lipid pegylated phospholipid, PEG-DSPE (1,2-distearoyl-sn-glycero-3-phosphoethanolamine-N-[methoxy-(poly(ethylene glycol))-2000]).[150] Pegylated lipids, because of their amphiphilic character, have frequently been used for stabilizing hydrophobic formulations in medicine, where the PEG chains help to provide a hydrophilic coating on the particle surface. The co-introduction of a paramagnetic lipid (Gd-DTPA-bis(stearylamide) during hydrophillization was used to make the construct detectable by MRI. These multimodal probes were estimated to contain ~150 Gd-lipids and produced high MR relaxivity. The lipid coat surrounding the QDs was further doped with maleimide containing paramagnetic, which was used to conjugate to the cyclic RGD peptide to target angiogenic vessels and tumor cells overexpressing the $\alpha v \beta 3$ integrins.[211] The overall size of the construct after lipid modification was about 10 nm, which is significantly less than other nanoparticulate contrast agents.[40,212] The multimodal character and the biological specificity were demonstrated by specific labeling of human umbilical vein endothelial cells *in vitro*. A similar multimodal nanoparticle construct was bioconjugated with annexin A5 and used for cellular labeling. Annexin A5 binds with high affinity to phosphatidylserine, which is asymmetrically distributed PS at the outer monolayer of the cell membrane of apoptotic cells.[213] The specific targeting ability of these particles was demonstrated *in vitro* using a T lymphoma cell line. Strong signal enhancement from apoptotic labeled cells using MR and fluorescence was detected.

The use of biotin-avidin affinity for the synthesis of paramagnetic and bioconjugated QDs has been reported.[214] Biotinylated Annexin A5 was first bound to the streptavidin coated QDs in 1:1 molar ratio, followed by the binding of biotinylated Gd-DTPA to the remaining binding sites on streptavidin-coated QDs. It is emphasized that in biotinylated Annexin A5, the biotin is connected to the N-terminal tail that is away from the phosphatidylserine binding site. To increase the MR sensitivity of the multimodal construct, the number of Gd attached to each QDs needs to be increased. In order to achieve this, a lysine-based, wedge-shaped biotinylated construct capable of binding up to eight Gd-DTPA has been used. The mean particle size of the QDs after the surface modifications was observed to be ~7 nm. Selective labeling of apoptotic cells was observed with Annexin-coated particles while no labeling was observed for viable cells. MR experiments of labeled cells demonstrated stronger signal from the wedge constructs. The bimodal character of Annexin A5 coated conjugated Gd wedge particles was also demonstrated in a blood-clot system as the activated platelets are known to expose phosphatidylserine. 35–50% higher MR signal intensity over the control was observed with the particles in the incidence of Annexin A5 competing with the coagulation factors for binding to phosphatidylserine. Optical imaging of whole blood clots was performed using two-photon laser scanning microscopy.[215,216] Specific labeling of platelets was observed with the particles, similar to results achieved from FITC labeled Annexin A5. In an *ex vivo* study, the same particles were also used to label the vascular wall of the murine carotid artery and imaged by fluorescence and MR.

In another approach, micellization of the hydrophobic QDs was carried out using silica precursors to make them water dispersible as well as to introduce functional groups on the surface to enable bioconjugation.[217] CdSe/ZnS QDs prepared by hot-phase synthesis were first capped with short-chain, negatively charged detergents and then coated with successive layers of hydrophobic, amphiphilic, and hydrophilic silica precursors resulting in an 18-nm diameter particle. The final precursor helped in making the particles hydrophilic as well as by introducing amino groups on the silica shell. It is reported that this method enabled single QD nanocrystal encapsulation in each shell, which minimizes the possibility of self-quenching, providing a quantum yield improvement of 50% over the starting QDs. Hydrophobic paramagnetic Gd have been incorporated in the hydrophobic layer between the QD and the silica shell and some preliminary studies were performed. However, ~90% of the Gd compound leached out during the dialysis, indicating the need for using either a stronger hydrophobic compound or a covalent coupling of the paramagnetic-linked species. Addressing the issue of covalent coupling of the Gd chelate to the QD matrix, Yang and others have prepared CdS:Mn/ZnS QDs conjugated to Gd using the microemulsion route.[218] The QDs were coated with a silane precursor bearing chelating groups, which were used to capture the ~107 Gd ions per QD. The presence of the Gd on the outer surface provided rapid water exchange with the bulk giving rise to high relaxivity values of 20.5 and 151 mM^{-1} s^{-1} for R_1 and R_2, respectively. The 4–7 nm silica shell on the QDs was further expected to reduce the toxicity concerns due to QDs matrix degradation by capturing free Cd or Zn.

In general it is observed that the effect of surface modifications on the QDs results in (a) an effective increase in the particle diameter of up to 20 nm, (b) reduction in the quantum yield, and (c) enabling the introduction of paramagnetic entities for synthesizing bimodal nanoparticles. Some additional ways of preparing multimodal nanoparticles are described briefly later.

Fluorescent and paramagnetic liposomes bioconjugated with cyclic RGD have also been used for specific labeling of tumor cells *in vivo*. In a report by Mulder and others, the bioconjugated particles have been used to specifically target subcutaneously implanted LS174 human colon adenocarcinoma xenograft model in nude mice and image angiogenesis by detecting the expression of $\alpha v \beta 3$.[211] The liposomal particles were prepared with 0.75 mol% Gd-DTPA-bis (stearylamide) and 0.1 mol% lipid-bound chromophore such as 1,2-dioleoyl-sn-glycero-3-phosphoethanolamine-N-lissamine rhodamine B sulfonyl ammonium salt. The control particles were bioconjugated with another peptide RAD, which differs in one methyl group position as compared to RGD but does not show any specific binding affinity for the $\alpha v \beta 3$ integrins. Both MR and fluorescence imaging were used to determine the targeting of the liposomes. While MR experiments depicted an increase in the MR signal on the tumor sites with both RGD and

RAD conjugated liposomes, the *ex vivo* fluorescence experiments established RGD liposomes association with the activated tumor endothelium and RAD liposomes present in the extravascular compartment.

While dendrimers have been used for numerous drug delivery and imaging applications, few reports exist for its use as multimodal contrast agent for MRI and OI. Availability of various generatons of dendrimers with different functional groups has opened large possibilities for their applications in medical field. Talanov and coworkers have used a generation polyamidoamine ethylenediamine core dendrimer with 256 terminal amino groups to covalently attach amine-reactive DTPA derivative, 2-(4-isothiocyanatobenzyl)-6-methyl-diethylenetriaminepentaacetic acid and Cyanine 5.5 dye to the dendrimer.[219] The authors note a marginal increase in the quantum yield of the dye upon covalent conjugation to the dendrimer and no effect of chelated Gd on the dye showing robust fluorescent properties. The dendrimer construct conjugated to ~1–2 dyes and ~150 Gd molecules was used for *in vivo* imaging of sentinel lymph nodes.[219,220] It was observed that lymph nodes could be detected with only 0.5 nmol Cyanine 5.5 while ~750 n mole of Gd was necessary for consistent MR response.

MAGNETIC CORE—LUMINESCENT LABEL PARTICLES

The second subclassification of magnetofluorescent multimodal nanoparticles, that is, those having a magnetic core and modified with luminescent moieties, are described in this section. A major component of this class comprises an iron oxide magnetic core consisting of SPIO, USPIO, or MIONs and conjugated to fluorescent dyes such as FITC and cyanine dyes. The cyanine class of dyes such as Cy5.5, Cy 7, and others that fluoresce in the red and NIR regions are frequently employed due to lower absorption and scattering of the emitted light and poor autofluorescence in this region.

Magnetic iron oxide particles are generally synthesized, in an oxygen-free environment, by the coprecipitation of Fe^{2+} and Fe^{3+} ions with a base.[182] Particles prepared in this manner are typically hydrophobic and are unsuitable for biological applications. In aqueous biological environments, strong hydrophobic interactions between these particles lead to agglomeration and a subsequent change in their magnetic character—from superparamagnetic to ferromagnetic. Surface modification of these particles allows them to be rendered water dispersible and provides useful reactive groups for enabling bioconjugation and dye attachment. In the literature a number of different moieties have been attached to the surface of these particles, including targeting molecules for specific delivery, fluorescent dyes for optical imaging, and drug molecules for therapeutic applications. Dextran has been preferentially used for the surface modification of iron oxide particles; however, silanization and pegylation have also been applied. Table 16.3 provides a summary of some examples of multimodal iron oxide nanoparticles found in the literature. These examples are further discussed in the following sections.

Dextran-Coated Iron Oxide Nanoparticles

Dextran-coated iron oxide particles are preferred for biological application due to their high colloidal stability under harsh physiological conditions. Iron oxide particles

TABLE 16.3

Comparison of Surface-Modified Iron Oxide Nanoparticles

Coating	Dye	Targeting Agent	Size (nm)	R1 (mM^{-1} s^{-1})	R2 (mM^{-1} s^{-1})	Dye/ Particle	Bioconjugation/ Particle	Ref.
Dextran	FITC	TAT	41	22.3	71.9	6.7	6.7	40
Dextran	CY5.5	R4	62	29	92.5	1.79	15.5	42
Dextran	CY5.5	Annexin V	50	19	48	1.8	3.5	221
Dextran	CY5.5	EPPT	35.8	26.43	53.44	5	14	222
Dextran	CY5.5	RGD	28	a	a	7.2	27.2	223
PEG-Silane	CY5.5	Cltx	15	a	a	1.22	10.2	224

a: data not available

coated with dextran are synthesized by coprecipitation from aqueous solutions containing ferrous salt and dextran by controlled addition of a cold solution of a base.[225] These particles are then crosslinked with epichlorohydrin to form cross-linked iron oxide (CLIO) nanoparticles. CLIO particles are further reacted with ammonia to yield primary amine on their surface for attaching dye and targeting molecules.

Jaffer and colleagues[226] have used CLIO-Cy5.5 particles for multimodal imaging of cellular inflammation in apolipoprotein E-deficient (apo E-/-) mouse model. Preferential uptake of these particles by macrophages enabled *in vivo* imaging using MRI while *ex vivo* cellular inflammation in murine artherosclerosis was monitored using MR and OI.

Bioconjugation of the fluorescent CLIO nanoparticles has been carried out to study various cellular phenomena. For instance, conjugation of FITC-labeled membrane translocating signal peptide (HIV-1 tat) to the CLIO nanoparticle was used to demonstrate a 100-fold increase in internalization of the particles in lymphocytes.[40] In another application, CLIO-Cy5.5 particles were conjugated with a synthetic peptide EPPT (derived from the CDR3V$_h$ region of a monoclonal antibody (ASM2) raised against human epithelial cancer cells) to monitor changes in tumor volume caused by treatment with a chemotherapeutic agent (5-FU).[227] In this study, optical imaging provided information about the location and relative dimensions of the tumor while MR imaging was used to quantitatively evaluate the size of the tumor. In the mouse model of cancer, cyclic RGD peptide conjugated CLIO-Cy5.5 nanoparticles were selectively targeted to BT-20 tumor cells, which express $\alpha v \beta 3$ integrins.[223] By using MR and fluorescence imaging modalities, it was demonstrated that BT-20 cells internalize these particles more than 9 L tumor cells, which produce less $\alpha v \beta 3$ integrins. These particles were eventually cleared from the blood by liver and spleen macrophages and did not show any toxic effect. CLIO-Cy5.5 nanoparticles carrying a specific peptide sequence, containing the VHSPNKK motif that has homology to the alpha-chain of very late antigen (a known ligand for vascular adhesion molecule-1 (VCAM-1)), were successfully employed to target VCAM-1-expressing endothelial cells in a murine tumor necrosis factor-α induced inflammatory model.[228]

Schellenberger and others[221] have developed CLIO-Cy5.5 nanoparticles labeled with Annexin V, a human protein that binds with phosphatidylserine (a lipid that is present on the membrane of a cell during apoptosis) for determining apoptosis in organs and tissues. A single reactive sulfhydryl group was added to the Annexin V protein by optimizing its reaction with SATA in a manner that does not alter its ability to bind with apoptotic Jurkat T cells. This optimized SATAylated Annexin V protein was labeled on CLIO-Cy-5.5 dye to yield the multimodal nanoparticle 1.8 Cy5.5 molecules for optical detection and 2064 Fe atoms for MR detection. These particles were tested *in vitro* for both NIR flourescence and MR imaging modalities.

CLIO-Cy5.5 particles have also been employed to design "smart" nanoprobes whose position can be located by MR imaging, while their optical properties can be activated through cleavage of attached dye molecule by a specific enzyme.[42] Fluorescent quenching of Cy5.5 dye occurs due to its interaction with CLIO nanoparticles. Cy5.5 dye, when attached to the surface of a CLIO nanoparticle in such a fashion that it can be cleaved by the action of a specific enzyme, restores its fluorescence properties upon action of a specific enzyme. These probes were synthesized using a thioether linker between arginyl peptide and amino-CLIO particles and were utilized for *in vivo* MRI and NIR flourescence imaging of lymph nodes in mice to demonstrate their magnetic and activatable optical imaging capabilities, respectively. In these studies, fluorochrome was chosen in near infrared range to minimize autofluorescence.

Alternate Surface Modifications

Rather than using dextran, other researchers have applied alternative surface coatings enabling hydrophilization, fluorescent contrast, and biomolecular functionalization.

PEG-modified phospholipid micelles have been employed to coat superparamagnetic iron oxide particles.[229] Nanoparticles were synthesized by the reverse micelle technique and coated with PEG-phospholipid micelles [1:8 mixture of 1,2-distearoyl-sn-glycero-3-phosphoethanolamine-N-methoxy(polyethyleneglycol) 2000 and 1,2-distearoyl-sn-glycero-3-phosphoethanolamine-N-[amino(polyethyleneglycol) 2000] to render them water soluble and biocompatible. The amine groups available on the surface of the micelles were used for attaching Texas Red dye molecules and TAT peptides. The average size of MIONs was approximately 14.6 nm and each particle had 6070 iron atoms. *In vitro* OI and MR experiments were performed on primary human dermal fibroblast (HDF) and Madin-Darby bovine kidney (MDBK) derived cells labeled with the multimodal nanoparticles.

For targeting gliomas cells, iron oxide nanoparticles coated with a silane layer have been developed. Fe_3O_4 nanoparticles of 10.5 nm were synthesized using a coprecipitation process of iron chloride and sodium hydroxide, and coated with amine-terminated PEG silane. Cy5.5 molecules were attached to the surface amine groups to permit OI.[224] For specific targeting, sulfhydryl-modified chlorotoxin was conjugated with iodoacetate-derivatized nanoparticles. These particles were tested *in vitro* on 9 L glioma cells for their optical and magnetic properties using confocal fluorescence microscopy and MR imaging. Consistent results were obtained using both optical and MR imaging, and preferential uptake of multimodal nanoparticles was demonstrated on 9 L glioma cells.

In another example, silane-modified multimodal iron oxide nanoparticles were synthesized for the detection and selective lysing of cancer cells via magnetocytolysis.[230] Iron oxide particles were prepared by the colloidal reaction of iron dodecyl sulfate $(Fe(DS)_2)$ with methylamine. A two-photon dye ASPI-SH (1-methyl-4 -(E)-2-{4-[methyl(2-sulfanylethyl)-amino]phenyl}-1-ethenyl)pyridinium iodide) was attached to these particles and a thin silica layer was produced on the particle surface by the hydrolysis of TEOS in the presence of NH_4OH. 3-(Triethoxysilylpropyl-carbamoyl)butyric acid, which offers a COOH group for the coupling of targeting agent, was attached to the silica coating in the presence of DMF. A biotargeting group, Luteinizing Hormone-Releasing Hormone (LH-RH), was attached to the COOH group in the presence of BOP. These nanoparticle probes were tested on oral epithelial carcinoma cells (KB cells, LH-RH receptor positive). The selective uptake of the multimodal nanoprobe by KB cells was verified by two-photon laser scanning microscopy. Selective lysing of KB cells as a result of magnetocytolysis was also demonstrated by the application of a high-frequency ac magnetic field.

FUTURE PROSPECTS

Due to the complementary nature of information provided by MR and OI, significant progress has been made in the development of conventional and multimodal nanoparticulate contrast agents for use with these imaging modalities. Some of the important examples of these nanomaterials include dye-doped silica, quantum dots, iron oxides, lipids, and dendrimers as described in this chapter. These nanoparticulate platforms have and are continuing to be developed for achieving improved diagnostic and therapeutic efficacy in modern medicinal research. The development of targeted multimodal nanoparticulate probes are anticipated to lead to early disease detection technologies, thus enhancing treatment efficacy and patient care. However, many challenges need to be overcome before their full potential is realized. An understanding of the intricate interplay between biological and nanoparticulate materials will be necessary for the design of next-generation multimodal contrast agents with superior homing and clearance properties. As well, methods to leverage nanoconfinement for improved contrast from sequestered MRI and optical contrast agents are desired. For *in vivo* applications, the development of probes in the NIR window is essential. In addition, high sensitivity and photostability of probes under *in vivo* conditions need to be addressed, and their toxic potential minimized.

ACKNOWLEDGMENTS

The authors acknowledge the financial support of the Particle Engineering Research Center (PERC) at the University of Florida, the National Science Foundation (NSF Grant EEC-94-02989, NSF-NIRT Grant EEC-0506560), the National Institutes of Health (Grant 1-P20-RR020654-01), James & Esther King Biomedical Research

Program (Grant 06NIR-05), and the Industrial Partners of the PERC for support of this research. Any opinions, findings, and conclusions or recommendations expressed in this material are those of the author(s) and do not necessarily reflect those of the National Science Foundation.

REFERENCES

1. Massoud, T.F. & Gambhir, S.S. Molecular imaging in living subjects: Seeing fundamental biological processes in a new light. *Genes & Development* **17**, 545–580 (2003).

2. Weissleder, R. & Mahmood, U. Molecular imaging. *Radiology* **219**, 316–333 (2001).

3. Weissleder, R. Molecular imaging in cancer. *Science* **312**, 1168–1171 (2006).

4. Weissleder, R. Molecular imaging: Exploring the next frontier. *Radiology* **212**, 609–614 (1999).

5. Barentsz, J. et al. Commonly used imaging techniques for diagnosis and staging. *J Clin Oncol* **24**, 3234–3244 (2006).

6. Weissleder, R. & Ntziachristos, V. Shedding light onto live molecular targets. *Nat Med* **9**, 123–128 (2003).

7. Weissleder, R. Scaling down imaging: Molecular mapping of cancer in mice. *Nat Rev Cancer* **2**, 11–18 (2002).

8. Ntziachristos, V. & Weissleder, R. Charge-coupled-device based scanner for tomography of fluorescent near-infrared probes in turbid media. *Med Phys* **29**, 803–809 (2002).

9. Graves, E.E., Ripoll, J., Weissleder, R., & Ntziachristos, V. A submillimeter resolution fluorescence molecular imaging system for small animal imaging. *Med Phys* **30**, 901–911 (2003).

10. Ntziachristos, V. Fluorescence molecular imaging. *Annu Rev Biomed Eng* **8**, 1–33 (2006).

11. Graves, E.E., Weissleder, R., & Ntziachristos, V. Fluorescence molecular imaging of small animal tumor models. *Curr Mol Med* **4**, 419–430 (2004).

12. Ntziachristos, V., Bremer, C., & Weissleder, R. Fluorescence imaging with near-infrared light: New technological advances that enable *in vivo* molecular imaging. *Eur Radiol* **13**, 195–208 (2003).

13. Ohlerth, S. & Scharf, G. Computed tomography in small animals—Basic principles and state of the art applications. *Vet J* **173**, 254–271 (2007).

14. Ritman, E.L. Molecular imaging in small animals—Roles for micro-CT. *J Cell Biochem,* **87**, 116–124 (2002).

15. Paulus, M.J., Gleason, S.S., Kennel, S.J., Hunsicker, P.R., & Johnson, D.K. High resolution x-ray computed tomography: An emerging tool for small animal cancer research. *Neoplasia* **2**, 62–70 (2000).

16. Arrington, S.A., Schoonmaker, J.E., Damron, T.A., Mann, K.A., & Allen, M.J. Temporal changes in bone mass and mechanical properties in a murine model of tumor osteolysis. *Bone* **38**, 359–367 (2006).

17. Nahrendorf, M. et al. High-resolution imaging of murine myocardial infarction with delayed-enhancement cine micro-CT. *Am J Physiol—Heart Circ Physiol* **292**, H3172–H3178 (2007).

18. Walters, E.B., Panda, K., Bankson, J.A., Brown, E., & Cody, D.D. Improved method of *in vivo* respiratory-gated micro-CT imaging. *Phys Med Biol* **49**, 4163–4172 (2004).

19. Wirtzfeld, L.A. et al. A new three-dimensional ultrasound microimaging technology for preclinical studies using a transgenic prostate cancer mouse model. *Cancer Res* **65**, 6337–6345 (2005).

20. Coatney, R.W., Zhao, S.F., Woods, T., & Legos, J. Non-invasive imaging in animal research: principles and applications of ultrasound imaging in rodents. *J Pharmacol Sci* **91**, 18P–18P (2003).

21. Herschman, H.R. PET reporter genes for noninvasive imaging of gene therapy, cell tracking and transgenic analysis. *Crit Rev Oncol Hematol* **51**, 191–204 (2004).

22. Cherry, S.R. & Gambhir, S.S. Use of positron emission tomography in animal research. *Ilar J* **42**, 219–232 (2001).

23. Schmidt, K.C. & Smith, C.B. Resolution, sensitivity and precision with autoradiography and small animal positron emission tomography: Implications for functional brain imaging in animal research. *Nucl Med Biol* **32**, 719–725 (2005).

24. Guerra, P. et al. Performance analysis of a low-cost small animal PET/SPECT scanner. *Nucl Instr Meth Phys Res Sect A—Accel Spectrom Detect Assoc Equip* **571**, 98–101 (2007).

25. Hedlund, L.W. & Johnson, G.A. Morphology of the small-animal lung using magnetic resonance microscopy. *Proc Am Thorac Soc* **2**, 481–483, 501–502 (2005).

26. Mori, S., Zhang, J., & Bulte, J.W. Magnetic resonance microscopy of mouse brain development. *Methods Mol Med* **124**, 129–147 (2006).

27. Pirko, I., Fricke, S.T., Johnson, A.J., Rodriguez, M., & Macura, S.I. Magnetic resonance imaging, microscopy, and spectroscopy of the central nervous system in experimental animals. *NeuroRx* **2**, 250–264 (2005).

28. Tatlisumak, T., Strbian, D., Abo Ramadan, U., & Li, F. The role of diffusion- and perfusion-weighted magnetic resonance imaging in drug development for ischemic stroke: From laboratory to clinics. *Curr Vasc Pharmacol* **2**, 343–355 (2004).

29. Bae, K.T. et al. Gadolinium-enhanced computed tomography angiography in multi-detector row computed tomography: Initial observations. *Acad Radiol* **11**, 61–68 (2004).

30. Gierada, D.S. & Bae, K.T. Gadolinium as a CT contrast agent: Assessment in a porcine model. *Radiology* **210**, 829–834 (1999).

31. McDonald, M.A. & Watkin, K.L. Small particulate gadolinium oxide and gadolinium oxide albumin microspheres as multimodal contrast and therapeutic agents. *Invest Radiol* **38**, 305–310 (2003).

32. Watkin, K.L. & McDonald, M.A. Multi-modal contrast agents: a first step. *Acad Radiol* **9 Suppl 2**, S285–S289 (2002).

33. Alexiou, C. et al. Magnetic mitoxantrone nanoparticle detection by histology, X-ray and MRI after magnetic tumor targeting. *J Magnet Magnet Mater* **225**, 187–193 (2001).

34. Santra, S. et al. Synthesis and characterization of fluorescent, radio-opaque, and paramagnetic silica nanoparticles for multimodal bioimaging applications. *Advanced Materials* **17**, 2165–2169 (2005).

35. Santra, S., Yang, H.S., Holloway, P.H., Stanley, J.T., & Mericle, R.A. Synthesis of water-dispersible fluorescent, radio-opaque, and paramagnetic CdS: Mn/ZnS quantum dots: A multifunctional probe for bioimaging. *J Amer Chem Soc* **127**, 1656–1657 (2005).

36. Benamor, M. et al. PET/CT imaging: What radiologists need to know. *Cancer Imaging* **7 Spec No A**, S95–S99 (2007).

37. Otsuka, H., Morita, N., Yamashita, K., & Nishitani, H. FDG-PET/CT for cancer management. *J Med Invest* **54**, 195–199 (2007).

38. Nahmias, C. et al. Positron emission tomography/computerized tomography (PET/CT) scanning for preoperative staging of patients with oral/head and neck cancer. *J Oral Maxillofac Surg* **65**, 2524–2535 (2007).

39. Fritz, V. et al. Micro-CT combined with bioluminescence imaging: a dynamic approach to detect early tumor-bone interaction in a tumor osteolysis murine model. *Bone* **40**, 1032–1040 (2007).

40. Josephson, L., Tung, C.H., Moore, A., & Weissleder, R. High-efficiency intracellular magnetic labeling with novel superparamagnetic-tat peptide conjugates. *Bioconjugate Chem* **10**, 186–191 (1999).

41. Crich, S.G. et al. Improved route for the visualization of stem cells labeled with a Gd-/Eu-chelate as dual (MRI and fluorescence) agent. *Magn Reson Med* **51**, 938–944 (2004).

42. Josephson, L., Kircher, M.F., Mahmood, U., Tang, Y., & Weissleder, R. Near-infrared fluorescent nanoparticles as combined MR/optical imaging probes. *Bioconjugate Chem* **13**, 554–560 (2002).

43. Zhao, H. et al. Emission spectra of bioluminescent reporters and interaction with mammalian tissue determine the sensitivity of detection *in vivo*. *Journal of Biomedical Optics* **10**, – (2005).

44. Lyons, S.K. Advances in imaging mouse tumour models *in vivo*. *Journal of Pathology* **205**, 194-205 (2005).

45. Lyons, S.K. et al. Noninvasive bioluminescence imaging of normal and spontaneously transformed prostate tissue in mice. *Cancer Res* **66**, 4701–4707 (2006).

46. Billinton, N. & Knight, A.W. Seeing the wood through the trees: A review of techniques for distinguishing green fluorescent protein from endogenous autofluorescence. *Anal Biochem* **291**, 175–197 (2001).

47. Labas, Y.A. et al. Diversity and evolution of the green fluorescent protein family. *Proc Nat Acad Sci USA* **99**, 4256–4261 (2002).

48. Matz, M.V., Lukyanov, K.A., & Lukyanov, S.A. Family of the green fluorescent protein: Journey to the end of the rainbow. *Bioessays* **24**, 953–959 (2002).

49. Kennedy, M.D., Jallad, K.N., Thompson, D.H., Ben-Amotz, D., & Low, P.S. Optical imaging of metastatic tumors using a folate-targeted fluorescent probe. *J Biomed Opt* **8**, 636–641 (2003).

50. King, R.C., Mills, S.L., & Medina, J.E. Enhanced visualization of parathyroid tissue by infusion of a visible dye conjugated to an antiparathyroid antibody. *Head Neck* **21**, 111–115 (1999).

51. Ramjiawan, B. et al. Noninvasive localization of tumors by immunofluorescence imaging using a single chain Fv fragment of a human monoclonal antibody with broad cancer specificity. *Cancer* **89**, 1134–1144 (2000).

52. Ballou, B. et al. Tumor labeling *in vivo* using cyanine-conjugated monoclonal antibodies. *Cancer Immunol Immunother* **41**, 257–263 (1995).

53. Pelegrin, A. et al. Antibody fluorescein conjugates for photoimmunodiagnosis of human colon-carcinoma in nude-mice. *Cancer* **67**, 2529–2537 (1991).

54. Bugaj, J.E., Achilefu, S., Dorshow, R.B., & Rajagopalan, R. Novel fluorescent contrast agents for optical imaging of *in vivo* tumors based on a receptor-targeted dye-peptide conjugate platform. *Journal of Biomedical Optics* **6**, 122–133 (2001).

55. Montet, X., Ntziachristos, V., Grimm, J., & Weissleder, R. Tomographic fluorescence mapping of tumor targets. *Cancer Res* **65**, 6330–6336 (2005).

56. Ntziachristos, V. et al. Visualization of antitumor treatment by means of fluorescence molecular tomography with an Annexin V-Cy5.5 conjugate. *Proc Nat Acad Sci USA* **101**, 12294–12299 (2004).

57. Bremer, C., Ntziachristos, V., & Weissleder, R. Optical-based molecular imaging: contrast agents and potential medical applications. *Eur Radiol* **13**, 231–243 (2003).

58. Schellenberger, E.A. et al. Optical imaging of apoptosis as a biomarker of tumor response to chemotherapy. *Neoplasia* **5**, 187–192 (2003).

59. Montet, X. et al. Tomographic fluorescence imaging of tumor vascular volume in mice. *Radiology* **242**, 751–758 (2007).

60. De Grand, A.M. & Frangioni, J.V. An operational near-infrared fluorescence imaging system prototype for large animal surgery. *Technol Cancer Res Treat* **2**, 553–562 (2003).

61. Chen, Y., Intes, X., & Chance, B. Development of high-sensitivity near-infrared fluorescence imaging device for early cancer detection. *Biomed Instrum Technol* **39**, 75–85 (2005).

62. Gurfinkel, M., Ke, S., Wen, X., Li, C., & Sevick-Muraca, E.M. Near-infrared fluorescence optical imaging and tomography. *Dis Markers* **19**, 107–121 (2003).

63. Shang, H.M. et al. Optically transparent superhydrophobic silica-based films. *Thin Solid Films* **472**, 37–43 (2005).

64. Sharrna, P., Brown, S., Walter, G., Santra, S., & Moudgil, B. Nanoparticles for bioimaging. *Adv Colloid Interf Sci* **123**, 471–485 (2006).

65. Diaz, A. et al. Growth of hydroxyapatite in a biocompatible mesoporous ordered silica. *Acta Biomater* **2**, 173–179 (2006).

66. McLaughlin, J.K., Chow, W.H., & Levy, L.S. Amorphous silica: A review of health effects from inhalation exposure with particular reference to cancer. *J Toxicol Environ Health* **50**, 553–566 (1997).

67. Hulchanskyy, T.Y. et al. Organically modified silica nanoparticles with covalently incorporated photosensitizer for photodynamic therapy of cancer. *Nano Lett* **7**, 2835–2842 (2007).

68. Simovic, S. & Prestidge, C.A. Nanoparticle layers controlling drug release from emulsions. *Euro J Pharm Biopharm* **67**, 39–47 (2007).

69. Yang, X.L., Han, X., & Zhu, Y.H. (PAH/PSS)(5) microcapsules templated on silica core: Encapsulation of anticancer drug DOX and controlled release study. *Colloids and Surfaces A—Physicochem Eng Asp* **264**, 49–54 (2005).

70. Zhou, J. et al. Synthesis of porous magnetic hollow silica nanospheres for nanomedicine application. *J Phys Chem C* **111**, 17473–17477 (2007).

71. Santra, S., Zhang, P., Wang, K.M., Tapec, R., & Tan, W.H. Conjugation of biomolecules with luminophore-doped silica nanoparticles for photostable biomarkers. *Anal Chem* **73**, 4988–4993 (2001).

72. Schiestel, T., Brunner, H., & Tovar, G.E.M. Controlled surface functionalization of silica nanospheres by covalent conjugation reactions and preparation of high density streptavidin nanoparticles. *J Nanosci Nanotechnol* **4**, 504–511 (2004).

73. Qhobosheane, M., Santra, S., Zhang, P., & Tan, W.H. Biochemically functionalized silica nanoparticles. *Analyst* **126**, 1274–1278 (2001).

74. Santra, S. et al. Folate conjugated fluorescent silica nanoparticles for labeling neoplastic cells. *J Nanosci Nanotechnol* **5**, 899–904 (2005).

75. Santra, S. et al. TAT conjugated, FITC doped silica nanoparticles for bioimaging applications. *Chem Commun*, 2810–2811 (2004).

76. Santra, S., Dutta, D., & Moudgil, B.M. Functional dye-doped silica nanoparticles for bioimaging, diagnostics and therapeutics. *Food Bioprod Process* **83**, 136–140 (2005).

77. Santra, S., Wang, K.M., Tapec, R., & Tan, W.H. Development of novel dye-doped silica nanoparticles for biomarker application. *J Biomed Opt* **6**, 160–166 (2001).

78. Wang, L. et al. Watching silica nanoparticles glow in the biological world. *Anal Chem* **78**, 646–654 (2006).

79. Bagwe, R.P., Yang, C.Y., Hilliard, L.R., & Tan, W.H. Optimization of dye-doped silica nanoparticles prepared using a reverse microemulsion method. *Langmuir* **20**, 8336–8342 (2004).

80. Zhao, X.J., Bagwe, R.P., & Tan, W.H. Development of organic-dye-doped silica nanoparticles in a reverse microemulsion. *Adv Mater* **16**, 173–176 (2004).

81. Zhou, X.C. & Zhou, J.Z. Improving the signal sensitivity and photostability of DNA hybridizations on microarrays by using dye-doped core-shell silica nanoparticles. *Anal Chem* **76**, 5302–5312 (2004).

82. Carbonaro, C.M. et al. Photostability of porous silica—rhodamine 6G hybrid samples. *Mater Sci Eng C—Biomim Supramol Syst* **26**, 1038–1043 (2006).

83. Yan, J.L. et al. Dye-doped nanoparticles for bioanalysis. *Nano Today* **2**, 44–50 (2007).

84. Zhao, X.J., Tapec-Dytioco, R., & Tan, W.H. Ultrasensitive DNA detection using highly fluorescent bioconjugated nanoparticles. *J Amer Chem Soc* **125**, 11474–11475 (2003).

85. Lian, W. et al. Ultrasensitive detection of biomolecules with fluorescent dye-doped nanoparticles. *Anal Biochem* **334**, 135–144 (2004).

86. Zhao, X.J. et al. A rapid bioassay for single bacterial cell quantitation using bioconjugated nanoparticles. *Proc Nat Acad Sci USA* **101**, 15027–15032 (2004).

87. Deng, T., Li, J.S., Jiang, J.H., Shen, G.L., & Yu, R.Q. Preparation of near-IR fluorescent nanoparticles for fluorescence-anisotropy-based immunoagglutination assay in whole blood. *Adv Func Mater* **16**, 2147–2155 (2006).

88. Wang, L., Yang, C., & Tan, W. Dual-luminophore-doped silica nanoparticles for multiplexed signaling. *Nano Lett* **5**, 37–43 (2005).

89. Zhao, X., Tapec-Dytioco, R., & Tan, W. Ultrasensitive DNA detection using highly fluorescent bioconjugated nanoparticles. *J Am Chem Soc* **125**, 11474–11475 (2003).

90. Yao, G. et al. FloDots: luminescent nanoparticles. *Anal Bioanal Chem* **385**, 518–524 (2006).

91. Callan, J.F., De Silva, A.P., Mulrooney, R.C., & Mc Caughan, B. Luminescent sensing with quantum dots. *J Inclus Phenomena Macrocycl Chem* **58**, 257–262 (2007).

92. Huo, Q. A perspective on bioconjugated nanoparticles and quantum dots. *Colloids and Surfaces B—Biointerf* **59**, 1–10 (2007).

93. Jamieson, T. et al. Biological applications of quantum dots. *Biomaterials* **28**, 4717–4732 (2007).

94. Yang, D.Z., Xu, S.K., & Chen, Q.F. Applications of quantum dots to biological probes. *Spectrosc Spectral Anal* **27**, 1807–1810 (2007).

95. Zhong, Y., Kaji, N., Tokeshi, M., & Baba, Y. Nanobiotechnology: Quantum dots in bio-imaging. *Expert Rev Proteom* **4**, 565–572 (2007).

96. Hines, M.A. & Guyot-Sionnest, P. Synthesis and characterization of strongly luminescing ZnS-capped CdSe nanocrystals. *J Phys Chem* **100**, 468–471 (1996).

97. Chan, W.C.W. et al. Luminescent quantum dots for multiplexed biological detection and imaging. *Curr Opinion Biotechnol* **13**, 40–46 (2002).

98. Bao, H.B., Gong, Y.J., Li, Z., & Gao, M.Y. Enhancement effect of illumination on the photoluminescence of water-soluble CdTe nanocrystals: Toward highly fluorescent CdTe/CdS core-shell structure. *Chem Mater* **16**, 3853–3859 (2004).

99. Weng, J.F. & Ren, J.C. Luminescent quantum dots: A very attractive and promising tool in biomedicine. *Curr Medic Chem* **13**, 897–909 (2006).

100. Frangioni, J.V. *In vivo* near-infrared fluorescence imaging. *Curr Opinion Chem Biol* **7**, 626–634 (2003).

101. Peng, Z.A. & Peng, X.G. Formation of high-quality CdTe, CdSe, and CdS nanocrystals using CdO as precursor. *J Amer Chem Soc* **123**, 183–184 (2001).

102. Watson, A., Wu, X.Y., & Bruchez, M. Lighting up cells with quantum dots. *Biotechniques* **34**, 296–300 (2003).

103. Waggoner, A. Fluorescent labels for proteomics and genomics. *Curr Opin Chem Biol* **10**, 62–66 (2006).

104. Medintz, I.L., Uyeda, H.T., Goldman, E.R., & Mattoussi, H. Quantum dot bioconjugates for imaging, labelling and sensing. *Nat Mater* **4**, 435–446 (2005).

105. Smith, A.M. & Nie, S.M. Chemical analysis and cellular imaging with quantum dots. *Analyst* **129**, 672–677 (2004).

106. Yang, H. & Holloway, P.H. Efficient and photostable ZnS-Passivated CdS: Mn luminescent nanocrystals. *Adv Func Mater* **14**, 152–156 (2004).

107. Peng, X.G., Schlamp, M.C., Kadavanich, A.V., & Alivisatos, A.P. Epitaxial growth of highly luminescent CdSe/CdS core/shell nanocrystals with photostability and electronic accessibility. *J Amer Chem Soc* **119**, 7019–7029 (1997).

108. Gao, X.H. & Nie, S.M. Molecular profiling of single cells and tissue specimens with quantum dots. *Trends Biotechnol* **21**, 371–373 (2003).
109. Ness, J.M., Akhtar, R.S., Latham, C.B., & Roth, K.A. Combined tyramide signal amplification and quantum dots for sensitive and photostable immunofluorescence detection. *J Histochem Cytochem* **51**, 981–987 (2003).
110. Fu, A.H., Gu, W.W., Larabell, C., & Alivisatos, A.P. Semiconductor nanocrystals for biological imaging. *Curr Opin Neurobiol* **15**, 568–575 (2005).
111. Pinaud, F. et al. Advances in fluorescence imaging with quantum dot bio-probes. *Biomaterials* **27**, 1679–1687 (2006).
112. Gao, X.H., Cui, Y.Y., Levenson, R.M., Chung, L.W.K., & Nie, S.M. *In vivo* cancer targeting and imaging with semiconductor quantum dots. *Nat Biotechnol* **22**, 969–976 (2004).
113. Murray, C.B., Norris, D.J., & Bawendi, M.G. Synthesis and characterization of nearly monodisperse Cde (E = S, Se, Te) semiconductor nanocrystallites. *J Amer Chem Soc* **115**, 8706–8715 (1993).
114. Pradhan, N., Reifsnyder, D., Xie, R.G., Aldana, J., & Peng, X.G. Surface ligand dynamics in growth of nanocrystals. *J Amer Chem Soc* **129**, 9500–9509 (2007).
115. Munro, A.M., Plante, I.J.L., Ng, M.S., & Ginger, D.S. Quantitative study of the effects of surface ligand concentration on CdSe nanocrystal photoluminescence. *J Phys Chem C* **111**, 6220–6227 (2007).
116. Al-Salim, N., Young, A.G., Tilley, R.D., McQuillan, A.J., & Xia, J. Synthesis of CdSeS nanocrystals in coordinating and noncoordinating solvents: Solvent's role in evolution of the optical and structural properties. *Chem Mater* **19**, 5185–5193 (2007).
117. Asokan, S. et al. The use of heat transfer fluids in the synthesis of high-quality CdSe quantum dots, core/shell quantum dots, and quantum rods. *Nanotechnology* **16**, 2000–2011 (2005).
118. Santra, S. et al. Rapid and effective labeling of brain tissue using TAT-conjugated CdS: Mn/ZnS quantum dots. *Chem Commun*, 3144–3146 (2005).
119. Yang, H.S., Holloway, P.H., & Santra, S. Water-soluble silica-overcoated CdS: Mn/ZnS semiconductor quantum dots. *J Chem Phys* **121**, 7421–7426 (2004).
120. Gaponik, N. et al. Thiol-capping of CdTe nanocrystals: An alternative to organometallic synthetic routes. *J Phys Chem B* **106**, 7177–7185 (2002).
121. Zhang, H. et al. Hydrothermal synthesis for high-quality CdTe nanocrystals. *Adv Mater* **15**, 1712–1715 (2003).
122. Wang, Y. et al. Mechanism of strong luminescence photoactivation of citrate-stabilized water-soluble nanoparticles with CdSe cores. *J Phys Chem B* **108**, 15461–15469 (2004).
123. Chen, Q.F., Yang, D.Z., Xu, S.K., & Qu, Z. [Spectroscopy study of aqueous CdTe quantum dots synthesized by microwave irradiation method]. *Guang Pu Xue Yu Guang Pu Fen Xi* **27**, 650–653 (2007).
124. He, Y. et al. Microwave-assisted growth and characterization of water-dispersed CdTe/CdS core-shell nanocrystals with high photoluminescence. *J Phys Chem B* **110**, 13370–13374 (2006).
125. Li, L., Qian, H.F., & Ren, J.C. Rapid synthesis of highly luminescent CdTe nanocrystals in the aqueous phase by microwave irradiation with controllable temperature. *Chem Commun*, 528–530 (2005).
126. Qian, H.F., Li, L., & Ren, J.C. One-step and rapid synthesis of high quality alloyed quantum dots (CdSe-CdS) in aqueous phase by microwave irradiation with controllable temperature. *Mater Res Bull* **40**, 1726–1736 (2005).
127. Chan, W.C.W. & Nie, S.M. Quantum dot bioconjugates for ultrasensitive nonisotopic detection. *Science* **281**, 2016–2018 (1998).
128. Pathak, S., Choi, S.K., Arnheim, N., & Thompson, M.E. Hydroxylated quantum dots as luminescent probes for *in situ* hybridization. *J Amer Chem Soc* **123**, 4103–4104 (2001).

129. Wang, X.S. et al. Surface passivation of luminescent colloidal quantum dots with poly(dimethylaminoethyl methacrylate) through a ligand exchange process. *J Amer Chem Soc* **126**, 7784–7785 (2004).

130. Nann, T. Phase-transfer of CdSe@ZnS quantum dots using amphiphilic hyperbranched polyethylenimine. *Chem Commun*, 1735–1736 (2005).

131. Osaki, F., Kanamori, T., Sando, S., Sera, T., & Aoyama, Y. A quantum dot conjugated sugar ball and its cellular uptake on the size effects of endocytosis in the subviral region. *J Amer Chem Soc* **126**, 6520–6521 (2004).

132. Smith, A.M., Ruan, G., Rhyner, M.N., & Nie, S.M. Engineering luminescent quantum dots for *in vivo* molecular and cellular imaging. *Ann Biomed Eng* **34**, 3–14 (2006).

133. Goldman, E.R. et al. Avidin: A natural bridge for quantum dot-antibody conjugates. *J Amer Chem Soc* **124**, 6378–6382 (2002).

134. Mattoussi, H. et al. Self-assembly of CdSe-ZnS quantum dot bioconjugates using an engineered recombinant protein. *J Amer Chem Soc* **122**, 12142–12150 (2000).

135. Goldman, E.R. et al. Conjugation of luminescent quantum dots with antibodies using an engineered adaptor protein to provide new reagents for fluoroimmunoassays. *Anal Chem* **74**, 841–847 (2002).

136. Bruchez, M., Moronne, M., Gin, P., Weiss, S., & Alivisatos, A.P. Semiconductor nanocrystals as fluorescent biological labels. *Science* **281**, 2013–2016 (1998).

137. Wu, X.Y. et al. Immunofluorescent labeling of cancer marker Her2 and other cellular targets with semiconductor quantum dots. *Nat Biotechnol* **21**, 41–46 (2003).

138. Mitchell, G.P., Mirkin, C.A., & Letsinger, R.L. Programmed assembly of DNA functionalized quantum dots. *J Amer Chem Soc* **121**, 8122–8123 (1999).

139. Akerman, M.E., Chan, W.C.W., Laakkonen, P., Bhatia, S.N., & Ruoslahti, E. Nanocrystal targeting *in vivo*. *Proc Nat Acad Sci USA* **99**, 12617–12621 (2002).

140. Gao, X.H., Chan, W.C.W., & Nie, S.M. Quantum-dot nanocrystals for ultrasensitive biological labeling and multicolor optical encoding. *J Biomed Opt* **7**, 532–537 (2002).

141. Hanaki, K. et al. Semiconductor quantum dot/albumin complex is a long-life and highly photostable endosome marker. *Biochem Biophys Res Commun* **302**, 496–501 (2003).

142. Parak, W.J. et al. Cell motility and metastatic potential studies based on quantum dot imaging of phagokinetic tracks. *Adv Mater* **14**, 882–885 (2002).

143. Dabbousi, B.O. et al. (CdSe)ZnS core-shell quantum dots: Synthesis and characterization of a size series of highly luminescent nanocrystallites. *J Phys Chem B* **101**, 9463–9475 (1997).

144. Derfus, A.M., Chan, W.C.W., & Bhatia, S.N. Intracellular delivery of quantum dots for live cell labeling and organelle tracking. *Adv Mater* **16**, 961–966 (2004).

145. Derfus, A.M., Chan, W.C.W., & Bhatia, S.N. Probing the cytotoxicity of semiconductor quantum dots. *Nano Letters* **4**, 11–18 (2004).

146. Sukhanova, A. et al. Biocompatible fluorescent nanocrystals for immunolabeling of membrane proteins and cells. *Anal Biochem* **324**, 60–67 (2004).

147. Dahan, M. et al. Diffusion dynamics of glycine receptors revealed by single quantum dot tracking. *Biophys J* **86**, 363A–363A (2004).

148. Dahan, M. et al. Diffusion dynamics of glycine receptors revealed by single-quantum dot tracking. *Science* **302**, 442–445 (2003).

149. Lidke, D.S. et al. Quantum dot ligands provide new insights into erbB/HER receptor-mediated signal transduction. *Nat Biotechnol* **22**, 198–203 (2004).

150. Dubertret, B. et al. *In vivo* imaging of quantum dots encapsulated in phospholipid micelles. *Science* **298**, 1759–1762 (2002).

151. Tokumasu, F. & Dvorak, J. Development and application of quantum dots for immunocytochemistry of human erythrocytes. *J Microsc—Oxford* **211**, 256–261 (2003).

152. Jaiswal, J.K., Mattoussi, H., Mauro, J.M., & Simon, S.M. Long-term multiple color imaging of live cells using quantum dot bioconjugates. *Nat Biotechnol* **21**, 47–51 (2003).

153. Voura, E.B., Jaiswal, J.K., Mattoussi, H., & Simon, S.M. Tracking metastatic tumor cell extravasation with quantum dot nanocrystals and fluorescence emission-scanning microscopy. *Nat Med* **10**, 993–998 (2004).

154. Rozenzhak, S.M. et al. Cellular internalization and targeting of semiconductor quantum dots. *Chem Commun*, 2217–2219 (2005).

155. Vu, T.Q. et al. Peptide-conjugated quantum dots activate neuronal receptors and initiate downstream signaling of neurite growth. *Nano Lett* **5**, 603–607 (2005).

156. Weng, J. & Ren, J. Luminescent quantum dots: A very attractive and promising tool in biomedicine. *Curr Med Chem* **13**, 897–909 (2006).

157. Lagerholm, B.C. et al. Multicolor coding of cells with cationic peptide coated quantum dots. *Nano Lett* **4**, 2019–2022 (2004).

158. Mattheakis, L.C. et al. Optical coding of mammalian cells using semiconductor quantum dots. *Anal Biochem* **327**, 200–208 (2004).

159. Kaul, Z., Yaguchi, T., Kaul, S.C., & Wadhwa, R. Quantum dot-based protein imaging and functional significance of two mitochondrial chaperones in cellular senescence and carcinogenesis. *Ann N Y Acad Sci* **1067**, 469–473 (2006).

160. Hoshino, A. et al. Quantum dots targeted to the assigned organelle in living cells. *Microbiol Immunol* **48**, 985–994 (2004).

161. Chen, F.Q. & Gerion, D. Fluorescent CdSe/ZnS nanocrystal-peptide conjugates for long-term, nontoxic imaging and nuclear targeting in living cells. *Nano Lett* **4**, 1827–1832 (2004).

162. Hoshino, A., Hanaki, K., Suzuki, K., & Yamamoto, K. Applications of T-lymphoma labeled with fluorescent quantum dots to cell tracing markers in mouse body. *Biochem Biophys Res Commun* **314**, 46–53 (2004).

163. Frangioni, J.V. Self-illuminating quantum dots light the way. *Nat Biotechnol* **24**, 326–328 (2006).

164. Kim, S. et al. Near-infrared fluorescent type II quantum dots for sentinel lymph node mapping. *Nat Biotechnol* **22**, 93–97 (2004).

165. Soltesz, E.G. et al. Intraoperative sentinel lymph node mapping of the lung using near-infrared fluorescent quantum dots. *Ann Thorac Surg* **79**, 269–277 (2005).

166. Larson, D.R. et al. Water-soluble quantum dots for multiphoton fluorescence imaging in vivo. *Science* **300**, 1434–1436 (2003).

167. Kim, S., Fisher, B., Eisler, H.J., & Bawendi, M. Type-II quantum dots: CdTe/CdSe(core/shell) and CdSe/ZinTe(core/shell) heterostructures. *J Amer Chem Soc* **125**, 11466–11467 (2003).

168. Morgan, N.Y. et al. Real time *in vivo* non-invasive optical imaging using near-infrared fluorescent quantum dots. *Acad Radiol* **12**, 313–323 (2005).

169. Kircher, M.F. et al. *In vivo* high resolution three-dimensional imaging of antigen-specific cytotoxic T-lymphocyte trafficking to tumors. *Cancer Res* **63**, 6838–6846 (2003).

170. Gutierrez, F.R., Brown, J.J., & Mirowitz, S.A. *Cardiovascular Magnetic Resonance Imaging*, xviii, 233 p. (Mosby Year Book, St. Louis, 1992).

171. Nelson, K.L. & Runge, V.M. Basic Principles of MR Contrast. *Top Magnet Reson Imag* **7**, 124–136 (1995).

172. Hashemi, R.H., Bradley, W.G., & Lisanti, C.J. *MRI: The Basics*, xiii, 353 p. (Lippincott Williams & Wilkins, Philadelphia, 2004).

173. Lowe, M.P. MRI contrast agents: The next generation. *Austral J Chem* **55**, 551–556 (2002).

174. Rinck, P.A., Bjørnerud, A., & European Magnetic Resonance Forum. *Magnetic Resonance in Medicine: The Basic Textbook of the European Magnetic Resonance Forum*, iv, 245 p. (Blackwell Wissenschafts-Verlag, Berlin; Boston, 2001).

175. Bushong, S.C. *Magnetic Resonance Imaging: Physical and Biological Principles*, xi, 509 p. (Mosby, St. Louis, Mo., 2003).

176. Bellin, M.F. MR contrast agents, the old and the new. *Euro J Radiol* **60**, 314–323 (2006).

177. Arbab, A.S., Liu, W., & Frank, J.A. Cellular magnetic resonance imaging: Current status and future prospects. *Exp Rev Med Devices* **3**, 427–439 (2006).

178. Bulte, J.W.M. & Kraitchman, D.L. Iron oxide MR contrast agents for molecular and cellular imaging. *NMR in Biomed* **17**, 484–499 (2004).

179. Bjornerud, A. & Johansson, L. The utility of superparamagnetic contrast agents in MRI: Theoretical consideration and applications in the cardiovascular system. *NMR in Biomed* **17**, 465–477 (2004).

180. Corot, C. et al. Macrophage imaging in central nervous system and in carotid atherosclerotic plaque using ultrasmall superparamagnetic iron oxide in magnetic resonance imaging. *Investig Radiol* **39**, 619–625 (2004).

181. Simon, G.H. et al. MRI of arthritis: Comparison of ultrasmall superparamagnetic iron oxide vs. Gd-DTPA. *J Magnet Reson Imag* **23**, 720–727 (2006).

182. Gupta, A.K. & Gupta, M. Synthesis and surface engineering of iron oxide nanoparticles for biomedical applications. *Biomaterials* **26**, 3995–4021 (2005).

183. Gupta, A.K., Naregalkar, R.R., Vaidya, V.D., & Gupta, M. Recent advances on surface engineering of magnetic iron oxide nanoparticles and their biomedical applications. *Nanomed* **2**, 23–39 (2007).

184. Dunning, M.D., Kettunen, M.I., Ffrench Constant, C., Franklin, R.J., & Brindle, K.M. Magnetic resonance imaging of functional Schwann cell transplants labelled with magnetic microspheres. *Neuroimage* **31**, 172–180 (2006).

185. Jung, C.W. & Jacobs, P. Physical and chemical properties of superparamagnetic iron oxide MR contrast agents: Ferumoxides, ferumoxtran, ferumoxsil. *Magn Reson Imaging* **13**, 661–674 (1995).

186. Weissleder, R., Lee, A.S., Khaw, B.A., Shen, T., & Brady, T.J. Antimyosin-labeled monocrystalline iron oxide allows detection of myocardial infarct: MR antibody imaging. *Radiology* **182**, 381–385 (1992).

187. Smirnov, P. et al. Single-cell detection by gradient echo 9.4 T MRI: A parametric study. *Contr Media Molec Imag* **1**, 165–174 (2006).

188. Bottrill, M., Nicholas, L.K., & Long, N.J. Lanthanides in magnetic resonance imaging. *Chem Soc Rev* **35**, 557–571 (2006).

189. Aime, S., Botta, M., Fasano, M., & Terreno, E. Lanthanide(III) chelates for NMR biomedical applications. *Chem Soc Rev* **27**, 19–29 (1998).

190. Aime, S., Botta, M., & Terreno, E. Gd(III)-based contrast agents for MRI. *Adv Inorgan Chem—Includ Bioinorgan Stud* **57**, 173–237 (2005).

191. Caravan, P., Ellison, J.J., McMurry, T.J., & Lauffer, R.B. Gadolinium(III) chelates as MRI contrast agents: Structure, dynamics, and applications. *Chem Rev* **99**, 2293–2352 (1999).

192. Aime, S. et al. High sensitivity lanthanide(III) based probes for MR-medical imaging. *Coordin Chem Rev* **250**, 1562–1579 (2006).

193. Toth, E., Helm, L., & Merbach, A.E. Relaxivity of MRI contrast agents. *Contr Agents I* **221**, 61–101 (2002).

194. Caravan, P. Strategies for increasing the sensitivity of gadolinium based MRI contrast agents. *Chem Soc Rev* **35**, 512–523 (2006).

195. Thomsen, H.S. Nephrogenic systemic fibrosis: A serious late adverse reaction to gadodiamide. *Eur Radiol* **16**, 2619–2621 (2006).

196. Runge, V.M. Gd-Dpta—An IV contrast agent for clinical MRI. *Nucl Med Biol* **15**, 37–44 (1988).

197. Liu, Y.T. et al. Dynamic susceptibility contrast perfusion imaging of cerebral ischemia in nonhuman primates: Comparison of Gd-DTPA and NMS60. *J Magnet Reson Imag* **22**, 461–466 (2005).

198. Mohs, A.M. et al. Modification of Gd-DTPA cystine copolymers with PEG-1000 optimizes pharmacokinetics and tissue retention for magnetic resonance angiography. *Magnet Reson Med* **58**, 110–118 (2007).

199. Tweedle, M.F. The ProHance story: The making of a novel MRI contrast agent. *Euro Radiol* **7**, S225–S230 (1997).

200. Wedeking, P., Kumar, K., & Tweedle, M.F. Dissociation of gadolinium chelates in mice— Relationship to chemical characteristics. *Magnet Reson Imag* **10**, 641–648 (1992).

201. Gibby, W.A., Gibby, K.A., & Gibby, W.A. Comparison of Gd DTPA-BMA (Omniscan) versus GdHP-DO3A (ProHance) retention in human bone tissue by inductively coupled plasma atomic emission spectroscopy. *Invest Radiol* **39**, 138–142 (2004).

202. Bousquet, J.C. et al. Gd-DOTA: characterization of a new paramagnetic complex. *Radiology* **166**, 693–698 (1988).

203. Aime, S., Barge, A., Cabella, C., Crich, S.G., & Gianolio, E. Targeting cells with MR imaging probes based on paramagnetic Gd(III) chelates. *Curr Pharm Biotechnol* **5**, 509–18 (2004).

204. Aime, S. et al. Insights into the use of paramagnetic Gd(III) complexes in MR-molecular imaging investigations. *J Magn Reson Imaging* **16**, 394–406 (2002).

205. Santra, S., Dutta, D., Walter, G.A., & Moudgil, B.M. Fluorescent nanoparticle probes for cancer imaging. *Technol Cancer Res Treat* **4**, 593–602 (2005).

206. Rieter, W.J. et al. Hybrid silica nanoparticles for multimodal imaging. *Angew Chem Int Ed Engl* **46**, 3680–3682 (2007).

207. Lee, J.H., Jun, Y.W., Yeon, S.I., Shin, J.S., & Cheon, J. Dual-mode nanoparticle probes for high-performance magnetic resonance and fluorescence imaging of neuroblastoma. *Angew Chem-Int Ed* **45**, 8160–8162 (2006).

208. Perez, J.M., Josephson, L., O'Loughlin, T., Hogemann, D., & Weissleder, R. Magnetic relaxation switches capable of sensing molecular interactions. *Nat Biotechnol* **20**, 816–820 (2002).

209. Lin, Y.S. et al. Multifunctional composite nanoparticles: Magnetic, luminescent, and mesoporous. *Chem Mater* **18**, 5170–5172 (2006).

210. Mulder, W.J.M. et al. Quantum dots with a paramagnetic coating as a bimodal molecular imaging probe. *Nano Lett* **6**, 1–6 (2006).

211. Mulder, W.J.M. et al. MR molecular imaging and fluorescence microscopy for identification of activated tumor endothelium using a bimodal lipidic nanoparticle. *FASEB J* **19**, 2008–2110 (2005).

212. Mulder, W.J.M., Strijkers, G.J., van Tilborg, G.A.F., Griffioen, A.W., & Nicolay, K. Lipid-based nanoparticles for contrast-enhanced MRI and molecular imaging. *NMR in Biomed* **19**, 142–164 (2006).

213. van Tilborg, G.A.F. et al. Annexin A5-conjugated quantum dots with a paramagnetic lipidic coating for the multimodal detection of apoptotic cells. *Bioconj Chem* **17**, 865–868 (2006).

214. Prinzen, L. et al. Optical and magnetic resonance imaging of cell death and platelet activation using Annexin A5-functionalized quantum dots. *Nano Lett* **7**, 93–100 (2007).

215. van Zandvoort, M. et al. Two-photon microscopy for imaging of the (atherosclerotic) vascular wall: A proof of concept study. *J Vasc Res* **41**, 54–63 (2004).

216. Denk, W., Strickler, J.H., & Webb, W.W. 2-Photon laser scanning fluorescence microscopy. *Science* **248**, 73–76 (1990).

217. Bakalova, R. et al. Silica-shelled single quantum dot micelles as imaging probes with dual or multimodality. *Anal Chem* **78**, 5925–5932 (2006).

218. Yang, H.S., Santra, S., Walter, G.A., & Holloway, P.H. Gd-III-functionalized fluorescent quantum dots as multimodal imaging probes. *Adv Mater* **18**, 2890–2894 (2006).

219. Talanov, V.S. et al. Dendrimer-based nanoprobe for dual modality magnetic resonance and fluorescence imaging. *Nano Lett* **6**, 1459–1463 (2006).

220. Koyama, Y. et al. A dendrimer-based nanosized contrast agent, dual-labeled for magnetic resonance and optical fluorescence imaging to localize the sentinel lymph node in mice. *J Magnet Reson Imag* **25**, 866–871 (2007).

221. Schellenberger, E.A., Sosnovik, D., Weissleder, R., & Josephson, L. Magneto/optical annexin V, a multimodal protein. *Bioconj Chem* **15**, 1062–1067 (2004).

222. Moore, A., Medarova, Z., Potthast, A., & Dai, G.P. *In vivo* targeting of underglycosylated MUC-1 tumor antigen using a multimodal imaging probe. *Cancer Res* **64**, 1821–1827 (2004).

223. Montet, X., Montet-Abou, K., Reynolds, F., Weissleder, R., & Josephson, L. Nanoparticle imaging of integrins on tumor cells. *Neoplasia* **8**, 214–222 (2006).

224. Veiseh, O. et al. Optical and MRI multifunctional nanoprobe for targeting gliomas. *Nano Lett* **5**, 1003–1008 (2005).

225. Palmacci, S. & Josephson, L. Synthesis of polysaccharide covered superparamagnetic oxide colloids. U.S. Patent 5262176 (1993).

226. Jaffer, F.A. et al. Cellular imaging of inflammation in atherosclerosis using magneto-fluorescent nanomaterials. *Molec Imag* **5**, 85–92 (2006).

227. Medarova, Z., Pham, W., Kim, Y., Dai, G.P., & Moore, A. *In vivo* imaging of tumor response to therapy using a dual-modality imaging strategy. *Int J Cancer* **118**, 2796–2802 (2006).

228. Kelly, K.A. et al. Detection of vascular adhesion molecule-1 expression using a novel multimodal nanoparticle. *Circ Res* **96**, 327–336 (2005).

229. Nitin, N., LaConte, L.E.W., Zurkiya, O., Hu, X., & Bao, G. Functionalization and peptide-based delivery of magnetic nanoparticles as an intracellular MRI contrast agent. *J Biol Inorgan Chem* **9**, 706–712 (2004).

230. Levy, L., Sahoo, Y., Kim, K.S., Bergey, E.J., & Prasad, P.N. Nanochemistry: Synthesis and characterization of multifunctional nanoclinics for biological applications. *Chem Mater* **14**, 3715–3721 (2002).

Index